Lectures on Quantum Electronics

Advances
in
Science
and
Technology

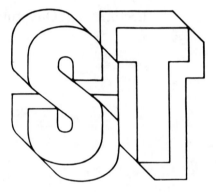

Physics Series

ADVANCES IN
SCIENCE AND TECHNOLOGY

Lectures on Quantum Electronics

N. V. KARLOV

Translated from the Russian by EUGENE YANKOVSKY

Mir Publishers
Moscow

CRC Press
Boca Raton   Ann Arbor   Tokyo   London

Library of Congress Cataloging-in-Publication Data

Karlov, N. V. (Nikolai Vasil'evich)
  [Lektsii po kvantovoi elektronike. English]
  Lectures on quantum electronics / N.V. Karlov; translated from the Russian by Eugene
    Yankovsky.
      p. cm. — (Advances in science and technology in the USSR)
  Translation of : Lektsii po kvantovoi elektronike.
  Includes bibliographical references and index.
  ISBN 0-8493-7538-X
  1. Lasers. 2. Quantum electronics. I. Title. II. Series.
  QC689.K3713   1992
  621.36′6—dc20   92-27128
                                                                                              CIP

© 1993 by CRC Press, Inc.

International Standard Book Number 0-8493-7538-X
Library of Congress Card Number 92-27128

Printed in the United States   1 2 3 4 5 6 7 8 9 0

Printed on acid-free paper

CONTENTS

6   Contents

8   Contents

Contents

9

PREFACE TO THE SECOND RUSSIAN EDITION

In preparing the second edition I have drawn on my experience in teaching the basics of quantum electronics to students of physical engineering, physical technology, and physics. I have also taken into account the methods of preparing for Ph.D. examinations in physical electronics, including quantum electronics. Basically this has meant adding simple problems and numerical examples. These are placed at the end of the appropriate lectures, and the answers to the problems are given at the end of the book.

For these problems and answers I am indebted to I.T. Sorokina and E.V. Sorokin, graduates of the Division of Wave Processes of the Department of Physics at Moscow State University and practicing researchers at the General Physics Institute of the USSR Academy of Sciences. I also feel that I must single out a fine book by Svelto,[1] which should be used in getting acquainted with the basics of quantum electronics. Finally, I am grateful to L.A. Savushkina for her assistance in preparing the manuscript for publication.

N.V. Karlov

---

[1] O. Svelto, *Principles of Lasers* (3d ed., New York: Plenum Press, 1989).

PREFACE TO THE FIRST RUSSIAN EDITION

This book is a collection of lectures on the basics of quantum electronics delivered at the Moscow Institute of Physics and Technology.

Working with undergraduates, graduates, and practicing researchers has convinced me of the need for a course of lectures on quantum electronics setting forth the basic physical ideas of this branch of science in condensed form and describing the principles of operation of the most important devices. The aim is to give future experimental physicists, engineering physicists, and radio engineers who intend to work in laser research and applications of laser light the necessary minimum of initial information on quantum electronics. Preparing such a course is extremely complicated and perhaps cannot even be done comprehensively. Therefore, I have chiefly relied on my experience of working in a research laboratory of a general physical nature that widely employs laser light. This approach may have made the manner of presentation somewhat one-sided.

The course begins by explaining the physical bases of quantum electronics, which reduce essentially to applying Einstein's theory of radiation to thermodynamically nonequilibrium systems with discrete energy levels. I have paid great attention to the concept of feedback realized in stimulated emission of radiation in cavities and discussed in detail the ways in which a laser cavity forms a beam of light of high directivity and how the mode composition of laser light can be controlled. Part One concludes with a list of the basic formulas of quantum electronics.

Part Two describes the methods of creating an active medium and the properties of the most widely known lasers, such as the helium-neon, argon, copper-vapor, carbon dioxide, carbon monoxide, dye, color center, ruby, and neodymium-doped glass lasers. Great attention is paid to gasdynamic, chemical, excimer, semiconductor, and free-electron lasers. I also consider the problem of tuning the wavelength of laser light. Part Two concludes with a summary of the characteristic features of the most widely known lasers and a discussion of the trends in quantum electronics and the various applications of lasers.

A distinctive feature of quantum electronics is the heavy penetration of the ideas of the theory of oscillations and the methods of microwave electronics into optics. At the same time spectroscopy and physical optics form the groundwork of quantum electronics. Hence, I have tried to show

how the combination of radiophysics, spectroscopy, and wave optics in quantum electronics made it possible to create exceptionally high concentrations (reaching the limit) of the energy of electromagnetic radiation in space, time, and a frequency interval. In view of the vigorous development of quantum electronics I have focused not on specific achievements in establishing laser parameters but on the principles of operation and the potentials of the lasers considered. I call attention again and again to the similarity of processes taking place in different lasers, at the same time stressing their distinctive features.[2]

I am deeply indebted to my teacher Alexander M. Prokhorov, communication with whom over the years has made the writing of this course of lectures possible. I wish to thank my friends and colleagues, F.V. Bunkin, V.G. Veselago, and P.P. Pashinin, for carefully reading the manuscript and making numerous helpful remarks. I am also grateful to the reviewers in the Division of Optics and Spectroscopy of the Department of Physics at Moscow State University and to Yu.A. Kravtsov for many constructive remarks, which I was pleased to accept when working on the final version of the manuscript. Finally, I must thank I.I. Goryachko and L.A. Farforina for their great help in preparing the manuscript and M.N. Andreyeva for help in preparing the illustrations.

As one of the first students in the Department of Physics and Technology at Moscow State University way back in the forties, I wish to dedicate these lectures to the founders of the Moscow Institute of Physics and Technology.

---

[2] In beginning to study the basics of quantum electronics I would recommend using the following books: *Vibrations and Waves: Introduction to Acoustics, Radiophysics, and Optics* (Moscow: Fizmatgiz, 1959; in Russian) by G.S. Gorelik, *Optics* (Moscow: Nauka, 1976; in Russian) by G.S. Landsberg, *Principles of Lasers* (New York: Plenum Press, 1976) by O. Svelto, and *Handbook of Lasers* (Moscow: Sovetskoe radio, 1978, vols. I and II: in Russian) by A.M. Prokhorov (ed.).

PART ONE

Basics of Laser Physics

LECTURE 1

Einstein's Coefficients

Definition of quantum electronics. Stimulated and spontaneous transitions and Einstein's coefficients. The coherence of stimulated radiation

We start this course of lectures on the basics of quantum electronics by defining the topic. According to the definition given in the *Large Soviet Encyclopedia*, "quantum electronics is the branch of physics that studies the methods of generating and amplifying electromagnetic radiation by using the effect of stimulated emission of radiation in thermodynamically nonequilibrium quantum systems, the properties of the respective generators and amplifiers, and the application of these devices." The best-known devices of quantum electronics are masers and lasers. Hence, in the narrow sense one can speak of quantum electronics as the science of masers and lasers, bearing in mind that masers are quantum amplifiers and generators of coherent electromagnetic radiation in the microwave range, while the term "lasers" refers to such devices in the optical (light) range.

The high concentration of luminous energy in an extremely narrow solid angle and a small spectral interval, that is, highly directional and mono-chromatic radiation, is the main feature of lasers that distinguishes them from ordinary sources of light. Another feature, closely linked to this, is the ability of lasers to pack large amounts of energy into extremely narrow time intervals. If we turn to masers, we discover that as generators they differ from ordinary sources of radio waves by their high stability of frequency, while as amplifiers they have an extremely low noise level compared with ordinary electronic radio-amplifiers.

All these properties are explained by the fact that amplification in quantum electronics is achieved through the so-called phenomenon of stimulated emission of radiation, a fact reflected in the firmly established terminology. The term "maser" ("laser") is formed from the first letters of *m*icrowave (*l*ight) *a*mplification by *s*timulated *e*mission of *r*adiation.

We see that there is no great difference between masers and lasers, and the difference practically vanishes as we move from millimeter to submillimeter waves. Lasers attract the greatest interest, however, owing to their ability to concentrate large amounts of energy in narrow spatial, temporal, and frequency intervals. This ability of lasers can serve as their fullest functional definition. By luminous energy, of course, one must understand the energy of the optical radiation in the infrared (IR), visual, and ultraviolet (UV) ranges.

In view of what has been said, a course in the basics of quantum electronics must be understood as a course in the basics of laser physics augmented with a discussion of the operational principles of the more interesting types of lasers. The very important question of application of lasers can be treated here only by way of reference to the most representative examples of application.

The foundation of quantum electronics as a science is the phenomenon of stimulated emission of radiation, whose existence was postulated in 1916 by Albert Einstein (1879-1955). In quantum systems possessing discrete energy levels there are three types of transition between energy states, namely, transitions induced by an electromagnetic field, or stimulated transitions, spontaneous transitions, and nonradiative relaxation transitions. The properties of stimulated radiation determine the coherence of emission and amplification of radiation in quantum electronics. Spontaneous emission determines the presence of noise, serves as a trigger for amplification and excitation of oscillations, and, together with nonradiative relaxation transitions, plays an important role in generating and maintaining a thermodynamically nonequilibrium radiant state.

In stimulated transitions a quantum system may transfer from one energy state to another (Figure 1.1) with simultaneous absorption of energy from the electromagnetic field (in a transition from a lower energy level to a higher) or with emission of electromagnetic energy (in a transition from a higher energy state to a lower).

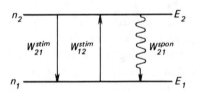

Figure 1.1. The diagram of two energy levels $E_2 > E_1$ (populations n_2 and n_1, respectively). The stimulated transitions from the upper level to the lower with a probability W_{21}^{stim} and from the lower to the upper with a probability W_{12}^{stim} are shown by straight arrows, and the spontaneous transition with a probability W_{21}^{spon} is shown by a wavy arrow.

Stimulated transitions possess the following important properties.

First, the probability of stimulated transitions is nonzero only if the external electromagnetic field oscillates with a resonance frequency, that is, the energy of a quantum of this field, $h\nu$ ($\hbar\omega$), coincides with the difference in energies of the two isolated states considered (two energy levels E_2 and E_1, where the subscripts "2" and "1" stand for the higher and lower energies, respectively). This condition is equivalent to Bohr's postulate

$$h\nu = \hbar\omega = E_2 - E_1. \tag{1.1}$$

Second, the electromagnetic field quanta, or photons, emitted in stimulated transitions are identical to the photons of the field that induced these transitions. This means that the external electromagnetic field and the field generated as a result of stimulated transitions have the same frequency, phase, polarization, and direction of propagation, that is, the fields are identical.

Third, the probability of stimulated transitions per unit time, or the transition rate, is proportional to the energy density of the external field per unit spectral interval, or simply the spectral volumetric density of the radiation present, ϱ_ν [J/cm$^3 \cdot$ Hz]:

$$W_{12}^{\text{stim}} = B_{12}\,\varrho_\nu, \tag{1.2}$$

$$W_{21}^{\text{stim}} = B_{21}\,\varrho_\nu, \tag{1.3}$$

where B_{12} and B_{21} are Einstein's coefficients for stimulated absorption and emission, respectively, with the order of subscripts "1" and "2" designating the direction of a transition.

Thus, stimulated radiation is induced by an external field. The complete equivalence of the stimulated (secondary) and stimulating (primary) radiations results in the coherence of amplification and emission in quantum electronics.

However, in addition to the emission of radiation stimulated by an external field there is spontaneous emission of radiation. Atoms (or molecules, ions, and electrons) that are in a higher energy state may perform spontaneous transitions to a lower state. Such transitions are called spontaneous. The decay of the upper energy state resembles the radioactive decay of an unstable nucleus. The probability of spontaneous transitions does not depend on an external electromagnetic field; the acts of spontaneous emission are in no way linked with the external field. Hence, spontaneous radiation is incoherent with respect to the external field and plays the role of internal noise. In addition, spontaneous emission depletes higher energy levels, assisting the return of an atom to a lower energy state.

Spontaneous emission is a purely quantum effect and does not allow for a classical interpretation. In classical mechanics a metastable state having a higher energy with respect to a ground stable state can be prolonged indefinitely in the absence of external perturbations. In the quantum world such a metastable state decays spontaneously at a nonzero average rate.

Let us now study in greater detail the properties of stimulated and spontaneous emissions. The relationship between the probabilities of spontaneous and stimulated transitions can be established thermodynamically, along the lines put forward by Einstein.

Let us consider an ensemble of quantum particles placed in a thermostat with a temperature T. We wish to find the equilibrium conditions for such an ensemble in the field of its own radiation emitted and absorbed in transitions between the energy levels of the constituent particles.

Suppose that the quantum system has two energy levels, E_1 and E_2, with $E_2 > E_1$, with an energy quantum $h\nu$ emitted or absorbed in the transitions between these levels. In thermodynamic equilibrium the ensemble does not lose or acquire energy. Hence, the overall number of transitions from the higher energy state to the lower per unit time must be equal to the overall number of transitions from the lower energy state to the higher per unit time. The overall number of transitions is determined by the number of particles on the energy levels or, as is customarily said in quantum electronics, by the level populations.

In thermal equilibrium the level populations are linked by the Boltzmann formula

$$\frac{n_2}{g_2} = \frac{n_1}{g_1} \exp\left(-\frac{E_2 - E_1}{kT}\right), \tag{1.4}$$

where g_2 and g_1 are the degeneracy multiplicities (statistical weights) of levels "2" and "1", and k is the Boltzmann constant.

The total number of transitions $2 \rightarrow 1$ is equal to the product of the number n_2 of particles in state "2" by the probability, per unit time, of a $2 \rightarrow 1$ transition of a single particle.

We already know that a free particle in an excited state will finally give off its excess energy in the form of a radiation quantum, irrespective of an external perturbation. The probability of a spontaneous transition of a particle from the upper level to the lower is proportional to time. By assumption, in a time interval dt this probability grows by

$$dw^{\text{spon}} = A_{21} dt, \tag{1.5}$$

where A_{21} is Einstein's coefficient for spontaneous emission. We have, therefore, postulated that the probability of spontaneous emission per unit time or, which is the same, the spontaneous decay rate, is constant and equal by definition to the respective Einstein coefficient A_{21}:

$$W_{21}^{\text{spon}} = A_{21}. \tag{1.6}$$

Spontaneous emission describes the process of a spontaneous transition of a particle from the upper state to the lower. There can be no spontaneous transitions in the opposite direction, from the lower state to the upper. The upper level becomes populated in the situation considered here only through stimulated transitions accompanied by the absorption of radiation quanta.

The particles in the ensemble are in the field of their own radiation, whose energy density per unit spectral interval is ϱ_ν. This field induces transitions from the upper state to the lower and in the opposite direction. According to assumptions (1.2) and (1.3), the probabilities of these transitions are proportional to ϱ_ν. Combining (1.6), (1.4), (1.3), and (1.2) and employing the equilibrium condition

$$g_1 B_{12} \varrho_\nu e^{-E_1/kt} = g_2 (B_{21} \varrho_\nu + A_{21}) e^{-E_2/kT}, \tag{1.7}$$

we can find a formula that links A_{21}, B_{12}, and B_{21}. In Eq. (1.7) we have equated the total numbers of transitions from the lower level to the upper (on the left-hand side) with that from the upper level to the lower (on the right-hand side). This equation provides an easy way to find the radiative energy in the equilibrium quantum system considered here:

$$\varrho_\nu = \frac{A_{21}}{B_{21}} \left(\frac{g_1 B_{12}}{g_2 B_{21}} \exp \frac{E_2 - E_1}{kT} - 1 \right)^{-1}. \tag{1.8}$$

This leads to important corollaries. Einstein postulated that radiation emitted and absorbed in equilibrium transitions between the energy states of the two-level equilibrium quantum system considered here is described by Planck's formula for blackbody radiation. Then in free space

$$\varrho_\nu = \frac{8\pi\nu^2}{c^3} \frac{h\nu}{\exp(h\nu/kT) - 1},$$ (1.9)

where c is the speed of light.

If we compare these two formulas with Bohr's frequency condition (1.1), we see that Einstein's postulate agrees with Bohr's postulate. Further comparison leads to the important conclusion that Einstein's B_{12} and B_{21} coefficients for stimulated transitions are linked by the following formula:

$$g_1 B_{12} = g_2 B_{21}.$$ (1.10)

The relation expresses the important fact that stimulated emission and absorption have the same probability (per nondegenerate state).

The probability of spontaneous emission is proportional to Einstein's coefficient for stimulated emission:

$$A_{21} = \frac{8\pi\nu^2}{c^3} h\nu B_{21}.$$ (1.11)

Thus, to describe the thermodynamic equilibrium between a system of quantum particles (i.e. particles with discrete energy levels) and the field of their radiation Einstein introduced the concept of field-stimulated equally probable transitions (which takes into account the multiplicity of degeneracy) from the upper state to the lower and from the lower to the upper. The equilibrium condition leads to such a relationship between the spontaneous and the stimulated emission that for a single particle the probability of transition per unit time accompanied by emission of radiation quanta is

$$W^{em} = \left(\frac{8\pi\nu^2}{c^3} h\nu + \varrho_\nu\right) B_{21}.$$ (1.12)

It is important that W^{em} is proportional to B_{21} and, hence, where stimulated transitions are forbidden there can be no spontaneous emission, and vice versa, where there is no spontaneous emission there can be no stimulated emission.

The equilibrium radiation generated by the entire ensemble of particles acts, in relation to each particle, as an external electromagnetic field that stimulates absorption or emission of radiation by the particle, depending on the particle's state. Hence, Eqs. (1.10)-(1.12), obtained as a byproduct of equilibrium conditions, are also valid for the case of a quantum system in an external radiation field.

The relation linking the spontaneous and stimulated emission probabilities contains the factor $8\pi\nu^2/c^3$, which is the number of oscillators (types of waves, types of oscillations or modes) per unit spectral interval in free space. The probability of spontaneous emission is proportional to ν^3, with the result that it plays a minor role at radio frequencies but a major role in optics.

What is important for quantum electronics, however, is that the stimulated emission probability is proportional to the energy density of the stimulating field. At a high enough density of this field the (coherent) radiation produced by stimulated emission becomes predominant.

Now it is pertinent to discuss more thoroughly the coherence of radiation produced by stimulated emission. It is important to stress once more that while spontaneous emission consitutes a purely quantum effect (from the classical viewpoint an excited free atom may live indefinetely), stimulated emission can be interpreted classically.

For example, a harmonic oscillator performing natural oscillations and placed in a monochromatic radiation field tuned in resonance with the oscillator is driven by this external force. As is well known, the frequency and phase of these oscillations are determined by the frequency and phase of the driving force. At the same time, for a certain phase relationship between the initial natural oscillations and the driving force the power absorbed by the oscillator may be negative. This means that at some values of phase the oscillator transfers its energy to the external field owing to the field's action on the oscillator. This manifests itself in stimulated emission of radiation that, in view of the classical harmonic nature of the oscillator and force, is coherent.

One must bear in mind, however, that the classical effect of energy transfer from an oscillator to a driving field occurring in the case where the vibrations of the oscillator and the field are in opposite phases yields an expression for the radiant energy proportional to the first power of the field strength of the driving force. On the other hand, the probability of stimulated emission is proportional to the energy density of the stimulating field, that is, to the square of the field strength. Nevertheless, a classical theory of stimulated emission can be built, but only if we consider an ensemble of classical oscillators that group under the action of the external field. Stimulated radiative processes are brought about by the coherent radiation generated by the "blobs" formed, in which the distance between particles is much smaller than the wavelength.

Let us now return to the quantum domain.

The coherence criterion for oscillations is the existence of a constant-phase relationship between them. In quantum theory, in view of the number of particles vs. phase uncertainty relation,

$$\Delta n \Delta \varphi \gtrsim 1/2, \tag{1.13}$$

the phase of the electromagnetic field, φ, is exactly determinate only if the number of photons, n, is indeterminate. It is therefore senseless to speak of the phase of a single photon. But if the phase difference between two waves is known instead of the values of the individual phases, the uncertainty relation allows for determining the overall number of photons while leaving indeterminate the question of what photons belong to what waves. For this reason when coherent electromagnetic radiations add up to a single wave, we speak of the addition of identical photons.

There is no way in which we can distinguish between photons that have the same frequency, direction of propagation, phase, and polarization. The photons, or quanta of electromagnetic radiation, obey the Bose-Einstein statistics, that is, the number of photons per field oscillator (type of vibrations, mode) with the same value of frequency, phase, etc. can be unlimited. The state of the entire radiation field is determined by the number of photons in the mode. It is this fact that for a large number of identical photons makes it possible to resort to the classical interpretation of electromagnetic radiation, for which the principle of linear combination of oscillations, coherent oscillations included, is characteristic. In view of the characteristic properties of bosons (particles that obey the Bose-Einstein statistics), as the number of individual acts of stimulated emission of radiation per unit time increases, the intensity of the stimulating, or primary, wave grows while its frequency, phase, etc. remain unchanged.

A consistent quantum theory of emission and absorption of light was formulated in 1927 by Paul Adrien Maurice Dirac (1902-1984). The qualitative reasoning given above concerning spontaneous and stimulated radiation and the coherence of the stimulated radiation is rigorously discussed in §61 and §62 in his well-known book *The Principles of Quantum Mechanics* (4th ed., Oxford: Oxford Univ. Press, 1958).

However, important results can be obtained as well by a semiclassical treatment, whose application on the whole is characteristic of quantum electronics. In this approach the system consisting of a particle and the radiation field is split into two parts, the quantum particle and the classical radiation field. The particle is described by a wave function satisfying the Schrödinger equation whose constituent Hamiltonian describes the inter-action with the field, which is considered a perturbation. The equation describes the effect of the field on the particle and, at least in principle, allows calculating the dipole moment of the particle. It is then assumed that the dipole emits radiation in the classical manner, and this, in turn, makes it possible to allow for the effect of the particle on the field. Here the conservation of the coherence of the emitted photons is easily traced. The field brings to life an oscillating dipole moment that is coherent to the driving force. The oscillating dipole, in turn, emits a coherent field.

This coherence, as is known, produces all the many properties of lasers that distinguish them so markedly from ordinary light sources.

Problems to Lecture 1

1.1. Table 1.1 represents a wavelength and frequency scale for electromagnetic waves ranging from radio waves to X-rays. Fill in the empty spaces.

Table 1.1

Range	Radio emission	NH$_3$ maser	Lasers				UV emission	X-ray emission
			CO$_2$	Nd:YAG	He-Ne	KrCl		
λ, μm								
log ν								

1.2. Using the data from Table 1.1, find by how many times the probability of spontaneous emission in the case of an Nd : YAG laser is greater than in the case of an NH_3 maser.

1.3. Suppose that two excited levels are in thermodynamic equilibrium. Find the ratio of the upper-level population to the lower-level population if the energy gap between the levels corresponds to a maser transition frequency of 24 GHz at 300 K, 77 K, and 4.2 K.

1.4. Find the ratio of the same quantities if the energy gap between the levels corresponds to the lasing frequency of a CO_2 laser ($\lambda = 10.6$ μm) and the lasing frequency of an argon laser ($\lambda = 0.488$ μm).

1.5. Calculate the ratio of the equilibrium population difference in a two-level system at $T = 300$ K to the overall number of particles, assuming that the level degeneracy multiplicities are the same, for the cases where the energy gap between the levels is 10, 1, 0.1, 0.01, and 0.001 eV. What are the wavelength and frequency of the radiation in each case?

1.6. Basing your reasoning on the thermodynamic approach to Planck's formula, derive Eq. (1.10).

1.7. Using only the assumption about the discrete nature of energy and the fact that photons obey the Bose-Einstein statistics, derive Planck's formula for the energy density of thermal equilibrium blackbody radiation.

LECTURE 2

Linewidth

The time-energy uncertainty relation, natural lifetime, and spontaneous emission linewidth. Lorentzian lineshape. Probability of stimulated transitions in monochromatic radiation. Homogeneous and inhomogeneous line broadening. Gaussian lineshape in the case of Doppler broadening

Let us consider the problem of the width of absorption or emission lines.

Up till now we have spoken of a two-level system without taking into account that levels have finite widths. Any processes that reduce the lifetime of particles on levels broaden the lines of the respective transitions. Indeed, the energy of a state can be defined only by a measuring process shorter than the lifetime in this state, τ. Then, in accordance with the time-energy uncertainty relation,

$$\Delta E \Delta t \gtrsim \hbar, \tag{2.1}$$

the uncertainty in determining the energy cannot be smaller than $\hbar/\tau$. The uncertainty leads to an uncertainty in the transition frequency of $1/2\pi\tau$. The time constant τ is the measure of the time necessary for an excited system to give off its excess energy and is determined by the rates of spontaneous emission and nonradiative relaxation transitions.

In the absence of external forces spontaneous emission determines the lifetime of a state. Therefore, the smallest possible or, as it is sometimes called, natural linewidth $\Delta\nu_0$ is determined by the probability of a spontaneous transition A_{21} as follows:

$$\Delta\nu_0 = A_{21}/2\pi. \tag{2.2}$$

The natural linewidth usually plays a significant role only at high frequencies ($A_{21} \propto \nu^3$) and for superallowed transitions. Ordinarily we can ignore the effect of spontaneous emission on linewidths since, in reality, relaxation transitions shorten lifetimes more effectively.

As mentioned earlier, in systems with discrete energy levels, in addition to spontaneous and stimulated transitions, nonradiative relaxation transitions may play a significant role. These transitions occur as a result of the interaction of a quantum particle with its surroundings, and the mechanism of such interactions greatly depends on the type of system. For instance, this may be the interaction of an ion and the lattice in a crystal or collisions between molecules of a gas or liquid. In the final analysis, the result of relaxation processes is the exchange of energy between the subsystem of particles considered and the thermal motion in the entire system, an exchange that leads to thermodynamic equilibrium in the system.

Usually the time required for equilibrium to set in, the lifetime of a particle on a level, is denoted by T_1 and called the longitudinal relaxation time. Such terminology has a long history, dating from the days when nuclear magnetic resonance, NMR, and electron paramagnetic resonance, EPR, were studied. Longitudinal relaxation corresponds to the situation in which the high-frequency magnetization vector of the system moves along the direction of the constant external field. There is also a transverse relaxation time T_2, which corresponds to the motion of the magnetization vector in a plane perpendicular to the direction of the constant external field.

As a rule one is interested in the behavior of a large ensemble of particles, more precisely, the electric or magnetic dipole moment of this ensemble. The total moment is determined by the phase relations between the moments of individual particles. Let us assume that at time zero all particles of the ensemble have dipole moments that oscillate in phase. This phase feature of the oscillations will in time break down. Time T_2 is the measure of the time interval during which the particles will acquire random phases (with respect to each other).

Any process contributing to the relaxation time T_1, that is, any process in which the particles lose their energy, contributes to loss of phase. Hence,

$$T_2 \lesssim T_1. \tag{2.3}$$

But there are interactions that increase the randomness without changing the energy of the ensemble of particles, interactions that disrupt the phase relations between the particles. For instance, in gas-kinetic collisions one molecule undergoes a $2 \rightarrow 1$ transition while another undergoes a $1 \rightarrow 2$ transition. The system as a whole does not change its energy, but the phase information is lost, the phase memory is disrupted. It can be assumed that T_2 is the time in which the phase memory is intact, or the time it takes the energy to transfer from particle to particle along the level.

In all systems where the particle-to-particle interactions (spin-spin coupling, dipole-dipole interactions in general, elastic collisions) are essential, $T_2 < T_1$ is often $T_2 \ll T_1$. In molecular beams, where the flying particles do not collide with each other, $T_2 = T_1$, which in turn is equal to the time of transit through the region where a particle interacts with the radiation field. This is true, obviously, only if the natural lifetime is relatively long ($T_1 \ll 1/A_{21}$).

Since T_2 is the shortest relaxation time possible, it defines the linewidth of a transition, $\Delta\nu_{line}$:

$$\Delta\nu_{line} = 1/2\pi T_2. \tag{2.4}$$

In what follows it will be more convenient to use τ to denote the lifetime of a particle on a level and place a subscript "0" if we are speaking of the natural lifetime (the lifetime with respect to spontaneous emission), τ_0. The relaxation time, which determines the linewidth of a transition in any process, will be denoted by $1/2\pi\Delta\nu_{line} = 1/\Delta\omega_{line}$.

Thus, the fact that a particle stays in an excited energy state only for a finite time interval leads to energy-level broadening. Emission from broadened levels acquires a spectral width. The most general and fundamental mechanism that limits the lifetime of a particle on an excited level from above is spontaneous emission, which must, therefore, have a spectral width corresponding to the rate of spontaneous decay acts.

Quantum electrodynamics makes it possible to calculate the spectral distribution of spontaneous emission photons originating from a level of width

$$\Delta E = \hbar/\tau_0. \tag{2.5}$$

The shape of a spontaneous emission line proves to be Lorentzian with a spectral width

$$\begin{aligned} \Delta\nu_{line} &= \Delta E/h \\ &= 1/2\pi\tau_0. \end{aligned} \tag{2.6}$$

The Lorentzian lineshape is characterized by a form factor

$$q(\nu) = \frac{1}{2\pi} \frac{\Delta\nu_{line}}{(\nu - \nu_0)^2 + \Delta\nu^2_{line}/4} \tag{2.7}$$

and resembles a resonance curve with its peak at $\nu = \nu_0$ and falling off to half the peak value at frequencies $\nu = \nu_0 \pm \Delta\nu_{line}/2$. Obviously, the total width of the curve at half-height is $\Delta\nu_{line}$.

If we bear in mind that not only the upper of the two levels can decay spontaneously but the lower level can too if it is not the ground level, then $\Delta\nu_{line}$ in (2.7) is the sum of two terms, each determined by the decay rate of the level, $1/\tau_{02}$ and $1/\tau_{01}$:

$$\Delta\nu_{line} = \frac{1}{2\pi}\left(\frac{1}{\tau_{02}} + \frac{1}{\tau_{01}}\right). \tag{2.8}$$

The expression for the form factor, (2.7), can easily be obtained in the classical approximation.

The equation of motion of an oscillating dipole with radiative decay is reduced to the equation of a harmonic oscillator with viscous damping:

$$\ddot{x} + \gamma\dot{x} + \omega_{01}x = 0, \tag{2.9}$$

where ω_{01} is the circular frequency of the electron vibrating in the dipole, which consists of the electron and a nucleus, and γ the classical damping coefficient of the dipole's radiation. The solution to this equation is well-known:

$$x = C \exp\left(-\gamma t/2\right) \exp\left(j\omega_1 t\right), \tag{2.10}$$

where $\omega_1^2 = \omega_{01}^2 - (\gamma/2)^2$, C is an arbitrary constant amplitude, and j is the unit imaginary number $(j^2 = -1)$. Since the intensity of the radiation emitted by this Lorentzian oscillator is proportional to the square of the peak value of oscillations of the electron in the oscillator, we conclude that the radiation intensity falls off exponentially in time with a characteristic time constant

$$\tau = 1/\gamma. \tag{2.11}$$

This is the mean lifetime of the excited state.

The time dependence (2.10) corresponds to a spectral distribution function $g(\omega)$ easily obtainable through the Fourier transform of $x(t)$:

$$g(\omega) = \frac{1}{\sqrt{2\pi}} \int_{-\infty}^{\infty} x(t) \, e^{-j\omega t} \, dt$$

$$= \frac{jC/\sqrt{2\pi}}{\omega_1 - \omega + j\gamma/2}. \tag{2.12}$$

In calculating (2.12) we allowed for the fact that (2.10) was obtained for t nonnegative. Formula (2.12) gives the spectral distribution of the amplitudes, and the spectral distribution of the radiation intensity is the square of the absolute value of the spectral distribution function. Hence,

$$G(\omega) = g(\omega) g^*(\omega)$$

$$= \frac{C^2/2\pi}{(\omega_1 - \omega)^2 + \gamma^2/4}. \tag{2.13}$$

If we introduce the notation $\gamma = 2\pi\nu_{\text{line}}$, $\omega_1 = 2\pi\nu_0$, and $\omega = 2\pi\nu$ and set $C^2 = 4\pi\Delta\nu_{\text{line}}$, we arrive at formula (2.7) for the form factor $g(\nu)$.

These are the properties of spontaneous emission. Its intensity is frequency dependent. Hence, its probability is frequency dependent, too, and has a certain spectral density

$$W_\nu^{\text{spon}} = q(\nu) W^{\text{spon}}$$

$$= A_{21} q(\nu). \tag{2.14}$$

We must also require that

$$\int_\nu W_\nu^{\text{spon}} \, d\nu = W^{\text{spon}}. \tag{2.15}$$

All this leads to the following normalization condition for the form factor $q(\nu)$:

$$\int_0^\infty q(\nu) \, d\nu = 1, \tag{2.16}$$

which has already been taken into account in (2.7) on the assumption that $\nu_0 \gg \Delta\nu_{\text{line}}/2$.

Next, as we have already established, the probabilities of spontaneous emission and stimulated emission are linked. This was proved on the grounds of extremely general thermodynamic considerations. Hence the probability of stimulated emission is also frequency dependent and has a spectral density

$$
\begin{aligned}
W_\nu^{\text{stim}} &= q(\nu) W^{\text{stim}} \\
&= q(\nu) B_{21} \varrho_\nu .
\end{aligned}
\tag{2.17}
$$

Here

$$
W^{\text{stim}} = \int_\nu q(\nu) \varrho_\nu B_{21} d\nu .
\tag{2.18}
$$

If the stimulated radiation is monochromatic,

$$
\varrho_\nu = \varrho \delta(\nu - \nu_0),
\tag{2.19}
$$

where ϱ is the radiative energy density, and $\delta(x)$ the Dirac delta function. The integral (2.18) can then be easily evaluated, and

$$
\begin{aligned}
W^{\text{stim}} &= q(\nu_0) B_{21} \varrho \\
&= 2 B_{21} \varrho / \pi \Delta \nu_{\text{line}} .
\end{aligned}
\tag{2.20}
$$

A reduction in the lifetime, which results in a finite linewidth $\Delta \nu_{\text{line}}$, decreases the probability of stimulated transitions caused by a monochromatic radiation field and is inversely proportional to the linewidth.

Line broadening associated with the finiteness of the lifetime of states linked by a transition is said to be homogeneous. Each atom in such a state emits a line, as it goes from the upper state to the lower, with a total width $\Delta \nu_{\text{line}}$ and a spectral shape $q(\nu)$. Similarly, each atom in the corresponding lower state absorbs radiation, as it moves from the lower state to the upper, represented by a line of total width $\Delta \nu_{\text{line}}$ and spectral shape $q(\nu)$. It is impossible to assign a definite spectral component in $q(\nu)$ to a single atom. In homogeneous broadening, irrespective of its nature, the form factor is a unique spectral characteristic of both a single atom and the entire ensemble of atoms. A variation in this broadening, possible in principle in an external perturbation of the ensemble, occurs simultaneously and in the same manner for all the atoms in the ensemble. Examples of homogeneous broadening are the natural linewidth and collision broadening in gases.

Inhomogeneous broadening originates in quite different circumstances. The spectral lines observed in experiments may be a structureless combination of several spectrally unresolvable homogeneously broadened lines. In such cases each particle emits or absorbs radiation not within the entire line experimentally observed. Such a line is said to be inhomogeneously broadened. The reason for inhomogeneous broadening may be any process that leads to a difference in the conditions of emission (or absorption) for a fraction of the otherwise identical atoms in the ensemble being studied, or the presence in the ensemble of atoms with close but different spectral properties (a hyperfine structure of any kind), atoms whose homogeneously broadened spectral lines overlap only partially. The term "inhomogeneous broadening" originated in NMR spectroscopy, in which broadening of this type occurs because of inhomogeneities of the external magnetizing field within the sample studied. Of similar origin is the inhomogeneous broadening in impure luminescent crystals, in which the inhomogeneity of the intracrystalline electric field leads to a difference in the Stark shifts of frequencies of the radiation emitted by impurity centers positioned in different parts of the crystal sample. A classical example of inhomogeneous broadening is Doppler broadening, characteristic of gases at low pressures and/or high frequencies.

The atoms (molecules, ions) of a gas are always in thermal motion. The linear Doppler effect results in a shift in the frequency of the radiation emitted by particles flying toward the observer at a speed u by $\nu_0 u/c$, where ν_0 is the frequency of the radiation emitted by a particle at rest, and c the speed of light. Natural broadening transforms the radiation at frequency ν_0 into a spectral line, but this broadening is homogeneous and the entire line experiences a shift in frequency by $\nu_0 u/c$. Since the atoms (molecules, ions) move in a gas with different velocities, the frequency shifts in their radiation are different and the overall shape of the spectral line is determined, on the whole, by the velocity distribution of the particles. Strictly speaking, this is true only if the natural linewidth is considerably smaller than the Doppler shifts in frequency, which is usually the case. If by $p(u)$ we denote the velocity distribution function of the particles, then $p(u)$ is linked to the form factor $q(\nu)$ of the Doppler line by the following simple formula:

$$q(\nu)\,d\nu = p(u)\,du. \tag{2.21}$$

The observed frequency is

$$\nu = \nu_0(1 + u/c). \tag{2.22}$$

Hence, $u = c(\nu - \nu_0)/\nu_0$ and $du = (c/\nu_0)\,d\nu$, with the result that

$$q(\nu) = \frac{c}{\nu_0} p\left(c\frac{\nu - \nu_0}{\nu_0}\right). \tag{2.23}$$

If the velocity distribution is Maxwellian,

$$p(u) = \frac{1}{u_0\sqrt{\pi}} \exp\left[-\left(\frac{u}{u_0}\right)^2\right], \tag{2.24}$$

where the mean thermal velocity

$$u_0 = \sqrt{2kT/m}. \tag{2.25}$$

Here k is the Boltzmann constant, T the temperature of the gas, m the mass of an atom (molecule) of the gas. Strictly speaking, the Maxwellian distribution is valid only when the system is in thermal equilibrium, but deviations from this state are usually minor even for excited (emitting radiation) particles. In any case, quantitative estimates that use the Maxwellian distribution have proved sufficiently accurate. Combining (2.25), (2.24), and (2.23), we easily find that

$$q(\nu) = \frac{c}{u_0\nu_0\sqrt{\pi}} \exp\left[-\frac{c^2}{u_0^2}\left(\frac{\nu - \nu_0}{\nu_0}\right)^2\right]$$

$$= \frac{1}{\Delta\nu_T\sqrt{\pi}} \exp\left[-\left(\frac{\nu - \nu_0}{\Delta\nu_T}\right)^2\right], \tag{2.26}$$

where we have introduced $\Delta\nu_T$ to designate the characteristic linewidth equal to the Doppler frequency shift at the mean thermal velocity of the radiating particle:

$$\Delta\nu_T = \nu_0 u_0/c. \tag{2.27}$$

A line whose shape is determined by the form factor (2.26) is known as the Doppler broadened line. Its shape, as (2.26) implies, is described by the Gaussian function and is symmetric with respect to the central frequency ν_0. The q vs. ν curve represented by (2.26) falls off much more rapidly for strong detunings from ν_0 than for the Lorentzian lineshape (2.7). Near ν_0 the Gaussian curve is flatter than the Lorentzian. Obviously, the width of the Gaussian curve is determined by parameter $\Delta\nu_T$, since the value of q decreases e-fold as we move away from the curve's center by $\Delta\nu_T$. If we define, as is customary, the width of a line (usually called the full width half magnitide, or FWHM) as the separation between two such points (in our case symmetric points) from the central frequency, points at which the intensity is half the peak value, in the case of Doppler broadening we can easily find, using (2.26), that this width is

$$\Delta\nu_D = \Delta\nu_T 2\sqrt{\ln 2}$$

$$= 2\nu_0 \sqrt{\frac{2kT}{mc^2} \ln 2}. \tag{2.28}$$

Note that the Gaussian lineshape in the form (2.26) is normalized to unity, $\int_0^\infty q(\nu)\,d\nu = 1$, on the assumption that $\nu_0 \gg \Delta\nu_T$. Figure 2.1 shows

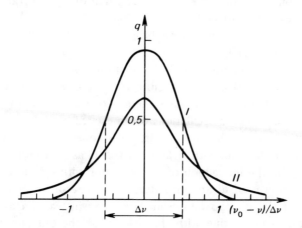

Figure 2.1. Gaussian (*I*) and Lorentzian (*II*) lineshapes (normalized to unity) for the same width at half-height. This width is marked $\Delta\nu$ along the horizontal axis. The scale along the vertical axis is chosen in units of $1/\Delta\nu$.

the lineshapes for homogeneous (2.7) and inhomogeneous (2.26) broadening in the case where $\Delta\nu_{\text{line}} = \Delta\nu_{\text{D}}$.

Doppler broadening becomes more prominent as the frequency grows. In the visual range of the spectrum at medium temperatures (300-600 K) the value of $\Delta\nu_{\text{D}}$ is about 0.8-1.5 GHz. For purposes of reference it is convenient to use formula (2.28) recalculated for wavelengths:

$$\Delta\lambda_{\text{D}}/\lambda_0 = 7 \times 10^{-7}\sqrt{T/M}, \tag{2.29}$$

with M the molecular weight of the gas. Since $d\lambda/\lambda = -d\nu/\nu$, we can write an equivalent formula in terms of frequencies:

$$\Delta\nu_{\text{D}}/\nu_0 = 7 \times 10^{-7}\sqrt{T/M}. \tag{2.30}$$

The fact that this dependence involves a square root implies that for rough estimates $\Delta\nu_{\text{D}}/\nu_0$ (or $\Delta\lambda_{\text{D}}/\lambda_0$) can be assumed constant and equal to $(2\text{-}3) \times 10^{-6}$ in conditions, say, of glow discharge in a gas.

Here are two examples that illustrate the above reasoning. For a CO_2 laser (CO_2 molecules emitting radiation at about 400 K on a wavelength of 10 μm) $\Delta\nu_{\text{D}}$ is about 60 MHz, and for a helium-neon laser (neon atoms emitting radiation at about 400 K on a wavelength of 0.63 μm) the Doppler linewidth reaches 1.35-1.40 GHz.

Now let us return to homogeneous broadening. What follows will make it clear that the natural linewidth can usually be ignored up to the very high frequencies corresponding to UV radiation. For gases homogeneous broadening is determined in real conditions by collision processes. Among the various types of collisions in gases there is one in which collision leads to a change in phase of the oscillations of an excited particle. In the classical approximation, where the excited particle is considered an oscillator vibrating with a constant amplitude and a fixed phase, variations in phase lead to a change in the interaction of the oscillator with the electromagnetic field. This variation is of a random nature. If as a result of collisions (one or several) the initial phase relationship breaks down, we can assume that a new oscillator is interacting with the field while the old one has disappeared. Hence, the mean free-transit time of the oscillator between two phase-disrupting collisions is the mean lifetime of gas particles with respect to collisions of this type. This time, τ_{col}, determines, in accordance with formula (2.4), the collision-broadened linewidth $\Delta\nu_{col}$:

$$\Delta\nu_{col} = 1/2\pi\tau_{col}. \tag{2.31}$$

Since gas-kinetic collisions constitute a random process acting, on the average, in the same way on all gas molecules (atoms) of the same type (provided the type of collision partner does not change), collision broadening is homogeneous.

The value of τ_{col} is determined by the collision gas-kinetic cross section of the phase-disrupting process, σ_{col}, the thermal-motion velocity of a gas particle, u, and the gas density n:

$$\tau_{col} = 1/n\langle u\sigma_{col}\rangle, \tag{2.32}$$

where the angle brackets denote averaging over velocities. In simple estimates we can assume that u in (2.32) is the mean thermal velocity u_0, with the result that $1/\tau_{col} = nu_0\sigma_{col}$. The value of σ_{col} requires a special experimental study in each case.

The fact that τ_{col} is inversely proportional to the gas density is extremely important for quantum electronics. Such dependence ensures that the collision-broadened linewidth $\Delta\nu_{col}$ is directly proportional to the gas pressure and may reach substantial values. For instance, in the case of the CO_2 laser already mentioned the rate of linewidth growth due to the increase in collision frequency with pressure is 4-8 MHz/torr. At a total pressure of 10-15 torr for a CO_2 laser, the homogeneous collision-broadened linewidth exceeds the inhomogeneous (Doppler) linewidth.

To conclude this lecture we note that in some cases it is necessary to allow for what is known as the transit linewidth. In atomic (molecular) beam spectroscopy the time of transit of the beam of particles through the region of space occupied by an electromagnetic field may be short. Obviously, the mean time interval in the course of which the beam interacts with the field, or transit time τ_{tr}, corresponds to the following homogeneous broadening:

$$\Delta\nu_{tr} = 1/2\pi\tau_{tr}. \tag{2.33}$$

Problems to Lecture 2

2.1. Imagine a substance in which identical atoms are rigidly fixed and do not interact with each other. In this case natural line broadening plays the dominant role. The lifetime of an allowed electric-dipole transition in the middle of the visible part of the spectrum is known to be 10 ns. What is the linewidth of such a transition in such a substance?

2.2. Generation of X-rays was recently achieved for the first time. The lasing wavelength was 10 nm. Using the results obtained in the previous problem, estimate the natural linewidth in this X-ray laser.

2.3. The natural linewidth of the radiation of a CO_2 laser ($\lambda = 10.6$ μm) produced by transitions between the vibrational levels of the ground electronic state is 50 MHz. What is the spontaneous lifetime of the upper laser level?

2.4. What mechanism of spectral line broadening is predominant in cubic crystals? In glasses? What is the cause of inhomogeneous broadening of spectral lines in impurity crystals?

2.5. The nonuniformity of internal electrostatic fields leads to strong inhomogeneous broadening. The line of luminescence of an Nd ion in glass corresponding to the $^4F_{3/2} \rightarrow {}^4I_{11/2}$ ($\lambda = 1.06$ μm) has a half-width $\Delta\lambda \approx 30$ nm. What is the effective transverse-relaxation time of an Nd ion?

2.6. Using the data of the previous problem, determine the width of the upper laser level $^4F_{3/2}$ of Nd ions in glass if the lower laser level decays at a rate $1/\tau_{02} = 10^8$ s^{-1}.

2.7. Suppose that a highly symmetric crystal is placed in a medium whose temperature is that of liquid helium. What is the nature of the broadening of the R line (the $^2E \rightarrow {}^4A_2$ transition) of chromium ions Cr^{3+} in such a crystal?

2.8. Suppose that nitrogen is being cooled to $T = 20$ K in a supersonic jet and that transverse electric pulses are sent through the nitrogen flow at certain intervals. Even without mirrors laser radiation with $\lambda = 337.1$ nm may be generated in such a system owing to the extremely high gain. Estimate the Doppler linewidth in such a superradiant nitrogen laser.

2.9. Estimate the collision-broadened linewidth of radiation in a gas kept at a pressure of 10 torr and a temperature of 300 K, assuming that the collision cross section is equal to the gas-kinetic one.

2.10. When the nitrogen pressure in a nitrogen laser exceeds 100 torr at $T = T_{room}$, the collision-broadened linewidth becomes greater than the Doppler linewidth. To what temperature must the gas be heated so that the Doppler linewidth at the same pressure is twice the collision-broadened?

2.11. Suppose that in a CO_2 laser the lasing wavelength is 10.6 μm and the Doppler linewidth (50 MHz) is greater than the collision-broadened (36 MHz). Does the Doppler linewidth become smaller in this case as a result of collisions? (*Hint*: Compare the mean free path with the radiation's wavelength.)

LECTURE 3

Amplification

Absorption and amplification. Active medium. Absorption cross section. Saturation effect. Saturation intensity. Pulsed mode. Saturation energy

An equilibrium quantum system absorbs the energy of external radiation, that is, if probability of stimulated transitions per particle from higher levels to lower levels (accompanied by energy emission) is equal to that of stimulated transitions per particle from lower levels to higher levels (accompanied by energy absorption), the overall number of transitions from the lower levels to the higher exceeds the number of transitions in the opposite direction because there are more particles on the lower levels than on the upper.

Indeed, the variation of the energy of an external radiation field per unit volume of the quantum system is determined by the difference in emitted and absorbed energies in individual upward and downward transitions. Since the emitted power is $n_2 W_{21} h\nu$ and the absorbed is $n_1 W_{12} h\nu$, then, in accordance with Eqs. (1.2), (1.3), (1.10), and (2.20), the rate of the energy density variation is

$$\frac{d\varrho}{dt} = \left(\frac{n_2}{g_2} - \frac{n_1}{g_1} \right) \frac{g_1 2B_{12} h\nu\varrho}{\pi \Delta\nu_{\text{line}}}, \tag{3.1}$$

which, in accordance with the Boltzmann formula (1.4) valid in thermodynamic equilibrium, is negative. This means that the energy of the external field is absorbed. Later the reader will see that when the energy absorption plays a noticeable role, the populations n_1 and n_2 change, and consequently the energy absorption rate changes as well. However, for the present we will not consider so strong an influence of the field on matter, assuming that relaxation processes return particles to the lower levels at a rate much higher than $d\varrho/dt$, which ensures the continuity of absorption of the energy of external electromagnetic radiation and transformation of this energy into heat.

Thus, in thermodynamic equilibrium $d\varrho/dt < 0$. To increase the radiation energy we must ensure that

$$n_2/g_2 > n_1/g_1. \tag{3.2}$$

In the absence of degeneracy this means that the population of the upper level must be greater than that of the lower. If degeneracy is present, the number of particles per nondegenerate state of the upper level must exceed the population of each nondegenerate state of the lower level.

Thus, an increase in the energy density of the external radiation field occurs in the quantum system when the equilibrium distribution of populations is disrupted in such a manner that the upper states are populated more heavily than the lower, that is, in the event of population inversion.

Systems of quantum particles in which of at least two energy levels the higher is populated more heavily than the lower are known as systems with population inversion. Sometimes such a system is said to be a negative temperature system. Such terminology formally stems from the application of the Boltzmann distribution to nonequilibrium systems with population inversion. Indeed, from the Boltzmann formula (1.4) it follows that for $E_2 > E_1$ the condition $n_2/g_2 > n_1/g_1$ is met automatically if one assumes that $T < 0$.

For our purposes, however, it is more important that systems with population inversion are systems with negative absorption, or amplification. For radiation propagating in the form of a travelling wave in the direction z at a speed c, the absorption coefficient α is defined thus:

$$\alpha = -\frac{1}{I}\frac{dI}{dz}, \tag{3.3}$$

where I [W/cm^2] is the irradiation intensity, or simply intensity (the energy flux per unit surface). Since $I = \text{const} \times \varrho$ and $dz = c\,dt$, we have

$$\alpha = -\frac{1}{c\varrho}\frac{d\varrho}{dt}. \tag{3.4}$$

Equation (3.1) then yields

$$\alpha = \left(\frac{n_1}{g_1} - \frac{n_2}{g_2}\right)\frac{g_1 2B_{12}h\nu}{c\pi\Delta\nu_{\text{line}}}. \tag{3.5}$$

If the condition $n_2/g_2 > n_1/g_1$ is met, the absorption coefficient α is negative and, hence, negative absorption corresponds to amplification, as expected. The absorption coefficient can then be called the amplification factor, or gain.

In view of the properties of stimulated emission, the amplification obtained as a result of population inversion is coherent. When such radiation propagates in a medium with negative absorption, its amplitude increases exponentially with an amplification increment equal to $\alpha/2$, or

$$\left(\frac{n_2}{g_2} - \frac{n_1}{g_1}\right) \frac{g_1 B_{12} h\nu}{c\pi\Delta\nu_{\text{line}}}. \tag{3.6}$$

For the sake of reference, here are formulas that relate the electromagnetic field strength E to intensity I:

$$E\,[\text{V/cm}] = 27\sqrt{I\,[\text{W/cm}^2]} \tag{3.7a}$$

for a linearly polarized wave, and

$$E\,[\text{V/cm}] = 19\sqrt{I\,[\text{W/cm}^2]} \tag{3.7b}$$

for a circularly polarized wave.

Thus, for amplification, that is, for inversion of the sign of absorption, there must be population inversion. To create this population inversion an external agent is required. Irrespective of the concrete mechanism of population inversion, this external agent must overcome processes aimed at restoring equilibrium in the population difference. To prevent such restoration processes one must expend energy, the pumping energy supplied by an external power source.

A collection of quantum particles that possesses the property of population inversion, that is, a medium with negative losses of energy of the radiation propagating in it, is known in quantum electronics as the active media.

Let us consider in greater detail the absorption coefficient in the equilibrium case. We will again restrict our discussion to a two-level system. In accordance with the Boltzmann distribution for a two-level system kept at a constant temperature T,

$$\alpha = n \frac{g_1 [1 - \exp(-h\nu/kT)] \, 2B_{12} h\nu}{g_1 + g_2 \exp(-h\nu/kT) \, c\pi\Delta\nu_{\text{line}}}, \tag{3.8}$$

where $n = n_1 + n_2$ is the total number density of the particles on both energy levels. At radio frequencies, as a rule, $h\nu \ll kT$ and

$$\alpha = n \frac{g_1}{g_1 + g_2} \frac{(h\nu)^2}{kT} \frac{2B_{12}}{c\pi\Delta\nu_{\text{line}}}. \tag{3.9}$$

In the optical range, $h\nu \gg kT$ and

$$\alpha = nh\nu \frac{2B_{12}}{c\pi\Delta\nu_{\text{line}}}. \tag{3.10}$$

For the sake of reference we note that at 300 K the equality $h\nu = kT$ sets in at $\lambda = 48~\mu m$ (which corresponds to a frequency of 6250 GHz or 207 cm^{-1}).

We can express Eq. (3.10) differently. The reader will recall that $g_1 B_{12} = g_2 B_{21}$ and $A_{21} = 8\pi\nu^2 h\nu B_{21}/c^3$, which transforms (3.10) into

$$\alpha = n \frac{g_2}{g_1} \frac{c^2}{4\pi\nu^2} \frac{A_{21}}{\pi\Delta\nu_{\text{line}}}. \tag{3.11}$$

Since $c/\nu = \lambda$ and $A_{21} = 1/\tau_0$, with τ_0 the natural (radiative, spontaneous) lifetime of a particle on the upper level, we have

$$\alpha = n \frac{g_2}{g_1} \frac{\lambda^2}{4\pi} \frac{1}{\pi\Delta\nu_{\text{line}}\tau_0}. \tag{3.12}$$

The product $\Delta\nu_{\text{line}}\tau_0$ is dimensionless. Hence, it is possible to characterize the absorbing properties of a particle by an effective cross section of the particle's interaction with the resonant electromagnetic field. This quantity is known as the absorption cross section and is denoted by σ. By definition,

$$\alpha = n\sigma. \tag{3.13}$$

Then Eq. (3.12) implies (see also Eq. (3.10)) that

$$\sigma = \frac{g_2}{g_1} \frac{\lambda^2}{2\pi} \frac{1}{2\pi\Delta\nu_{\text{line}}\tau_0}$$

$$= h\nu \frac{2B_{12}}{c\pi\Delta\nu_{\text{line}}}. \tag{3.14}$$

Since $2\pi\Delta\nu_{\text{line}}$ always exceeds $1/\tau_0$, the absorption cross section is always smaller than $\lambda^2/2\pi$ (in the optical range it is usually much smaller). The characteristic values of σ lie within a wide range of values $(10^{-12}\text{-}10^{-24}$ cm^2), depending on the spectral range and the type of quantum particle.

The above discussion concerned the so-called linear absorption coefficient, or the low-signal absorption coefficient, with α not depending on signal strength. The fact that α is independent of the intensity of the absorbed radiation corresponds to the well-known Bouguer-Lambert-Beer law. Here we arrive at this law by assuming that the absorbed radiation does not induce any deviations of the energy-level distribution of particle numbers from the thermodynamically equilibrium distribution.

However, the radiation absorbed by a system of particles always disrupts the thermal equilibrium in the system. When the probability of transitions induced by an external field becomes comparable with the probability of relaxation transitions, the equilibrium distribution of the populations is noticeably disrupted. The fraction of energy absorbed by the system decreases in the process, and so does the absorption coefficient, with the result that saturation finally sets in. Obviously, at the limit where the field strength is so high that the probability of stimulated transitions exceeds the probability of relaxation transitions, complete saturation sets in, with

$$n_1/g_1 - n_2/g_2 = 0. \tag{3.15}$$

At $n_1/g_1 = n_2/g_2$ the system becomes transparent for the resonant radiation, there is no absorption or amplification, $\alpha = 0$.

In the long history of optics the Bouguer-Lambert-Beer law and equivalent assumptions were considered axioms until the Soviet physicist Sergei I. Vavilov (1891-1951), long before lasers appeared, put forward and substantiated the idea of possible reduction in absorption accompanying an increase in the radiation intensity. The nonlinear nature of the absorption of light of high intensity prompted Vavilov to introduce the term "nonlinear optics". This term gained wide acceptance after the lasers appeared and initiated rapid development of this new branch of physics.

Let us consider the changes in populations in a two-level system brought about by a resonant electromagnetic field and relaxation and spontaneous transitions. We know that the result of any relaxation mechanisms is energy exchange between the system of particles considered and the thermal vibrations, an exchange that establishes thermal equilibrium throughout the system. As noted earlier, it is the relaxation interactions (together with spontaneous emission when the latter is substantial) that establish an equilibrium distribution of populations and provide the conditions for continued absorption of radiation energy.

Here are the rate (kinetic) equations for the populations n_1 and n_2 of the two energy levels. First we have the conservation law

$$n_1 + n_2 = n. \tag{3.16}$$

Next, the variation of the particle number density on the upper level, n_2, is given by the equation

$$\frac{dn_2}{dt} = -\left(w_{21} + \frac{1}{\tau_0}\right) n_2 + w_{12} n_1 - W_{21} n_2 + W_{12} n_1. \tag{3.17}$$

Here the first term indicates the particles leaving the upper level due to spontaneous decay (probability $1/\tau_0$) and relaxation (probability w_{21}), the second term indicates the particles filling the upper level due to relaxation from the lowel level (probability w_{12}), and the third and fourth terms describe stimulated transitions $1 \rightleftarrows 2$.

Writing W_{12} in the form $2B_{12} \varrho / \pi \Delta \nu_{\text{line}}$, recalling that $g_1 B_{12} = g_2 B_{21}$, and substituting $n - n_2$ for n_1, we get

$$\frac{dn_2}{dt} = -\left(\frac{1}{\tau} + \frac{g_1 + g_2}{g_2} 2B_{12} \frac{\varrho}{\pi \Delta \nu_{\text{line}}}\right) n_2$$

$$+ \left(w_{12} + 2B_{12} \frac{\varrho}{\pi \Delta \nu_{\text{line}}}\right) n, \tag{3.18}$$

where we have introduced the notation

$$\tau = \frac{1}{w_{21} + w_{12} + 1/\tau_0} \tag{3.19}$$

for the effective population relaxation time. In the absence of an external field, as Eq. (3.18) shows, the system relaxes with time τ.

In steady-state conditions ($dn_2/dt = 0$)

$$n_2 = \frac{w_{12}\tau + 2B_{12} \varrho\tau/\pi \Delta \nu_{\text{line}}}{1 + (g_1 + g_2) 2B_{12} \varrho\tau/g_2 \pi \Delta \nu_{\text{line}}} n \tag{3.20}$$

and

$$n_2 = \frac{g_2}{g_1 + g_2} n \tag{3.21}$$

as $\varrho \to \infty$. Accordingly,

$$n_1 = \frac{g_1}{g_1 + g_2} n \tag{3.22}$$

if we recall that $n_1 = n - n_2$, that is, to within the multiplicity of degeneracy the populations of the upper and lower levels even out. Complete saturation sets in.

In the absence of a radiation field ($\varrho = 0$)

$$n_2^0 = w_{12} \tau n$$

$$= \frac{w_{12}}{w_{12} + w_{21} + 1/\tau_0} n \tag{3.23}$$

and, respectively,

$$n_1^0 = \frac{w_{21} + 1/\tau_0}{w_{12} + w_{21} + 1/\tau_0} n. \tag{3.24}$$

Since equilibrium population differences obey the Boltzmann formula (1.4), the probabilities of upward and downward relaxation transitions must satisfy the following condition:

$$\frac{w_{12}}{w_{12} + 1/\tau_0} = \frac{g_2}{g_1} \exp\left(-\frac{E_2 - E_1}{kT}\right). \tag{3.25}$$

If the energy levels E_1 and E_2 have the same degeneracy multiplicities, the downward transitions are always more probable than the upward. For optical frequencies, when $E_2 - E_1 \gg kT$, the probability of an upward relaxation process is extremely low. At radio frequencies ($E_2 - E_1 \ll kT$) w_{12} is slightly lower than w_{21} (the rate $1/\tau_0$ can usually be ignored at radio frequencies).

Let us now return to the case of saturation. Usually we must deal not with the energy density ϱ but with intensity I, or the energy flux per unit surface. Since $I = \varrho c$, it is convenient to write Eq. (3.20) in the form

$$n_2 = \frac{w_{12}\tau + I2B_{12}\tau/c\pi\Delta\nu_{\text{line}})}{1 + (g_1 + g_2)I2B_{12}\tau/g_2 c\pi\Delta\nu_{\text{line}})}\, n, \qquad (3.26)$$

which makes it possible to introduce an effective characteristic of saturation,

$$I_{\text{sat}} = \frac{1}{4}\ \frac{c\pi\Delta\nu_{\text{line}}}{B_{12}\tau}, \qquad (3.27)$$

meaning the saturation intensity or the saturation energy flux per unit surface. Formula (3.14) links I_{sat} with resonance absorption cross section σ:

$$I_{\text{sat}} = h\nu/2\sigma\tau. \qquad (3.28)$$

In accordance with (3.5), the value of resonance absorption (amplification) is given by the formula

$$z = n_2/g_2 - n_1/g_1. \qquad (3.29)$$

Then, allowing for the fact that $n_2 = n - n_1$ and, hence, $z = n_2(g_1 + g_2)/g_1 g_2 - n_1/g_1$, we see that from (3.26) we can easily proceed to a simple formula for z:

$$z = \frac{z_0}{1 + (g_1 + g_2)I/2g_2 I_{\text{sat}}}, \qquad (3.30)$$

where z_0 stands for the population difference in the absence of an external field, that is, at $I = 0$.

We will repeatedly return to (3.30) in our narrative. Its meaning is clear. As the irradiation intensity grows, the initial population difference drops, and the rate of this change is specified by I_{sat}. When the intensity reaches I_{sat}, the initial population difference is reduced by a half. For $I \ll I_{\text{sat}}$ the saturation effect can be ignored.

The quantity I_{sat} (formula (3.28)) allows for a simple physical interpretation, namely, the product of intensity and the absorption cross section measured in units of $h\nu$, or $I\sigma/h\nu$, represents, for continuous irradiation, the mean rate of stimulated absorption acts. When this rate reaches (at

$g_1 = g_2$), as I grows, the value of the rate of relaxation decay of the population of the upper level, $1/\tau$, saturation becomes noticeable.

The above discussion is valid for a homogeneously broadened absorption line saturated as a whole as the intensity increases. Inhomogeneous broadening, for instance a Doppler line, requires a far more complicated analysis.

Note also that sometimes instead of the saturation intensity the inverse quantity is employed

$$S = 1/I_{sat}, \tag{3.31}$$

known as the saturation factor. However, I_{sat} has a more graphic physical interpretation.

The saturation effect plays an important role in quantum electronics. Saturation lowers the absorption coefficient of noninverted resonance absorption systems, reducing them to a transparent state, which often proves extremely useful. But it also lowers the gain of inverted systems, which is usually highly undesirable. Saturation is the nonlinearity that restricts the lasing intensity. Finally, in systems with many energy levels saturation of one resonance transition may lead to population inversion of another transition. We will discuss this last feature in greater detail later.

The order of magnitude of the saturation intensity is determined by the parameters of the transition of the quantum particle considered. In the visible part of the spectrum the value of I_{sat} amounts to 1-2 kW/cm^2 at $\tau = 10^{-6}$ s and $\sigma = 10^{-16}$ cm^2.

Up to this point we have dealt only with the continuous, or continuous-wave (cw), mode. However, the pulsed mode is also of great importance in quantum electronics. A lasing mode can be considered pulsed when the duration of the action of radiation on a quantum system is short compared to the characteristic relaxation time of the system. More precisely, the cw mode (of lasing or irradiation) takes up a time interval noticeably longer than the relaxation time. All the rest is the pulsed mode. In the pulsed mode the saturation effect is characterized by the so-called saturation energy, or saturation fluence.

Let us now go back to Eq. (3.18). Its solution for a pulsed function $\varrho = \varrho(t)$ yields the appropriate n_2 vs. t dependence. For simplicity we take the optical case: $w_{12} = 0$ and $n_2(0) = 0$. We also immediately introduce intensity $I = \varrho c$ and allow for formula (3.27), which gives the steady-state value of I_{sat}. As a result we arrive at the simple formula

$$\frac{dn_2}{dt} = \frac{1}{2} \frac{I}{I_{sat}} \frac{n}{\tau} - \left(1 + \frac{g_1 + g_2}{2g_2} \frac{I}{I_{sat}}\right) \frac{n_2}{\tau} \qquad (3.32)$$

with $I = I(t)$. The solution to this first-order linear differential equation is known. Introducing the notation

$$Q(t) = \left(1 + \frac{g_1 + g_2}{2g_2} \frac{I(t)}{I_{sat}}\right) \frac{1}{\tau}, \qquad (3.33)$$

$$R(t) = \frac{I(t)n}{2I_{sat}\tau} \qquad (3.34)$$

and recalling that $n_2(0) = 0$, we can write $n_2(t)$ in the form

$$n_2(t) = \exp\left[-\int_0^t Q(z)\, dz\right]$$

$$\times \int_0^t R(z) \exp\left[\int_0^t Q(x)\, dx\right] dz. \qquad (3.35)$$

Also,

$$\int_0^t Q(z)\, dz = \frac{t}{\tau} + \frac{1}{2} \frac{g_1 + g_2}{g_2} \int_0^t \frac{I(z)}{I_{sat}\tau}\, dz. \qquad (3.36)$$

The integral $\int_0^t I(z)\, dz$ represents the fraction of the radiation energy received by the system (in our case a two-level system) by time t, or fluence, the irradiation energy density by time t, and is measured in joules per square centimeter, which sets it apart from intensity (the energy flux per unit surface), measured in watts per square centimeter.

Let us introduce the notation

$$F(t) = \int_0^t I(z)\, dz, \qquad (3.37)$$

$$F_{sat} = I_{sat}\tau, \qquad (3.38)$$

where F_{sat} is obviously the saturation fluence. Allowing for Eqs. (3.36)-(3.38), we can write (3.35) in the form

$$n_2(t) = \frac{n}{2} \exp\left[-\frac{t}{\tau} + \frac{g_1 + g_2}{2g_2} \frac{F(t)}{F_{sat}} \right]$$

$$\times \int_0^t \exp\left[\frac{z}{\tau} + \frac{g_1 + g_2}{2g_2} \frac{F(z)}{F_{sat}} \right] \frac{I(z)}{F_{sat}} dz. \tag{3.39}$$

Further analysis requires specifying $I(t)$. If we are interested in the case of short pulses, whose duration τ_{pulse} is much shorter than τ, we can assume in calculating $n_2(\tau_{pulse})$ that the intensity is constant and equal to I_0. Then $F(\tau_{pulse}) = F_{pulse} = I_0 \tau_{pulse}$ means the pulse energy density, $F(z) = I_0 z$, and $I(z) = I_0$ for $0 \le z \le \tau_{pulse}$. As a result,

$$n_2(\tau_{pulse}) = n_2^{pulse}$$

$$= n I_0 \left\{ 1 - \exp\left[-\frac{g_1 + g_2}{2g_2} \frac{F_{pulse}}{F_{sat}} - \frac{\tau_{pulse}}{\tau} \right] \right\}$$

$$\times \left[2 I_{sat} \left(1 + \frac{g_1 + g_2}{2g_2} \frac{I_0}{I_{sat}} \right) \right]^{-1}. \tag{3.40}$$

This result differs from the one obtained in the steady-state case, (3.26)-(3.30), in the term with the exponential function as a factor, which characterizes the effect of the absence of effective relaxation during the irradiation pulse. As the energy of the pulse grows, saturation sets in in the sense that no further increase in n_2 is brought on by an increase in F_{pulse} simply because all particles have been transferred from level "1" to level "2" and no inverse relaxation occurs in the course of τ_{pulse}.

Comparison of (3.40) and (3.26) shows that in the pulsed mode the condition of complete saturation,

$$F_{pulse} \gg F_{sat}, \tag{3.41}$$

differs drastically from that in the cw mode,

$$I \gg I_{sat}. \tag{3.42}$$

It is these considerations that made it possible to introduce the concept of the saturation fluence in the pulsed mode via formula (3.38):

$$F_{sat} = I_{sat} \tau$$
$$= h\nu/2\sigma. \tag{3.43}$$

When the signal is strong, that is, when condition (3.42) has been met, there is a simple formula for n_2:

$$n_2^{pulse} = n \left[1 - \exp\left(-\frac{g_1 + g_2}{2g_2} \frac{F_{pulse}}{F_{sat}} \right) \right] \frac{g_2}{g_1 + g_2}. \tag{3.44}$$

If we know n_2, we can find n_1 and $n_2 - n_1$. If (3.44) holds true,

$$z = \frac{n_2}{g_2} - \frac{n_1}{g_1}$$

$$= z_0 \exp\left(-\frac{g_1 + g_2}{2g_2} \frac{F_{pulse}}{F_{sat}} \right), \tag{3.45}$$

where, as in the case (3.30), z_0 stands for the equilibrium population difference in the absence of an external field, $F_{pulse} = 0$. The marked difference between (3.45) and (3.31) can be explained by the different nature of saturation in the pulsed and cw modes.

Problems to Lecture 3

3.1. Derive formula (3.7a).

3.2. Can quantum amplification be maintained in a two-level system? Consider two cases: (a) steady-state and (b) time-dependent.

3.3. The most widespread crystal laser is the ND^{3+} : YAG laser, whose lasing transition $^4F_{3/2} \rightarrow {}^4I_{11/2}$ ($\lambda = 1.06 \ \mu m$) is well described by a Lorentzian curve with a width of 195 GHz at $T = 300$ K. The luminescence quantum yield η ($\eta = \tau/\tau_0$) from the $^4F_{3/2}$ level amounts to 0.42, and the lifetime on the upper laser level is 0.23 ms. Calculate the $^4F_{3/2} \rightarrow {}^4I_{11/2}$ transition cross section at the center of the line.

3.4. Suppose that the gas in a CO_2 laser is under such pressure that the collision mechanism of broadening of the line of the transition between the vibrational states $00°1$ and $10°0$ ($\lambda = 10.6$ μm) is predominant. The linewidth is 1 GHz. Estimate the spontaneous lifetime of the upper laser level if the transition cross section for this system is 9.5×10^{-18} cm^{-1}.

3.5. Find the absorption cross section for Cr^{3+} in a ruby sample with an ion concentration of 5×10^{18} cm^{-3} if at $\lambda = 0.54$ μm the intensity of the light passing through a plate 0.5-cm thick decreases 22 000-fold.

3.6. Estimate the lifetime of the upper laser level of a ruby laser ($\lambda = 0.69$ μm) if the saturation intensity is 2 kW/cm^2 at an effective cross section of the lasing transition of 2.5×10^{-20} cm^2.

LECTURE 4

Einstein's Coefficients and the Matrix Element of the Transition Operator

Wave functions of stationary states. The Schrödinger equation in the presence of perturbations. First-order perturbation theory. Superposition of wave functions of stationary states. Transition probability. Calculation of Einstein's coefficients for stimulated transitions in a two-level system. The matrix element of the dipole-moment transition operator. The oscillations of the upper level population. Rabi frequency

In the previous lectures we linked the absorption cross section, the amplification coefficient, and the saturation intensity with an Einstein coefficient introduced thermodynamically. However, it is highly important to make a quantum-mechanical study of this problem so as, first, to understand what is most important in quantum mechanics from the standpoint of quantum electronics and, second, to know the tools and methods used in determining the transition probabilities, absorption cross sections, etc.

We have already mentioned the fact that a customary approach to analyzing a system consisting of a particle and radiation field is to divide the system into two parts, the quantum particle and the classical field. Further, the energy of such a system is split into three terms, the internal energy of the particle, the energy of the radiation field, and the energy of the interaction between the two parts. The interaction is considered the perturbation with which one part of the system acts on the other, the perturbation of the radiation field on the particle.

As is known, micro-objects are described by Ψ functions obeying the Schrödinger equation

$$jh \frac{\partial}{\partial t} \Psi = \hat{H}\Psi, \tag{4.1}$$

where $\hat{H}$ is the energy operator.

The Hamiltonian of the system has the form

$$H = H_0 + H', \tag{4.2}$$

where H_0 is the total energy of the particle and field taken separately, and H' the energy of the interaction between the two (the energy reflects the perturbation of the internal energy of the particle by the radiation field). Such decomposition of H can always be done. We will consider the important specific case where the interaction energy H' is small compared to the total energy H_0 of the unperturbed components of the system. Since H_0 is determined by the strength of the interactions between the components of our molecule (atom, ion, etc.), this assumption agrees well with the case where microparticles interact with fields that are not very strong.

The results obtained on the assumption that H' is small compared to H_0 are valid in the so-called first-order perturbation theory.

Let us first consider an unperturbed system possessing energy levels E_n and wave functions Ψ_n. Since both the particle and the radiation field in the absence of any interaction between them can remain indefinitely in a given state, such states are stationary (time-independent). The wave functions of stationary states are special in that they can be represented in the form

$$\Psi_n = \psi_n \exp(-jE_n t/\hbar), \tag{4.3}$$

where the ψ_n depend only on spatial coordinates and satisfy the so-called stationary (time-independent) Schrödinger equation

$$\psi_n(x)E_n = \hat{H}_0(x)\psi_n(x). \tag{4.4}$$

As is well known, for finite quantum systems this equation has a solution only for certain energy values of E_n, which leads to quantization of the energy levels of the system for cases where the system cannot extend to infinity. In other words, the quantization of energy follows automatically from the stationary Schrödinger equation for finite systems.

Certain wave functions ψ_n correspond to the energy levels E_n. The functions are continuous, smooth (differentiable), and orthonormal,

$$\int \psi_m^* \psi_n \, dx \, dy \, dz = \begin{cases} 1 & \text{if } m = n, \\ 0 & \text{if } m \neq n, \end{cases} \tag{4.5}$$

the last property reflecting the fact that the stationary states are independent of each other. An unperturbed system can occupy only one specific state at each moment in time.

The situation changes when there is interaction in the system. The corresponding wave function must then satisfy the Schrödinger equation

$$jh \frac{\partial}{\partial t} \Psi' = (\hat{H}_0 + \hat{H}')\Psi'. \tag{4.6}$$

In general this equation cannot be solved analytically. However, because the interaction energy H' is small compared to the energy H_0 of the unperturbed system, the solution can be found by expanding the wave function Ψ' of the perturbed system in a series whose terms are solutions of the unperturbed Schrödinger equation

$$j\hbar \, \frac{\partial}{\partial t} \, \Psi = \hat{H}_0 \Psi, \qquad (4.7)$$

i.e., they are Ψ functions of stationary states.

In accordance with the approach taken in previous lectures, we restrict our analysis to the case where the particles have only two energy levels. Of course, it must be borne in mind that in reality there can be no systems having only two energy levels. But when the interaction with the field is of a resonant nature and the lines are sufficiently narrow, the two-level system proves to be a fairly good approximation.

A two-level system has two stationary states, that is, two Ψ functions, say Ψ_1 and Ψ_2. Hence, we must look for the solution to Eq. (4.6) in the form

$$\Psi' = a\Psi_1 + b\Psi_2. \qquad (4.8)$$

If prior to perturbation the particle was on the lower level E_1, we have $a = 1$ and $b = 1$; otherwise, we have $a = 0$ and $b = 1$. In general the expansion coefficients a and b are functions of time: $a = a(t)$ and $b = b(t)$. In the presence of interaction the particle and the field exchange energy and the energy state of the particle is time-dependent. After this interaction has continued for some time, a nonzero probability appears that the particle is in a stationary state differing from the initial one. For example, suppose that the system that prior to interaction was in state "1" is now described not by Ψ_1 but by Ψ_2. This means there was a transition from level E_1 to level E_2, and $|b(t)|^2 = b(t)b(t)^*$ is the probability of this transition.

Thus, we substitute Eq. (4.8) into Eq. (4.6). After collecting like terms and allowing for the fact that Ψ_1 and Ψ_2 satisfy Eq. (4.7), we get

$$j\hbar\Psi_1 \frac{\partial a}{\partial t} + j\hbar\Psi_2 \, \frac{\partial b}{\partial t} = a(t) \, \hat{H}'\Psi_1 + b(t)\hat{H}'\Psi_2. \qquad (4.9)$$

Now we multiply this equation into Ψ_2^* and integrate the product over the entire space. Allowing for the orthonormality property (4.5) and the fact that in view of (4.3)

$$\Psi_1 = \psi_1 \exp\left(-\frac{jE_1t}{h}\right), \quad \Psi_2 = \psi_2 \exp\left(-\frac{jE_2t}{h}\right), \quad (4.10)$$

we get

$$jh\frac{db}{dt} = a(t)\int \psi_1\hat{H}'\psi_2^* \exp\left(-\frac{jE_1t}{h} + \frac{jE_2t}{h}\right) dx\ dy\ dz$$

$$+ b(t)\int \psi_2\hat{H}'\psi_2^*\ dx\ dy\ dz. \quad (4.11)$$

If we had multiplied Eq. (4.9) into Ψ_1^*, we would have arrived at a similar equation for da/dt.

Now suppose that at time $t = 0$ the particle was on level E_1 and the perturbation was switched on, that is, let $a(0) = 1$ and $b(0) = 0$. Then, if the perturbation acts for a short time, Eq. (4.11) can be solved by assuming that the functions $a(t)$ and $b(t)$ in it are replaced by their initial values. Hence,

$$jh\frac{db}{dt} = \int \psi_1\hat{H}'\psi_2^* \exp\left(-\frac{j(E_1 - E_2)t}{h}\right) dx\ dy\ dz. \quad (4.12)$$

The right-hand side of this equation contains the integral $\int \psi_1\hat{H}'\psi_2^*\ dx\ dy\ dz$ taken over the entire space. This integral is known as the matrix element of the interaction operator for the $1 \rightarrow 2$ transition.

Let the perturbation energy H' reflect the dipole interaction of the particle with an oscillating electromagnetic field. If the dipole moment is μ and the field is $\mathbf{E}[\exp(j\omega t) + \exp(-j\omega t)]$, the interaction energy is the scalar product of these two vectors:

$$H' = -\mu\cdot\mathbf{E}[\exp(j\omega t) + \exp(-j\omega t)]. \quad (4.13)$$

We assume, for the sake of simplicity, that μ and $\mathbf{E}$ are parallel. In the dipole approximation the interaction operator transforms into the dipole moment operator $\hat{\mu}$. Its matrix element for the case of parallel μ and $\mathbf{E}$

can be expressed in the form

$$\int \psi_1 \hat{H}' \psi_2^* \, dx \, dy \, dz = -E \left(e^{j\omega t} + e^{-j\omega t} \right) \int \psi_1 \hat{\mu} \psi_2^* \, dx \, dy \, dz$$

$$= -\langle \mu \rangle E \left(e^{j\omega t} + e^{-j\omega t} \right), \tag{4.14}$$

where $\langle \mu \rangle$ defines the matrix element of the dipole-moment transition operator. In this way we arrive at the equation

$$\frac{db}{dt} = -\frac{1}{j\hbar} \langle \mu \rangle \, E \left(e^{j(\omega + \omega_0)t} + e^{j(\omega_0 - \omega)t} \right), \tag{4.15}$$

with $\omega_0 = (E_2 - E_1)\hbar$ the transition's resonance frequency. (It is customary here to use circular frequencies.) If the interaction occurs at a frequency ω close to the resonance frequency, the first exponential function oscillates too rapidly and can be ignored. Then

$$\frac{db}{dt} = -\frac{1}{j\hbar} \langle \mu \rangle \, E e^{j(\omega_0 - \omega)t}. \tag{4.16}$$

This equation can easily be integrated from 0 to t. The result is

$$b(t) = \frac{\langle \mu \rangle E}{\hbar} \frac{e^{j(\omega_0 - \omega)t} - 1}{\omega_0 - \omega}. \tag{4.17}$$

The probability of the $1 \rightarrow 2$ transition is determined by

$$|b(t)|^2 = \left(\frac{\langle \mu \rangle E}{\hbar} \right)^2 \left(\frac{\omega_0 - \omega}{2} \right)^{-2} \sin^2 \left(\frac{\omega_0 - \omega}{2} t \right). \tag{4.18}$$

This expression implies, for one thing, that there is a sizable transition probability only when the external field has a frequency ω close to ω_0, that is, when the radiation is resonant.

To link the transition probability just obtained with Einstein's coefficient B_{12} for a stimulated transition we must apply this result to thermal radiation and allow for the spectral width of the transition. The quantity E^2 in formula (4.18) for $|b(t)|^2$ is related to the energy density of the corresponding field through the following simple and well-known equation: $\varrho = E^2/8\pi$. This refers, however, to the case of an electric field polarized along the direction of the dipole. In the case of isotropic thermal radiation, for which the thermodynamic analysis of Einstein's coefficients was carried out, the energy density of the field along a selected direction amounts to one-third of the total energy density of the field. Hence,

$$E^2 = 8\pi\varrho/3. \tag{4.19}$$

The energy density of thermal radiation is distributed over the frequency spectrum according to Planck's formula. But we have derived a formula for $|b(t)|^2$ for the case of a monochromatic external force. The total transition probability in a thermal radiation field can be obtained by integrating (4.18) over all the frequencies present in the field, bearing in mind that the spectral energy density $E_\nu^2 = 8\pi\varrho_\nu/3$ enters (4.18) with ϱ_ν given by Planck's formula (1.9). Thus, the total transition probability Π can be written in the form

$$\Pi = \int_{-\infty}^{\infty} |b(t)|^2 d\nu = \int_{-\infty}^{\infty} \frac{\langle\mu\rangle^2}{\hbar^2} \frac{8\pi\varrho_\nu}{3} \frac{\sin^2[\pi(\nu_0 - \nu)t]}{\pi^2(\nu_0 - \nu)^2} d\nu. \tag{4.20}$$

The function ϱ_ν is extremely smooth, while the function $\sin^2[\pi(\nu_0 - \nu)t]/(\pi^2 - \nu^2)$ has a marked resonance peak, with the result that ϱ_ν can be taken outside the integral:

$$\Pi = \frac{8\pi}{3} \frac{\langle\mu\rangle^2}{\hbar^2} \varrho_\nu \int_{-\infty}^{\infty} \frac{\sin^2[\pi(\nu_0 - \nu)t]}{\pi^2(\nu_0 - \nu)^2} d\nu. \tag{4.21}$$

By introducing the variable $x = \pi(\nu_0 - \nu)t$ we can reduce the integral to tabular form: $\int_{-\infty}^{\infty} x^{-2}\sin^2x \, dx = \pi$. As a result,

$$\Pi = \frac{8\pi}{3} \frac{\langle\mu\rangle^2}{\hbar^2} \varrho_\nu t. \tag{4.22}$$

An important feature of this formula is that the probability of a transition occurring because of a perturbation is proportional to the time during which this perturbation acts, t. This is in complete agreement with Einstein's postulate concerning the probabilities of stimulated transitions.

Introducing the transition rate, or the transition probability per unit time,

$$W_{12} = \frac{\Pi}{t} = \frac{8\pi}{3} \frac{\langle\mu\rangle^2}{h^2} \varrho_\nu, \tag{4.23}$$

and comparing this expression with Einstein's postulate (1.2),

$$W_{12} = B_{12}\varrho_\nu, \tag{4.24}$$

we arrive at the final expression for Einstein's coefficient B_{12} in the form

$$B_{12} = \frac{8\pi}{3} \frac{\langle\mu\rangle^2}{h^2}. \tag{4.25}$$

A similar expression can be obtained for B_{21}, which in our two-level case without degeneracy proves to be equal to B_{12}. However, this method does not give us a formula for the probability of a spontaneous transition.

Thus, the values of Einstein's coefficients B_{12} and B_{21} are determined by the dipole matrix elements $\langle\mu\rangle$, whose calculation for the majority of simple configurations is quite possible via quantum mechanical methods. In many cases, however, one is forced to rely on the experimental data.

For the sake of reference we give the formulas for B_{12} and B_{21} for the case of degenerate levels:

$$B_{12} = \frac{8\pi}{3} \frac{1}{h^2} \frac{F_{12}}{g_1}, \quad B_{21} = \frac{8\pi}{3} \frac{1}{h^2} \frac{F_{21}}{g_2}, \tag{4.26}$$

where the so-called line strength $F_{12} = F_{21} = F$ of a transition is the sum of the squares of the matrix elements of the dipole-moment transition between nondegenerate states α and β, which comprise levels "1" and "2", respectively:

$$F = \sum_{\alpha, \beta} \langle\mu_{\alpha\beta}\rangle^2. \tag{4.27}$$

The above derivation of the connection between Einstein's coefficient B_{12} and the quantum mechanical characteristic of a transition, $\langle \mu \rangle^2$, is valid, strictly speaking, outside of resonance, when the resonance denominators of the $\omega_0 - \omega$ type (see Eq. (4.17)) are not too small. Otherwise, the assumption that the correction term in perturbation theory is small breaks down. However, after obtaining in first-order perturbation theory the expression (4.17) for $b(t)$ we carried out integration, (4.20), over all the frequencies in Planck's radiation, which intuitively justifies our calculations. But if we are interested in the resonance case of a field of frequency $\omega \approx \omega_{21} = \omega_0$, that is, the case where

$$\omega_{21} - \omega = \delta \ll \omega_{21}, \omega, \tag{4.28}$$

we must reconsider the problem from the very beginning. Once more we represent the Ψ functions of the mixed state generated by the radiation field acting in the two-level system in the form (4.8). After substituting Ψ into the Schrödinger equation (4.6), multiplying by the respective complex conjugates of Ψ_1 and Ψ_2, integrating the products, and collecting like terms, we arrive at exact equations for $a(t)$ and $b(t)$:

$$jh \frac{da}{dt} = V_{12}(t)b, \tag{4.29}$$

$$jh \frac{db}{dt} = V_{21}(t)a, \tag{4.30}$$

where the matrix elements of the perturbation operator $\hat{V}$ are (cf. (4.12))

$$V_{12}(t) = \int \Psi_1^* \hat{V} \Psi_2 dxdydz = V \exp[j(E_1 - E_2)t/h], \tag{4.31}$$

$$V_{21}(t) = \int \Psi_2^* \hat{V} \Psi_1 dxdydz = V \exp[-j(E_1 - E_2)t/h]. \tag{4.32}$$

If the interaction is of the dipole type (4.13), that is,

$$\hat{V} = -\hat{\mu}E(e^{j\omega t} + e^{-j\omega t}), \tag{4.33}$$

the matrix elements assume the form

$$V_{12}(t) = -\langle\mu\rangle E(e^{j(\omega + \omega_{21})t} + e^{j(\omega_{21} - \omega)t}), \tag{4.34}$$

$$V_{21}(t) = -\langle\mu\rangle E(e^{j(\omega + \omega_{12})t} + e^{j(\omega_{12} - \omega)t}). \tag{4.35}$$

Here, as in (4.14),

$$\langle\mu\rangle = \langle\mu_{12}\rangle = \langle\mu_{21}\rangle = \int \Psi_2^* \hat{\mu} \Psi_1 \, dx\,dy\,dz \tag{4.36}$$

and, in addition, we have introduced the frequencies

$$\omega_{21} = \frac{E_2 - E_1}{\hbar} = -\omega_{12} = -\frac{E_1 - E_2}{\hbar}. \tag{4.37}$$

As we did earlier, we assume that only the frequency difference δ plays a significant role (see (4.28)). Then, introducing the notation

$$-\langle\mu\rangle E = F, \tag{4.38}$$

we can transform Eqs. (4.34) and (4.35) into the following system of equations:

$$jh \frac{da}{dt} = Fbe^{j\delta t}, \quad jh \frac{db}{dt} = Fae^{-j\delta t}. \tag{4.39}$$

Substituting c for $be^{j\delta t}$ and canceling out a, we arrive at a second order linear differential equation with constant coefficients, that is, an equation of the harmonic oscillator type:

$$\frac{d^2c}{dt^2} - j\delta \frac{dc}{dt} + \frac{F^2}{\hbar^2} c = 0. \tag{4.40}$$

The solution to this equation is widely known. Introducing the notation

$$\Omega = \sqrt{\delta^2/4 + F^2/\hbar^2}, \tag{4.41}$$

we find that

$$b_1 = A \exp\left[-j\left(\frac{\delta}{2} - \Omega\right)t\right], \quad b_2 = B \exp\left[-j\left(\frac{\delta}{2} + \Omega\right)t\right],$$

$$\tag{4.42}$$

$$a_1 = \frac{\hbar}{F} A \left(\frac{\delta}{2} - \Omega\right) \exp\left[j\left(\frac{\delta}{2} + \Omega\right)t\right],$$

(4.43)

$$a_2 = \frac{\hbar}{F} B \left(\frac{\delta}{2} + \Omega\right) \exp\left[j\left(\frac{\delta}{2} - \Omega\right)t\right].$$

Hence, the sought Ψ function of the mixed state is

$$\Psi = \Psi_1 \frac{\hbar}{F} \left\{ A\left(\frac{\delta}{2} - \Omega\right) \exp\left[j\left(\frac{\delta}{2} + \Omega\right)t\right] \right.$$
$$+ B\left(\frac{\delta}{2} + \Omega\right) \exp\left[j\left(\frac{\delta}{2} - \Omega\right)t\right] \right\}$$
$$+ \Psi_2 \left\{ A \exp\left[-j\left(\frac{\delta}{2} - \Omega\right)t\right] + B \exp\left[-j\left(\frac{\delta}{2} + \Omega\right)t\right] \right\}.$$

(4.44)

The constants A and B can be found from the condition that $\Psi = \Psi_1$ at $t = 0$. Then $A = -B = F/2\hbar\mathbf{I}$. As a result,

$$\Psi = \Psi_1 \left\{ \frac{\delta - 2\Omega}{4\Omega} \exp\left[j\left(\frac{\delta}{2} + \Omega\right)t\right] \right.$$
$$- \frac{\delta + 2\Omega}{4\Omega} \exp\left[j\left(\frac{\delta}{2} - \Omega\right)t\right] \right\}$$
$$+ \Psi_2 \frac{F}{2\hbar\Omega} \left\{ \exp\left[-j\left(\frac{\delta}{2} - \Omega\right)t\right] - \exp\left[-j\left(\frac{\delta}{2} + \Omega\right)t\right] \right\}.$$

(4.45)

If at $t = 0$ we have $\Psi = \Psi_1$, the square of the modulus of the coefficient of Ψ_2 is the probability of finding the particle on the second level after the field was switched on. This probability is

$$\Pi = \frac{F^2}{2\hbar^2\Omega} [1 - \cos(2\Omega t)].$$

(4.46)

We see that Π oscillates with a period of 2Ω between zero and $F^2/\hbar^2\Omega^2$.

At exact resonance (see (4.41)) $\Omega^2 = F^2/\hbar^2$ and we have oscillations between 0 and 1:

$$\Pi = \frac{1}{2} \left[1 - \cos \left(2 \frac{F}{\hbar} t \right) \right]. \tag{4.47}$$

Returning to the initial notation (4.38) and (4.41), we see that the irradiation intensity and the dipole moment of the transition determine the oscillation frequency:

$$\Omega^2 = \frac{\delta^2}{4} + \left\langle \frac{\mu E}{\hbar} \right\rangle^2 = \frac{\delta^2}{4} + \Omega_R^2, \tag{4.48}$$

where Ω_R stands for the Rabi frequency,

$$\Omega_R = \left\langle \frac{\mu E}{\hbar} \right\rangle. \tag{4.49}$$

We write (4.46) in the new notation:

$$\Pi = \frac{2\Omega_R^2}{\delta^2 + 4\Omega_R^2} \left[1 - \cos \left(\delta^2 + 4\Omega_R^2 \right)^{1/2} t \right]. \tag{4.50}$$

The limiting cases are especially pictorial. At a small detuning (high intensities), $\delta \ll \Omega_R$,

$$\Pi = \frac{1}{2} [1 - \cos(2\Omega_R t)]. \tag{4.51}$$

For a large detuning (low intensities), $\delta \gg \Omega_R$,

$$\Pi = 2(1 - \cos \delta t) \frac{\Omega_R^2}{\delta^2}. \tag{4.52}$$

The meaning of (4.51) and (4.52) is clear. At high intensities, when the Rabi frequency considerably exceeds the detuning of the field from the exact resonance, the particle oscillates between the upper and lower levels with the Rabi frequency. In a weak field, when the corresponding Rabi frequency does not overlap the detuning of the field's frequency from the exact resonance, the probability of finding the particle on the upper level never becomes equal to 1 and oscillates with the detuning frequency. Nevertheless, at exact resonance the particle must reach the upper level even in a weak field, but only after an extremely long time interval, determined in this case by the low rate of the Rabi oscillations.

Here are some numerical estimates. For spectroscopically well resolved lines of resonance absorption, the typical value of $\langle \mu \rangle$ amounts to 1 debye, or 10^{-18} esu of electric dipole moment. According to formula (3.7), an intensity of, say, 1 MW/cm^2 corresponds to a field strength of 270×10^2 V/cm, or 10^2 esu of field strength. The value of $\hbar$ is well-known, approximately 10^{-27} erg·s. Then the Rabi frequency is 10^{11} Hz or, in units commonly used in spectroscopy, 3.3 cm^{-1}.

We also note that according to (4.25) an electric dipole moment of 1 debye corresponds to Einstein's coefficient B_{12} equal to 8×10^{18} esu and for a wavelength of 0.5 μm, to Einstein's coefficient A_{21} equal to 10^7 s^{-1} (see (1.11)), a value that agrees with a natural lifetime τ_0 of 0.1 μs.

All the material of this lecture corresponds only to the first-order perturbation theory. The reader can find a rigorous treatment of perturbations depending on time in § 40 of the well-known book *Quantum Mechanics* by L.D. Landau and E. M. Lifshitz (3rd ed., Oxford: Pergamon Press, 1977) and in the problem relating to that section.

Finally, an important feature of this lecture that sets it apart from Lecture 3 is that we took no account of either relaxation processes or the spontaneous decay of the upper level. Hence, everything stated here is valid only for time intervals that are short if compared with the lifetime of the upper state. For longer time intervals the coherence of states breaks down and we must use transport equations.

Problems to Lecture 4

4.1. For what values of time t is formula (4.18) valid?

4.2. Formula (4.25) for B_{12} has been derived for the special case but is universal. Prove its general validity.

4.3. The expression for the transition probability has been derived on the assumption that ϱ_ν is a sufficiently slowly varying function of the frequency. How does this formula look when this condition is not met?

4.4. What is the physical meaning of the fact that the results of calculations performed according to formula (4.20) depend on parameter t standing under the integral sign, that is, is t sufficiently great to justify carrying ϱ_ν outside the integral? (*Hint*: Refer to Lecture 2.)

4.5. What characteristics of the exciting radiation and the system considered are needed to measure the frequency of population oscillations experimentally?

4.6. Estimate what values of the parameters contained in the answer to the previous problem are necessary for directly measuring Ω_R in the example cited at the end of the lecture.

4.7. Estimate the values of the minimal intensity necessary for observing coherent population oscillations and the corresponding Rabi frequency for the example cited at the end of the lecture.

LECTURE 5

Laser Amplifiers

Amplification and generation. Passband of a travelling-wave amplifier. The noise generated by a quantum amplifier. Maximum output power. Pulsed mode. Maximum output energy. Change of pulse shape in nonlinear amplification

In quantum electronics the stimulated emission of radiation by the active medium is used for coherent amplification of electromagnetic waves and the creation of quantum amplifiers and quantum generators. The difference between quantum amplifiers and quantum generators is considerable.

Quantum amplifiers or, as we will often call them, laser amplifiers are designed to increase the electromagnetic field strength of their input signal. In this sense quantum amplifiers in the radio-frequency and optical ranges are similar to their predecessors, vacuum-tube and semiconductor amplifiers.

Quantum generators must be sources of radiation generated directly in the device and exiting to the ambient space. They are similar to ordinary radio-frequency generators and require positive feedback for their operation, just as radio-frequency generators do. In other words, quantum generators are self-oscillating systems in which the generation of electromagnetic oscillations occurs in the process of coherent amplification of the oscillations with the appropriate feedback. In accordance with the theory of ordinary self-oscillating systems, quantum generators must produce monochromatic radiation. And it is extremely important that under stimulated emission in the active medium the secondary photons not only have the same frequency but also propagate in the same direction as the primary photons. For this very reason laser radiation has sharp directivity and the laser beam is formed in the generator (oscillator) automatically.

It must be noted, however, that not all laser generators are self-oscillating systems. An important exception to this general rule will be discussed much later in connection with so-called superfluorescence lasers, with reference to specific lasers.

The feedback usually necessary for generation is achieved by placing the active medium in a cavity in which a system of standing electromagnetic waves can be generated. When an inevitable spontaneous transition from an upper level to a lower occurs, radiation is emitted spontaneously at some point in the cavity. If the cavity is tuned to the frequency of this transition and if the emitted photon lands in one of the standing waves, the radiation in the standing wave builds up and acts on the active media, producing stimulated radiation. If the power of this radiation is greater than the losses due to heating of the cavity walls, dissipation of the radiation, etc., as well as due to useful emission of radiation into the ambient space, that is, if the so-called self-excitation conditions are met (we will discuss these conditions later), then sustained oscillations are maintained in the cavity. In view of the properties of stimulated emission, these oscillations are highly monochromatic. All particles of the active media operate in phase because of the positive feedback carried out in the acts of stimulated emission via the radiation stored in the cavity, that is, radiation reflected by the cavity walls.

What has been said applies to an ideal laser generator, of course. In subsequent lectures we will discuss these questions more thoroughly.

At this point a small digression is in order.

A characteristic feature of quantum electronics and, probably, electronics in general is the following remarkable fact. The development of each new region and the mastery of every new wavelength band begin with the emergence of generators (oscillators). Only after generators have established themselves do amplifiers appear. The reason may be that in generators the new effect manifests itself more vividly and possibly amplifiers become necessary only after there is something to amplify. Be it as it may, rapid development of a new field of quantum electronics always begins with invention of new generators.

In the microwave range the first to appear was the ammonia maser generator and only then were paramagnetic maser amplifiers built.

In optics the first experiments in amplifying remained unnoticed until the appearance of the ruby and helium-neon laser generators gave an impetus to development of the quantum electronics of the optical range. Later powerful optical amplifiers appeared.

Nevertheless, we begin our analysis by considering amplifiers, not only because an amplifier is simpler than a generator but also because a generator is a self-excited amplifier with appropriate feedback.

We start with the passband of an amplifier in the linear mode. Formula (3.5) gives the expression for the amplifier gain at the center of the resonance line (for $n_2/g_2 > n_1/g_1$). We denote the linear gain, that is, the gain corresponding to small signals, at the center of an inverted resonance absorption line by α_0, which is measured in reciprocal centimeters and in the literature is sometimes called the amplification factor. We will follow tradition and call α_0 simply the gain.

The frequency dependence of the gain can be taken into account by introducing the form factor $q(\nu)$ of the line if we write $\alpha(\nu)$ in the form

$$\alpha(\nu) = \alpha_0 q(\nu)/q(\nu_0). \tag{5.1}$$

In the travelling wave mode the power gain of the entire amplifier is

$$G(\nu) = \exp\{[\alpha_0 q(\nu)/q(\nu_0) - \beta]l\}, \tag{5.2}$$

where l is the length of the amplifier, and β, the nonresonance loss factor. The bandwidth proves to be dependent on the value of the achieved gain and reduces as the gain grows. Indeed, if we define, as is commonly done, the bandwidth of an amplifier as the frequency band in which the gain exceeds half of the maximum gain, the equation

$$G(\nu) = G(\nu_0)/2 \tag{5.3}$$

provides the means for calculating the bandwidth:

$$q(\nu)/q(\nu_0) = 1 - \ln 2/(\alpha_0 l). \tag{5.4}$$

Specifying the shape of $q(\nu)$, we can obtain the values of ν that determine the bandwidth. For homogeneous broadening, that is, for the Lorentzian lineshape (2.7), simple calculations lead to the following expression for the bandwidth of a travelling-wave amplifier:

$$\Delta\nu = \Delta\nu_{\text{line}}(\ln 2)^{1/2}(\ln G_0 + \ln L - \ln 2)^{-1/2}, \tag{5.5}$$

where $G_0 = G(\nu_0) = \exp[(\alpha_0 - \beta)l]$ is the net gain at the center of the line and $L = \exp(\beta l)$ the loss factor.

We see that under inversion, that is, under amplification, the line narrows. This narrowing in the travelling wave mode is slow but can reach sizable values if there is considerable amplification. The reason is obvious. In view of the exponential dependence of gain on the amplifier length l, the spectral components corresponding to the center of the line are amplified more strongly. In the limit of long amplifier (high G_0) only the central component is amplified. Actually, this situation is realized in super-luminescence lasers, to be discussed much later.

Obviously, formula (5.5) is valid only if $G_0 > 2$.

Let us now consider the noise generated by a travelling-wave quantum amplifier. We will ignore thermal noise and focus only on the noise produced by spontaneous emission. At room temperature this corresponds to the optical range, while for radio-frequencies we move into the region of liquid helium temperatures. The optical case corresponds to free space and spontaneous emission into all modes. At radio-frequencies it is fairly easy to isolate a single type of waveguide propagation.

The equation for the rate of increase of energy per unit volume of active medium has the form

$$\frac{d\varrho}{dt} = (n_2 - n_1) \frac{h\nu B_{21}\varrho}{\pi \Delta \nu_{\text{line}}} + n_2 \frac{h\nu 8\pi \nu^2 h\nu B_{21}}{c^3 \pi \Delta \nu_{\text{line}}}. \tag{5.6}$$

Here we have assumed, for simplicity of notation, that $g_1 = g_2$. One must also bear in mind that the second term describes the emission of radiation into a solid angle of 4π sr.

Let us now turn to the intensity per unit spectral interval, that is, the amount of energy travelling through a unit surface per unit time per unit spectral interval along the amplifier axis z. Since $dz = c\, dt$ and $\varrho = I/c$, we have

$$\frac{dI}{dz} = (n_2 - n_1) \frac{h\nu B_{21}I}{c\pi \Delta \nu_{\text{line}}} + n_2 \frac{h\nu 8\pi h\nu B_{21}}{\lambda^2 c\pi \Delta \nu_{\text{line}}}. \tag{5.7}$$

We can easily find the equation for the total energy flux per unit spectral interval, $P = AI$, with A the surface area of the amplifier's aperture, namely,

$$\frac{dP}{dz} = (n_2 - n_1) \frac{h\nu B_{21}P}{c\pi \Delta \nu_{\text{line}}} + 2n_2 \frac{(h\nu)^2 \Omega(z)B_{21}A}{\lambda^2 c\pi \Delta \nu_{\text{line}}}, \tag{5.8}$$

if in the second term on the right-hand side of Eq. (5.7) we allow for the fact that the spontaneous emission gets amplified only in the direction of the output end of the amplifier into a solid angle $\Omega(z)$ and not into 4π sr. Hence, in this term we must allow only for the part that is equal to $\Omega(z)/4\pi$.

For a long amplifier with a fairly high gain we can approximately assume that Ω is the angle at which the input aperture of the amplifier is seen from the output aperture. Then

$$\Omega(z) \approx \Omega = A/(4\pi l^2).\tag{5.9}$$

As a result,

$$\frac{dP}{dz} = (n_2 - n_1)\frac{h\nu B_{21}}{c\pi\Delta\nu_{\text{line}}}P + 2n_2\frac{(h\nu)^2 B_{21}A^2}{\lambda^2 4\pi l^2 c\pi\Delta\nu_{\text{line}}}.\tag{5.10}$$

Integration of (5.10) at $P_{\text{in}} = P(z = 0) = 0$ yields

$$P_{\text{out}} = P(z = l)$$

$$= 2h\nu\,\frac{n_2}{n_2 - n_1}\frac{A}{\lambda^2}\frac{A}{4\pi l^2}$$

$$\times\left\{\exp\left[(n_2 - n_1)\frac{h\nu B_{21}}{c\pi\Delta\nu_{\text{line}}}l\right] - 1\right\}.\tag{5.11}$$

Since

$$\exp\left[(n_2 - n_1)\frac{h\nu B_{21}}{c\pi\Delta\nu_{\text{line}}}l\right] = \exp[\alpha_0 l] = G_0,\tag{5.12}$$

we have

$$P_{\text{out}} = 2h\nu\,\frac{n_2}{n_2 - n_1}\frac{A}{\lambda^2}\frac{A}{4\pi l^2}(G_0 - 1).\tag{5.13}$$

Recalculating for input, that is, finding the effective noise power in a unit spectral interval at the amplifier's input, yields

$$P_{\text{in}}^{\text{eff}} = 2h\nu\,\frac{n_2}{n_2 - n_1}\frac{A}{\lambda^2}\frac{A}{4\pi l^2}\frac{G_0 - 1}{G_0}.\tag{5.14}$$

The "two" reflects the two possible polarizations of the wave. $A/\lambda^2 \gg 1$ and $A/4\pi l^2 \ll 1$, but the condition $\sqrt{A}/l > \lambda/\sqrt{A}$ for the validity of the geometrical optics approximation always results in $(A/\lambda^2)(A/4\pi l^2) \gg 1$.

One can easily see, however, that in the event of one polarization and single-mode waveguide propagation (rather than in the geometrical optics approximation) formula (5.14) assumes the form

$$P_{in}^{eff} = h\nu \frac{n_2}{n_2 - n_1} \frac{G_0 - 1}{G_0},$$ (5.15)

which at high inversion and gain yields

$$P_{in}^{eff} = h\nu.$$ (5.16)

Formulas (5.15) and (5.16) correspond to the case of amplification when the radiation propagates, say, in an active medium placed in a microwave waveguide. They are also valid for optical waveguides manufactured in the form of single-mode dielectric fibers.

Hence, the minimum effective strength of input noise of a quantum amplifier in a unit spectral interval amounts to $h\nu$. It can be demonstrated that every coherent amplifier, that is, an amplifier that conserves the phase of the input signal while increasing the signal's strength, carries with it, in view of the uncertainty relation $\Delta n \Delta \varphi \geqslant 1/2$, unremovable input noise with a strength of $h\nu$ in a unit spectral interval. The first to notice this remarkable feature was the American physicist Charles H. Townes.

Now let us consider the energy characteristics of laser amplifiers or, more precisely, the power output of a travelling-wave quantum amplifier.

Quantum amplifiers are rarely used in optics to amplify weak signals in order to achieve a higher sensitivity of receivers in this frequency band of electromagnetic waves because there are good conventional receivers designed for this purpose. The use of quantum amplifiers for registering weak signals is justified in the far IR region and the microwave range. An interesting application of quantum amplifiers in information technology is their use as image intensifiers, say, in laser microscopy, where the relatively weak light reflected by the object or passing through the object without damaging it is greatly amplified by a laser amplifier so that projection of the image onto large screens becomes possible. Obviously, in this case the amplifier must be multimode since only a nonplanar wave carries the information about the spatial distribution of the characteristic features of the transmitted image. Image intensifiers usually operate at input signal levels considerably higher than the sensitivity threshold level.

An even higher input signal level is encountered when laser amplifiers are used to increase the output of laser generators for obtaining the highest possible values of power or energy output combined with high quality of input radiation. It is well-known that all types of radiation processing (tuning and stabilization of frequency, amplitude, frequency, phase and pulse modulation, formation of radiation pulses, etc.) are achieved most easily at moderate power outputs. If higher signal levels are needed, further amplification is necessary. In quantum electronics quantum amplifiers serve this purpose.

In analyzing the problem of the power output of laser amplifiers one must allow for saturation. Formula (3.30) for the population difference $n_2 - n_1$ was derived by employing rate equations for homogeneously broadened lines. We will restrict our further analysis to this case. Since $n_2 - n_1$ determines the gain, the rate equation for radiation propagating in a medium with a nonresonance loss factor β and a linear gain α_0 can be written, with due regard for (3.30) and the simplifying assumption $g_1 = g_2$, as

$$\frac{dI}{dz} = -\beta I + \frac{\alpha_0 I}{1 + I/I_{sat}}. \tag{5.17}$$

If we introduce the dimensionless intensity $J = I/I_{sat}$ and carry out simple transformations, we can write this equation as

$$dz = \frac{1 + J}{J} \frac{dJ}{\alpha_0 - \beta - \beta J}, \tag{5.18}$$

which can easily be integrated. For an amplifier of length l we get

$$(\alpha_0 - \beta)l = \ln \frac{J_2}{J_1} - \frac{\alpha_0}{\beta} \ln \frac{\alpha_0 - \beta(1 + J_2)}{\alpha_0 - \beta(1 + J_1)}, \tag{5.19}$$

with J_1 and J_2 the input and output (dimensionless) intensities. In such a general form this transcendental equation is too complicated to easily investigate although, of course, there is always the possibility of constructing the J_2 vs. J_1 curves for different values of α_0 and β as parameters. The following particular cases are of interest, however.

If $\beta J_2/(\alpha_0 - \beta)$ and $\beta J_1/(\alpha_0 - \beta)$ are small, then Eq. (5.19) is transformed into

$$(\alpha_0 - \beta)l = \ln \frac{J_2}{J_1} + \frac{\alpha_0}{\alpha_0 - \beta} (J_2 - J_1). \tag{5.20}$$

At low signal levels ($J_2 \ll 1$ and $J_1 \ll 1$) the first term on the right-hand side is predominant and we arrive at an exponential growth of the output intensity (linear amplification):

$$J_2 = J_1 \exp\left[(\alpha_0 - \beta)l\right]. \tag{5.21}$$

In the absence of energy losses ($\beta = 0$) but at high saturation levels ($J_1 \gg 1$) exponential growth is replaced by linear growth. Indeed, for $J_1 \gg 1$ we have

$$\ln \frac{J_2}{J_1} = \ln\left(1 + \frac{J_2 - J_1}{J_1}\right) = \frac{J_2 - J_1}{J_1},$$

which when combined with (5.20) yields

$$\alpha_0 l = \frac{J_2 - J_1}{J_1} + J_2 - J_1, \tag{5.22}$$

and

$$J_2 = J_1 + \alpha_0 l \frac{J_1}{1 + J_1} \approx J_1 + \alpha_0 l. \tag{5.23}$$

In the absence of energy losses and at high saturation levels each elementary segment of the amplifier adds energy to the common energy flux. If losses are taken into account, the situation changes drastically. For small but finite values of β/α_0 and for $J_1, J_2 \gg 1$ the initial equation (5.19) can be written as

$$\frac{J_2^{\beta/\alpha_0}}{\alpha_0 - \beta J_2} = \frac{J_1^{\beta/\alpha_0}}{\alpha_0 - \beta J_1} \exp \frac{(\alpha_0 - \beta)\beta l}{\alpha_0}. \tag{5.24}$$

Ignoring the difference between the values of J_2^{β/α_0} and J_1^{β/α_0} and the value of β/α_0 in comparison to unity, we find from (5.24) that

$$J_2 = \frac{\alpha_0}{\beta}\left(1 - e^{-\beta l}\right) + J_1 e^{-\beta l}. \tag{5.25}$$

At large amplifier lengths ($\beta l \gg 1$) the input signal attenuates and the output signal reaches its stationary value (in units of $I = JI_{sat}$)

$$I_{max} = \frac{\alpha_0}{\beta} I_{sat}. \tag{5.26}$$

This leads us to the important conclusion that in a travelling-wave laser amplifier the output intensity is determined in the end by the saturation intensity, the linear gain, and the nonresonance loss factor. The stationary intensity of the radiation propagating in the amplifier establishes itself when all the radiation that can be emitted by a unit length of the active medium in the complete saturation mode is absorbed due to nonresonance losses in the same unit segment. The balance between the absorbed and emitted energies results in no further amplification as the radiation travels along the amplifier.

The above results were obtained from the general solution to the energy transport equation. However, the analysis of these limiting cases can be made much more graphic by studying the initial equation (5.17). For instance, for $I/I_{sat} \gg 1$ Eq. (5.17) assumes the form

$$\frac{dI}{dz} = -\beta I + \alpha_0 I_{sat}. \tag{5.27}$$

If the intensity reaches its limit value I_{max}, no further amplification is possible. Hence, $dI/dz = 0$, which occurs at $I_{max} = (\alpha_0/\beta)I_{sat}$, that is, in accordance with Eq. (5.26). A direct solution of the simplified equation (5.27) yields in this case the same result as would the appropriate transformation of the general solution to Eq. (5.17). The same is true of the cases where $\beta = 0$ and the I are small.

The above analysis was carried out for the case of continuous amplification of continuous signals. The pulsed mode, that is, the mode whose characteristic times are shorter than the population relaxation time of the amplifier's active medium, requires special treatment.

A simple energy estimate can be made quite easily. In the case of strong signals corresponding to complete pulsed saturation of amplification an amount of energy equal to $(nh\nu/2)dz$ is emitted by an amplifier segment dz long, where n is the population inversion factor per unit amplifier length. On the same segment an amount of energy equal to $\beta F, dz$ is absorbed due to nonresonance losses, with F, the energy flux, passing through the cross section of the amplifier (the fluence).

As a result the balance equation assumes the form

$$n \frac{h\nu}{2} - \beta F = \frac{dF}{dz},$$

(5.28)

which is, in fact, equivalent to Eq. (5.27). The fluence reaches its maximum value F_{max} when dF/dz vanishes. Hence,

$$F_{max} = n \frac{h\nu}{2\beta}.$$

(5.29)

Turning to Eq. (3.43) for the saturation energy density in the pulsed mode, $F_{sat} = h\nu/2\sigma$, and Eq. (3.13) for the linear gain, $\alpha_0 = n\sigma$, we can easily find that

$$F_{max} = \frac{\alpha_0}{\beta} F_{sat}.$$

(5.30)

This formula is similar in appearance to (5.26) but not equivalent because it cannot be obtained by simply multiplying the right- and left-hand sides by the pulse length. Pulsed signals saturate a two-level quantum system differently than continuous signals do. (This was discussed at the end of Lecture 3.)

In the case of pulsed signals an important role is played not only by the energy characteristics of the amplifier but also by the shape and length of the pulses of the radiation being amplified. In the essentially nonlinear interaction of a powerful pulse of radiation with the amplifying medium, when the pulse propagating in the medium drives down the population inversion and thus forces emission of the energy stored in the media, the shape of the pulse changes in the course of amplification. The point is that for high intensities it is the leading edge of the pulse that drives down the population inversion considerably, with the result that the leading edge gets amplified more strongly than the trailing edge. This leads to a shift of the "center of gravity" in the energy distribution in the pulse toward the pulse's leading edge and, for sufficiently sharp edges to reduction in pulse length.

A digression is in order. When a short pulse of coherent radiation travels through an amplifying resonant medium, the coherent nature of the interaction of radiation with matter can come into play. In such interaction the dipole moment induced by the radiation field in the particles of the medium does not decay spontaneously in the course of the interaction. This means that in coherent interactions the pulse must be much shorter than the shortest relaxation time for the polarization of the active substance (see Lecture 4). When the radiation pulse length is longer than the phase memory of the substance, the interaction is incoherent.

In ordinary circumstances the interaction in high-power laser amplifiers is incoherent. The condition of incoherence of the interaction of a pulse of coherent radiation with a substance is the loss of coherence in the state of the substance during the pulse length:

$$\tau_{\text{pulse}} \gg \frac{1}{2\pi\Delta\nu_{\text{line}}}. \tag{5.31}$$

Since an amplifier cannot amplify signals whose duration is smaller than the inverse of the amplifier's bandwidth, condition (5.31) is practically always met for long amplifiers with a high gain.

In conditions of incoherent interaction we can use the ordinary equations of radiation transfer, which follow from the law of energy conservation. Since pulse amplification is a time-dependent process, an ordinary differential equation of the (5.17) type must be replaced by a partial differential equation,

$$\frac{1}{c}\frac{\partial I}{\partial t} + \frac{\partial I}{\partial z} = (\alpha - \beta)I, \tag{5.32}$$

where gain α is intensity-dependent. In the pulsed mode the saturation effect, which determines the dependence of gain on signal power, is expressed by formula (3.45). Then the energy transfer equation for a short pulse ($\tau_{\text{pulse}} \ll \tau$) assumes the form

$$\frac{1}{c}\frac{\partial I}{\partial t} + \frac{\partial I}{\partial z} =$$

$$\left[\alpha_0 \exp\left(-\frac{1}{2}\frac{g_1 + g_2}{g_2}\frac{1}{F_{\text{sat}}}\int_{-\infty}^{t} I(t', z)dt'\right) - \beta\right]I, \tag{5.33}$$

where α_0 is the linear gain (low-signal gain), $F_{\text{sat}} = h\nu/2\sigma$, and we have allowed for the possibility that the shape of $I(t, z)$ changes as the pulse propagates along the z axis.

Equation (5.33) cannot be solved in general form. The case of zero losses ($\beta = 0$), which is far from reality, leads to a solution according to which the energy grows linearly without limit when F_{sat} is exceeded. A similar situation exists for the continuous mode.

The variation of the pulse shape during amplification can be analyzed by solving Eq. (5.33) numerically. A simple picture of the reduction in length of an amplified signal is most easily studied if we consider a step-like rectangular input pulse. For pulses with a smooth leading edge the picture is different. Preferential amplification of the leading edge of the pulse leads to a gradual shift, during amplification, of the peak of the pulse along the leading edge in the direction of propagation of the radiation being amplified. The magnitude of this shift is determined by the shape of the leading edge of the initial pulse. The shift prevents the pulse from being compressed. Hence, in nonlinear amplification only pulses with sufficiently steep leading edges are compressed. Among these is the Gaussian-shaped pulse $I \propto \exp(-t^2/\tau_{pulse}^2)$. On the other hand, in the case of exponential growth of the leading edge there is no pulse compression. Hence, when nonlinear amplification is used to reduce the length of the pulse being amplified, the smooth component of the leading edge is chopped off by a fast shutter of some sort. At present this method of reducing the pulse length has not attained wide acceptance, although it could be used for preferable amplification of one short pulse in a train of such pulses.

Problems to Lecture 5

5.1. Estimate the bandwidth of a ruby laser with a 5-cm long active element, a population inversion density $n = 5 \times 10^{19}$ cm^{-3}, and a transition cross section $\sigma = 2.5 \times 10^{-20}$ cm^2. The transition linewidth $\Delta\nu$ is 330 GHz (at $T = 300$ K).

5.2. Does the solution to the previous problem depend on the input signal power? Estimate the input intensity at which the bandwidth will again begin to increase by employing the data of Problem 3.6.

5.3. By how many times will the effective strength of the input noise of a quantum amplifier based on the three-level scheme be greater than this quantity for a four-level quantum amplifier, all other things being equal?

5.4. Compare the minimum effective internal-noise levels of quantum amplifiers in the X-ray, optical, and microwave ranges.

5.5. In the saturation mode and the absence of losses the stimulated emission rate is much higher than the sum of rates of relaxation and spontaneous decay. Hence, practically all atoms in a unit volume shifted in one second to the upper laser level transfer to a lower level due to the stimulated emission. Calculate the maximum power output produced, via stimulated emission, by a unit volume of the active medium of an Nd^{3+}: YAG laser with an active particle number density equal to 5×10^{19} cm^{-3}, a lifetime on the upper laser level equal to 230 μs, and a gain equal to 0.5 cm^{-1}.

5.6. With what active media known to you can a 10-ns pulse of coherent radiation interact in an incoherent manner?

5.7. Suppose that you want to greatly reduce the length of a pulse of radiation with $\lambda = 1.06$ μm by nonlinear amplification. What laser amplifier would you select for this experiment, an Nd^{3+}-glass laser or an Nd^{3+}: YAG laser?

LECTURE 6

Generation

Open cavity and its Q-factor. Regeneration of a cavity by amplification. Pass cavity amplifier. Reflecting cavity amplifier. Self-excitation conditions. Resonance conditions. Oscillating frequency. Maximum power output

We have studied a travelling-wave amplifier. However, the interaction of the radiation being amplified and the active media may occur in the standing-wave mode as well. In this case the radiation passes many times through the sample of a substance placed in the cavity. The common cavity used in the radio frequency range has dimensions comparable with the length of the wave excited in it. Clearly, no such cavities can be built for waves with very short wavelengths. This prompted the Russian physicist Alexander M. Prokhorov to propose an open cavity instead of the closed volume resonator even for the IR range. Such a cavity is much larger than the wavelength of the electromagnetic waves generated in it. The use of open cavities in optics led to the invention of a laser.

The simplest open cavity consists of two parallel reflecting plates, or mirrors, separated by a distance $l \gg \lambda$. Often the mirrors are circular in shape, and their diameters are also much greater than the length λ of the wave generated in the cavity. It can be assumed that in the space between the mirrors a plane standing wave is generated if an integral number of half-waves fit into the space. If the mirrors are large enough, the energy losses due to diffraction can be ignored. If, in addition, the space between the mirrors is filled with a media that does not incur energy losses, the Q-factor of the cavity is determined by the energy losses during reflection or, more precisely, for ideal mirrors by energy losses due to radiation escaping to the ambient space through the mirrors.

At high values of the reflectivity R, the Q-factor of a cavity can be estimated from simple energy considerations. Let us assume that the distribution of the field in the cavity closely resembles that in a standing wave. A standing wave is equivalent to two travelling waves of equal intensity propagating in opposite directions. Suppose that the energy flux in each of these travelling waves is P. Then the flux lost as a result of reflection from two identical mirrors is $2P(1 - R)$, while the energy stored in the cavity is $2Pl/c$.

Physics of oscillations and electronics know many equivalent ways of defining the Q-factor of an oscillating system. The most general is the energy definition, which states that, by definition, the Q-factor of a cavity is

$$Q = 2\pi \frac{\text{energy stored by cavity}}{\text{energy lost by cavity per period}}. \tag{6.1}$$

Here

$$Q = \frac{2\pi l}{\lambda} \frac{1}{1 - R}, \tag{6.2}$$

where $R \leqslant 1$. The effect of the mirrors can be considered as an increase in the path l travelled by a plane wave in a cavity by a factor of $1/(1 - R)$.

Obviously, laser theory requires a more thorough analysis of open cavities and the radiation-field distribution in them. Here, however, it is important only to note that a system of two parallel reflecting disks, that is, two parallel mirrors, is a high-Q cavity. Actually, an open disk cavity is the Fabry-Perot interferometer, well-known in optics. The Q-factor of an interferometer considered to be a cavity is equal to the interferometer's resolving power, defined as the ratio of the wavelength to the width $\delta\lambda$ of the interference fringe at half-maximum intensity. The resolving power of a Fabry-Perot interferometer in transmitted light is calculated in optics courses. If the two mirrors are identical,

$$Q = \frac{\lambda}{\delta\lambda} = \frac{2\pi l}{\lambda} \frac{\sqrt{R} \, \exp{(-\alpha l)}}{1 - R \, \exp{(-\alpha l)}}, \tag{6.3}$$

where $\exp{(-\alpha l)}$ characterizes the energy losses per pass. At $\alpha = 0$ and $R \approx 1$ formulas (6.3) and (6.2) coincide.

If we assume that α is nonzero only because of resonance losses, then when population inversion is achieved the loss factor $\exp{(-\alpha l)}$ becomes the power gain per pass, $K = \exp{(\alpha_0 l)}$. As a result the cavity is regenerated, that is, its Q-factor grows and the passband narrows. We get

$$Q = \frac{2\pi l}{\lambda} \frac{\sqrt{RK}}{1 - RK}. \tag{6.4}$$

Obviously, this formula is valid only as long as RK is less than unity.

For a better understanding of what happens during regeneration let us consider a pass optical amplifier with a Fabry-Perot cavity, which is a travelling-wave amplifier placed inside a Fabry-Perot cavity. Suppose that the input signal is incident on the left semitransparent mirror of the cavity. The space between the mirrors is filled with an active media. Let us study the amplified signal emerging from the cavity through the right semitransparent mirror identical to the left mirror. In this situation the gain can be calculated by adding the amplitudes of the rays emerging from the interferometer after multiple reflections.

For an input signal of unit amplitude, the output field is given by the following sum:

$$E = (1 - R)\sqrt{K} \, \exp[-j\omega(t - l/c)]$$
$$\times \{1 + KR \, \exp(-2j\omega l/c) + K^2 R^2 \, \exp(-4j\omega l/c)$$
$$+ K^3 R^3 \, \exp(-6j\omega l/c) + ... \}. \tag{6.5}$$

Assuming that RK is less than unity and summing the above geometric progression, we arrive at the following formula for the complex-valued field-strength gain:

$$\Gamma = (1 - R)\sqrt{K} \, \frac{\exp(j\omega l/c)}{1 - RK \, \exp(-2j\omega l/c)}. \tag{6.6}$$

Hence, the power gain is determined by the formula

$$G = |\Gamma|^2$$
$$= \frac{(1 - R)^2 K}{1 - 2RK \, \cos(4\pi l/\lambda) + R^2 K^2}. \tag{6.7}$$

At resonance, that is, at $\cos(4\pi l/\lambda) = 1$, we get

$$G_0 = \frac{(1 - R)^2 K}{(1 - RK)^2}. \tag{6.8}$$

At $R = 0$ the travelling-wave mode is maintained and $G_0 = K$. At $R = 1$ the input signal is not admitted into the amplifier and $G_0 = 0$. At $K = 1$ there is neither amplification nor losses, and G acquires the meaning of the transmission coefficient of the Fabry-Perot etalon, which is equal to unity at resonance. As $K \to 1/R$ the gain G_0 tends to infinity, which means that generation sets in.

The above reasoning shows that positive feedback leads to generation. The condition

$$RK = 1 \tag{6.9}$$

is, in essence, the energy condition of self-excitation. Its meaning is simple: the energy losses per pass must be compensated for by amplification in the same pass.

We have considered the regeneration of a pass cavity filled with an active medium. This amplifier, which has identical semitransparent mirrors at input and output, becomes an oscillator when the self-excitation condition $RK = 1$ is met. The generator is connected in the same manner with the ambient space through its two opposite ends. Since this is not always convenient, a laser oscillator usually has only one semitransparent mirror, the output one, while the other is made a "dead-end," that is, is not transparent for radiation. In this case, when population inversion sets in, the reflecting amplifier is regenerated rather than the pass amplifier, that is, a device whose reflectivity exceeds unity.

If we now sum up the field amplitudes at the output of the system, as we did for the pass amplifier, that is, sum up the amplitudes of the fields undergoing multiple reflections in the amplifier, we find that

$$\Gamma = \frac{K \exp(-2j\omega l/c) - \sqrt{K}}{1 - K\sqrt{R} \exp(-j\omega l/c)}. \tag{6.10}$$

This implies that

$$G = |\Gamma|^2$$

$$= \frac{R - 2K\sqrt{R} \cos(4\pi l/\lambda) + K^2}{1 - 2K\sqrt{R} \cos(4\pi l/\lambda) + K^2 R}. \tag{6.11}$$

At resonance, that is, at $\cos(4\pi l/\lambda) = 1$, we have

$$G_0 = \left(\frac{K - \sqrt{R}}{1 - K\sqrt{R}}\right)^2. \tag{6.12}$$

At $R = 1$ there is no amplification, and at $R = 0$ we have $G_0 = K^2$, that is, amplification of a travelling wave that passes through the active medium twice. Generation sets in if the self-excitation condition is met, that is, if

$$RK^2 = 1. \tag{6.13}$$

Many different combinations of mirrors and active media can be devised, and for each the appropriate self-excitation condition can be developed. Common to all is the requirement that the gain per (effective) pass of radiation through the system of mirrors of the cavity and the active medium exceed the energy losses in the cavity during the same pass. The general energy balance must include both the energy losses due to useful emission outwards and the parasitic energy losses due to light scattering, absorption, etc.

Let us return to the resonance denominators in (6.7) and (6.11). The resonance condition

$$\cos (4\pi l/\lambda) = 1 \tag{6.14}$$

is equivalent to the well-known standing-wave condition (an integral number of half-waves)

$$l/\lambda = m/2, \quad m = 1, 2, 3, \dots . \tag{6.15}$$

In the optical range the value of m for open cavities is 10^5-10^6. This leads to a high density of resonance peaks, since the separation between adjacent peaks is

$$\delta\lambda = \lambda/2m. \tag{6.16}$$

It may so happen, and this is usually the situation in the majority of cases, that the inverted resonance line is broader than the separation between resonances:

$$\Delta\nu_{\text{line}}/\nu_{\text{line}} > \delta\lambda/\lambda = 1/2m. \tag{6.17}$$

This leads to a number of specific features, to be discussed later, in the properties of lasers.

The self-excitation condition $RK = 1$ (or equivalent conditions) provides the necessary balance of amplitudes. But for the self-oscillating mode to set in there must also be a certain balance in phases. The resonance conditions of the (6.14) type are met for radiation of a frequency for which all phase shifts cancel out. Hence, the phase balance condition yields the value of the oscillating frequency. Let us study this question in greater detail.

The dispersive properties of the active medium determine the phase shift in the active substance. As is well known, in resonantly absorbing (amplifying) media anomalous absorption is accompanied by anomalous dispersion. The fact that the phase shift depends on frequency follows from the frequency dependence of the dielectric susceptibility of the medium. The index of refraction is defined as

$$n = \sqrt{\varepsilon} = \sqrt{1 + 4\pi\chi}, \tag{6.18}$$

with the dielectric susceptibility χ written in the form

$$\chi = \chi' - j\chi''. \tag{6.19}$$

For small χ' and χ'' we have

$$n = 1 + 2\pi\chi' - j2\pi\chi''. \tag{6.20}$$

This approximation is equivalent to the approximation of low losses (amplification) of radiation over distances of one wavelength. Representing a plane wave of frequency ω (wavelength λ) that is propagating in the z direction in the form

$$\begin{aligned} E &= E_0 \exp\left[j(\omega t - 2\pi nz/\lambda)\right] \\ &= E_0 \exp\left(-4\pi^2\chi''z/\lambda\right) \\ &\quad \times \exp\left\{j[\omega t - 2\pi(1 + 2\pi\chi'z)/\lambda]\right\}, \end{aligned} \tag{6.21}$$

we see that the imaginary part of the dielectric susceptibility determines the losses (amplification) of radiation in the substance while the real part determines the phase shift:

$$\alpha = 8\pi^2\chi''/\lambda, \tag{6.22}$$

$$\varphi_{\text{line}} = -4\pi^2\chi'z/\lambda. \tag{6.23}$$

There are well-developed quantum mechanical methods for determining χ. But we will again resort to a classical analogy, in which a classical oscillator with damping represents the two-level quantum system with an upper level whose lifetime is finite. The equation

$$\ddot{x} + \gamma\dot{x} + \omega_{\text{line}}^2 x = A\exp(j\omega t) \tag{6.24}$$

has a well-known solution describing forced oscillations:

$$x = \frac{A\ \exp\ (j\omega t)}{\omega_{\text{line}}^2 - \omega^2 + j\gamma\omega}. \tag{6.25}$$

From the definition of dielectric susceptibility as the proportionality factor in the direct variation of the dipole moment per unit volume of substance with field strength there follows the direct proportionality of χ to the displacement x caused by the field. Then, if we write

$$\omega_{\text{line}}^2 - \omega^2 = (\omega_{\text{line}} + \omega)(\omega_{\text{line}} - \omega) \approx 2\omega_{\text{line}}\delta\omega_{\text{line}}, \tag{6.26}$$

where by definition $\delta\omega_{\text{line}} = \omega_{\text{line}} - \omega$ is the detuning of the radiation frequency ω from the central frequency ω_{line} of the resonance transition line, and reduce χ to (6.19), we get

$$\chi' = \frac{2\delta\omega_{\text{line}}}{\Delta\omega_{\text{line}}^2 + 4\delta\omega_{\text{line}}^2}B, \tag{6.27}$$

$$\chi'' = \frac{\Delta\omega_{\text{line}}}{\Delta\omega_{\text{line}}^2 + 4\delta\omega_{\text{line}}^2}B, \tag{6.28}$$

with B a constant, and $\Delta\omega_{\text{line}} = \gamma$. It has proved expedient to allow for the individual features of the active medium by assigning to it the quantity α via (6.28) and (6.22). This makes it possible to exclude B and represent the phase shift in the substance as follows:

$$\varphi_{\text{line}} = -\frac{\delta\omega_{\text{line}}}{\Delta\omega_{\text{line}}}\ \ln\ G, \tag{6.29}$$

where $G = \exp\ (-\alpha z)$.

Hence, a phase shift in the substance appears when there is a nonzero detuning of the radiation frequency from the center of the line. Since the active substance is inside the cavity, the cavity's dispersion curve and the corresponding phase shift should be taken into account. The passage of radiation through the cavity is described by the cavity's frequency transmission function. Within the framework of a single oscillation mode the phase characteristic of the cavity is completely equivalent to that of a single LRC circuit. As is known, and this can easily be demonstrated with the aid of the equivalent circuit of a resonance filter, the phase shift between the input and output signals can be represented in the form

$$\tan \varphi_{cav} = Q \frac{1 - \omega^2/\omega_{cav}^2}{\omega^2/\omega_{cav}^2}, \tag{6.30}$$

with Q the Q-factor of the cavity, and ω_{cav} the cavity's natural frequency. Introducing $\delta\omega_{cav} = \omega_{cav} - \omega$, which by definition is the detuning of the radiation frequency ω from ω_{cav}, and assuming that it is small, we find that

$$\varphi_{cav} = \arctan\left(Q \, 2\frac{\delta\omega_{cav}}{\omega_{cav}}\right) \approx Q \, 2\frac{\delta\omega_{cav}}{\omega_{cav}} = 2\frac{\delta\omega_{cav}}{\Delta\omega_{cav}}, \tag{6.31}$$

where $\Delta\omega_{cav}$ is the cavity's bandwidth (in a single mode). Then, equating the phase shifts (6.31) and (6.29), we get the phase balance condition, which specifies the oscillating frequency of the laser (maser).

The detunings of the oscillating frequency from the cavity's natural frequency and from the frequency at the center of the line are linked by the following formula:

$$\frac{\omega_{cav} - \omega}{\Delta\omega_{cav}} = \frac{\omega - \omega_{line}}{\Delta\omega_{line}} \ln K, \tag{6.32}$$

with $\ln K = (\ln G)/2$. In other words, the oscillating frequency

$$\omega = \frac{\omega_{cav}\Delta\omega_{line}/\ln K + \omega_{line}\Delta\omega_{cav}}{\Delta\omega_{line}/\ln K + \Delta\omega_{cav}} \tag{6.33}$$

differs from both the cavity's natural frequency and the line's central frequency only if the cavity is not tuned precisely to the line ($\omega_{cav} \neq \omega_{line}$). In the limiting case of an extremely narrow line or an extremely high gain ($\Delta\omega_{line}/\ln K \to 0$) does ω tend to ω_{line}. In the limiting case of an extremely broad line ($\Delta\omega_{line} \gg \Delta\omega_{cav}\ln K$) the oscillating frequency is determined by the cavity's natural frequency. These ideas are directly related to the problem of quantum frequency standards based on molecular oscillators, on the one hand, and on tunable lasers, on the other.

Formula (6.33) was derived by Prokhorov in 1954 in connection with an ammonia molecular oscillator (ammonia maser) and in a somewhat different form, but its general significance was noted even in those years. From the laser point of view, it is important to note that in the case of a broad line the oscillating frequency is determined by the tuning of the cavity. If the line is inhomogeneously broadened and overlaps several vibration modes of the cavity (see (6.17)), generation emerges at the frequencies corresponding to these modes.

Thus, when a quantum amplifier is regenerated in the feedback process that emerges when the amplifier is placed inside a cavity, it becomes self-excited, that is, transforms into a generator (oscillator). Analysis of the gain of a regenerated amplifier has made it possible to determine the self-excitation condition, that is, to find the minimum gain per pass necessary for generation to set in (the self-excitation threshold). Analysis of the phase relationships that exist in self-excitation has enabled finding the oscillating frequency. The linear theory cannot do more. The oscillation amplitude can be found only if we allow for the nonlinearity of the amplification process, only within the framework of the nonlinear theory.

Let us now turn to the question of the output power of laser oscillators. As in the case of quantum amplifiers, the predominant role here is played by the saturation effect and the presence of radiation losses.

We will consider a laser with a single completely reflecting ("dead-end") mirror and another mirror with transmittance $T = 1 - R$ (a partially transparent output mirror). As the radiation is amplified, it propagates in the active medium inside the laser cavity between the mirrors in the form of a sequence of building-up travelling waves assumed plane. We write the transport equation for waves travelling from left to right and from right to left:

$$\frac{dJ^+}{dz} = -\beta J^+ + \frac{\alpha_0 J^+}{1 + J^+ + J^-},$$

$$-\frac{dJ^-}{dz} = -\beta J^- + \frac{\alpha_0 J^-}{1 + J^+ + J^-}.$$

$$(6.34)$$

Here intensity is measured in units of saturation intensity, $J = I/I_{sat}$, and the superscripts $\pm$ stand for waves travelling in opposite directions. We also assume that in each cross section z the saturation effect is determined by the total averaged energy fluxes from left to right and from right to left, that is, the influence of the mode composition of the radiation in the cavity on the saturation effect is ignored.

The system of equations (6.34) corresponds to the diagram in Figure 6.1. For the values of J at the boundary $z = 0$ we introduce the notation $J^+(0) = J_1$ and $J^-(0) = J_2$. We also assume that $R(0) = R$ and $R(l) = 1$. Then the condition that the generation process be stationary leads us to the following boundary conditions:

$$J_1 = RJ_2, \quad J^-(l) = J^+(l) = J_0. \tag{6.35}$$

The output intensity is given by the formula

$$J_{out} = J_2 - J_1 = (1 - R) J_2 = \frac{1 - R}{R} J_1. \tag{6.36}$$

In Eqs. (6.34) we divide one equation by the other. The result is the equation

$$\frac{dJ^+}{dJ^-} = -\frac{J^+}{J^-}, \tag{6.37}$$

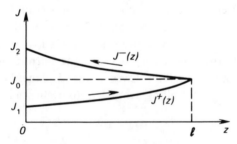

Figure 6.1. The intensity distribution in counterpropagating waves in a laser cavity of length l. The wave with intensity $J^+(z)$ propagates from left to right, and the wave with intensity $J^-(z)$ from right to left.

which is easily integrated:

$$J^+J^- = \text{const.} \tag{6.38}$$

Since $J^+ = J^- = J_0$ at $z = l$, we find that const $= J_0^2$. Hence, the solutions to Eqs. (6.34) must obey the following condition:

$$J^+J^- = J_0^2, \tag{6.39}$$

where J_0 is the intensity incident on the right (dead-end) mirror. We can show that J_0 is related through a simple formula to the output intensity. Indeed, since

$$J_1 = J_{\text{out}}R/(1 - R),$$
$$J_2 = J_{\text{out}}/(1 - R),$$

we have

$$J_1J_2 = J_0^2 = J_{\text{out}}^2 R/(1 - R)^2.$$

Hence,

$$J_0 = \frac{\sqrt{R}}{1 - R} J_{\text{out}}. \tag{6.40}$$

The solution now proceeds as follows. We write J^- in the form $J^- = J_0^2/J^+$, substitute it into the first equation in (6.34), drop the "+" in the superscript, and arrive at the equation

$$\frac{dJ}{dz} = - \left(\beta - \frac{\alpha_0 J}{J + J^2 + J_0^2} \right) J, \tag{6.41}$$

which is integrable in elementary functions.

What is interesting is the way in which the term determining the amplification and saturation varies owing to the presence of feedback and a return wave, in comparison with the case of a single travelling wave, $\alpha/(1 + J)$ (see (5.17)). Since this term in (6.41) is complicated, the solution to this equation is given by a transcendental algebraic equation that is difficult to comprehend and unwieldy for analysis.

In the particular case of strong saturation and a considerable excess of gain over losses ($\alpha_0/\beta \gg 1$, $J \gg 1$) we have

$$J_{\text{out}} = \frac{\alpha_0}{\beta} (1 - R) \frac{1 + \exp(-\beta l)}{1 - R \exp(-\beta l)}, \tag{6.42}$$

and

$$I_{\text{out}} \rightarrow (1 - R) \frac{\alpha_0}{\beta} I_{\text{sat}} \tag{6.43}$$

as $l \rightarrow \infty$. Since the oscillator is assumed infinitely long, Eq. (6.43) has no optimum in R. The highest possible intensity here is

$$I_{\text{out}}^{\text{max}} = \frac{\alpha_0}{\beta} I_{\text{sat}}, \tag{6.44}$$

which coincides, as expected, with the case of an amplifier (see (5.26)).

In the opposite particular case of a short oscillator ($\beta l \ll 1$) we can assume that both J^+ and J^- grow linearly with z. Then $J^+ + J^- = \text{const} = J$, and the initial equations (6.34) assume the form

$$\frac{dJ^+}{J^+} = \left(-\beta + \frac{\alpha_0}{1 + J} \right) dz,$$

$$\frac{dJ^-}{J^-} = \left(\beta - \frac{\alpha_0}{1 + J} \right) dz \tag{6.45}$$

and can easily be integrated.

As a result, the output intensity of our oscillator reaches its maximum value

$$I_{\text{out}}^{\text{max}} = \alpha_0 l (1 - \sqrt{\beta/\alpha_0})^2 I_{\text{sat}} \tag{6.46}$$

when the transmittance of the output mirror reaches its optimal value

$$T_{\text{opt}} = 1 - R_{\text{opt}} = 2\alpha_0 l (\sqrt{\beta/\alpha_0} - \beta/\alpha_0). \tag{6.47}$$

The feedback magnitude depends on the loss factor β to a greater extent than the maximum power output corresponding to this feedback. At

$\alpha_0 = 10^{-3}$ cm^{-1}, $l = 10^2$ cm, and $\beta = 10^{-4}$ cm^{-1} we have $T_{opt} = 0.042$ and $I_{out}^{max} = 0.07 I_{sat}$. This example shows what high demands are imposed on laser mirrors at the moderate gains characteristic, as a rule, of gas lasers. With such weak feedback only a small fraction of the energy flux escapes from the laser cavity. For the data of this example, the intensity incident on the dead-end mirror from inside the cavity is approximately 20 times higher than the output intensity (see (6.40)). At high gains the situation simplifies considerably.

Note that many medium-output lasers operate under conditions specified by (6.46) and (6.47).

It is also worth emphasizing that while the self-excitation conditions of a laser oscillator, (6.9) (or the equivalent formula (6.13)), were derived via the linear theory, the amplitude (strength) of the oscillations was determined by employing the results of the nonlinear theory. This conclusion is a particular manifestation of a more general idea of the theory of oscillations, namely, that the amplitude of steady-state oscillations in a self-oscillating system with the self-excitation conditions maintained is determined from the nonlinear theory, which allows for nonlinear effects in the initial equations. In quantum electronics the saturation effect is such a nonlinearity.

Thus, the intensity emitted by a laser oscillator with low β/α_0 and optimal feedback is $\alpha_0 l I_{sat}$, which means that all the energy stored in the laser's active medium is emitted into the ambient space. That this conclusion is of a general nature can be proved by estimating the power output of the laser cavity into the ambient space using the framework of an equivalent circuit with lumped parameters.

Let us represent the laser cavity in the form of a resonance LCR circuit (Figure 6.2). The active medium emitting radiation is equivalent to an oscillator with an e.m.f. E_{em} and an internal resistance R_{em} series-connected into an $L_0 C_0 R_0$ circuit. The emission of radiation into the ambient space is represented by the inductive coupling of the circuit with a matched long line. Let the load impedance of the line be Z_{load} and the coupling-loop mutual inductance be M.

The active-medium power output is related via the simple relationship

$$P_{em} = E_{em}^2/4R_{em} \tag{6.48}$$

to the e.m.f. and internal resistance of the oscillator. The following Kirchhoff equations correspond to the equivalent circuit of Figure 6.2 at the

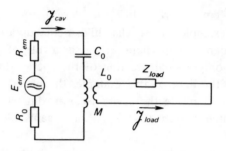

Figure 6.2. Equivalent circuit for a cavity with an active substance emitting radiation.

resonance frequency:

$$0 = Z_{\text{load}} \cdot \mathscr{I}_{\text{load}} + j\omega M \cdot \mathscr{I}_{\text{cav}},$$
$$E_{\text{em}} = \mathscr{I}_{\text{cav}}(R_{\text{em}} + R_0) + j\omega M \cdot \mathscr{I}_{\text{load}}. \tag{6.49}$$

Canceling out the cavity current $\mathscr{I}_{\text{cav}}$ in Eqs. (6.49), we arrive at the following expression for the load current $\mathscr{I}_{\text{load}}$:

$$\mathscr{I}_{\text{load}} = -j\frac{\omega M}{Z_{\text{load}}}\frac{1}{\omega_0 L}\frac{E_{\text{em}}}{1/Q_0 + 1/Q_{\text{cpl}} + 1/Q_{\text{em}}}, \tag{6.50}$$

where we have introduced the coupling Q-factor

$$Q_{\text{cpl}} = \frac{\omega L_0}{\omega^2 M^2/Z_{\text{load}}}, \tag{6.51}$$

the intrinsic Q-factor of the cavity

$$Q_0 = \omega L_0/R_0, \tag{6.52}$$

and the Q-factor related to the internal resistance of the e.m.f. oscillator

$$Q_{\text{em}} = \omega L_0/R_{\text{em}}. \tag{6.53}$$

The power that dissipates in the load impedance is

$$P_{load} = Z_{load} |\mathscr{I}_{load}|^2$$

$$= \frac{1/Q_{cpl}Q_{em}}{(1/Q_0 + 1/Q_{cpl} + 1/Q_{em})^2} 4P_{em}. \tag{6.54}$$

Expressing the relationship between the power output and the power generated in the e.m.f. source in terms of the Q-factors introduced by (6.51)-(6.53) and well known from the theory of resonance circuits but also possessing a general energy meaning (see (6.1)) attaches a general meaning to the derived relationship (6.54).

The power output can be optimized by varying the coupling with ambient space. Equation (6.54) makes it possible to find the maximum power output,

$$P_{load}^{max} = \frac{1/Q_{em}}{1/Q_0 + 1/Q_{em}} P_{em}, \tag{6.55}$$

achieved at optimal coupling

$$(1/Q_{cpl})_{opt} = 1/Q_0 + 1/Q_{em}. \tag{6.56}$$

When losses in the cavity are negligible ($1/Q_0 \ll 1/Q_{em}$), the entire radiative power at optimal coupling goes into the resistive load. Actually, the loop that couples the cavity with the line acts as an impedance transformer that matches the internal resistance of the e.m.f. source with load resistance. In the laser the transformer is a semitransparent output mirror that matches the active substance of the laser cavity with the free space.

From the standpoint of physics the generation process is enforced by the emission of radiation by the media with population inversion. Its intensity is determined by the population inversion density. To estimate the effect that an active media may have on this process, especially in the case of high-power lasers, the following simple line of reasoning has proved expedient. Suppose that in steady-state conditions the rate at which population inversion is created is Λ particles per unit time per unit volume. Then the maximum power output per unit volume is

$$P_1 = \Lambda h\nu. \tag{6.57}$$

This simple relation provides, say, for a CO_2 laser ($\lambda = 10.6\ \mu m$) an estimate from which it follows that for a mass flow rate M [kg/s] of the excited gas the maximum power output is $250M$ [kW], which corresponds to a power output of 250 kW at a mass flow rate of 1 kg/s.

For pulsed lasers the lase energy is determined by the amount of energy stored in the active substance by the time lasing begins, that is, by population inversion created up to the time of lasing. Since lasing ceases when population inversion vanishes, which happens when the populations of the upper and lower laser levels become equal, inversion with N particles corresponds to an emission energy

$$E_{em} = Nh\nu/2. \tag{6.58}$$

For a CO_2 laser this means 125 kJ for 1 kg of excited carbon dioxide.

Problems to Lecture 6

6.1. Compare the values of the Q-factor for open cavities in the optical and microwave ranges ($R \approx 0.99$). Determine the bandwidths of the cavities for the cases selected.

6.2. By how many times will the cavity bandwidth narrow under regeneration if the active medium is a ruby crystal, the reflectivity R of the output mirror is 0.5, the dead-end mirror has $R = 1$, and the power gain per pass of active medium is 1.3 dB? What must the gain of such a cavity be for oscillations to set in?

6.3. What is the greatest cavity length of a He-Ne laser with an oscillation linewidth $\Delta\nu_D = 1$ GHz at which the laser will still operate in a single mode.

6.4. Estimate the maximum possible energy yield of a unit volume of an Nd-glass active medium if the population inversion density is 10^{14} cm^{-3}.

6.5. How much energy is emitted by a ruby laser with a rod 7-cm long and 0.5 cm in diameter and a population inversion density of 3×10^{18} cm^{-3}?

6.6. The damage threshold of a ruby laser rod is 20 J/cm^2.

(a) What is the maximum population inversion density that we may expect to achieve in the active element described in the previous problem at which the rod is still intact?

(b) Can the given rod be damaged by its intrinsic radiation if the number density of the active particles is 1.6×10^{19} cm^{-3}?

(c) Find the maximum length of such a rod at which it is still intact?

6.7. What is the length of the pulses generated by a Nd^{3+}: YAG laser with an active element in the form of a rod 6-cm long and 0.5 cm in diameter, a population inversion density of 2×10^{17} cm^{-3}, and a radiative power output of 800 kW?

6.8. At optimal transmittance of the output mirror the output irradiance of the oscillator is 5% of the saturation irradiance. Find the reflectivity of the output mirror of this laser if the unsaturated gain is equal to two ($\alpha_0 = 0.1$ cm^{-1}).

LECTURE 7

Open Cavities

Cavities in electronics. Transition to short waves. Drop in the value of the Q-factor and the overlapping of resonances of closed volumes. Open cavities and spectrum purification. The Fresnel number. Modes. The lifetime of a mode of a passive cavity. Diffraction losses. The Fox-Lee method. The integral equation of an open cavity

The previous discussion suggests that at the base of quantum electronics there lies an inversely populated active media with positive feedback which is maintained by the stimulated emission of radiation in the cavity. In this combination of active media, stimulated emission, and cavity, the active media stores the energy and amplifies the generated radiation, the stimulated emission ensures that the amplification is coherent, and the cavity forms the spectral and spatial properties of the generated radiation.

Quantum electronics, at least in origin, is a part of electronics and at present constitutes the optical part of electronics. It is well-known that in the classical electronics of the long-wave and microwave ranges the most important characteristic of monochromatic radiation is its frequency, which is determined by the properties of the resonance circuit. The circuits used for long waves are quasi-stationary AC circuits, that is, circuits with lumped parameters, with the result that the size of the respective resonance circuits is much smaller than the radiation's wavelength. When we go over to microwave radiation, the circuits become essentially nonstationary, of the travelling wave type, owing to the sharp decrease of wavelength. Here various types of waveguides, such as coaxial, tubular and dielectric are used to channel the radiant energy. Sections of such waveguides, shorted in the proper manner and spatially organized in accordance with the electrodynamics of microwaves and the expected distribution of fields in these devices, serve as microwave cavities.

The best-known are hollow metal resonant cavities. The low losses in the walls at sizable reflectivities from a highly conductive metal surface lead to high values of the Q-factors of such cavities. The configuration and distribution of fields in microwave cavities differ considerably from the case of free space. The linear dimensions of such cavities are comparable to the wavelength. Hence, the natural oscillation spectrum of such an oscillator is considerably rarefied. As a rule it is comparatively easy in the microwave range to realize such configurations of resonant cavities for which the cavity possesses only one natural vibration mode within a broad frequency range. Resonant microwave systems determine the oscillating frequency of self-oscillating microwave systems.

Long waves are usually emitted into the ambient space in all directions, almost isotropically. As the wavelength gets shorter and especially as we move into the microwave range it becomes possible to form markedly anisotropic spatial distributions of radiation, which are known in electronics as directivity patterns. Formation of these patterns is carried out by antenna systems external with respect to the oscillator. Usually these are interference systems and in the short-wave region quasi-optical, but always with dimensions greater than the wavelength. Examples of these are the antenna arrays of radars and of satellite communication systems.

As we move to higher frequencies, that is, to the submillimeter or even the IR range, it becomes technologically impossible to manufacture resonant cavities with dimensions of the order of the wavelength. Thus, we are forced to go over to cavities whose dimensions are considerably greater than the wavelength. Briefly this aspect was discussed in connection with self-excitation conditions of lasers and, primarily, with the oscillating frequency. In a cavity whose dimensions considerably exceed the wavelength a set of directions of propagation of the radiation may realize itself. Feedback in such a cavity is maintained by stimulated emission of photons possessing the same frequency, polarization, and direction of propagation, that is, the same ω and $\mathbf{k}$. Hence, the cavity determines the phase balance in the four-dimensional space $(\omega t, \mathbf{k} \cdot \mathbf{r})$, with $\mathbf{r}$ a radius vector. In other words such a cavity determines the four-dimensional oscillating frequency or a set of such frequencies. This implies that in quantum electronics the cavity generates simultaneously the frequency of the oscillations and the direction of propagation of the generated radiation, that is, the radiation's temporal and spatial characteristics, which prove to be closely related.

Hence, cavities used in the optical range have dimensions that are much greater than the wavelength. It must be stressed that the drawback of cavities with dimensions comparable to wavelengths of the micrometer or sub-micrometer range is not only the technological difficulties encountered in manufacturing them or their small size and, hence, small power or energy of oscillations; as the dimensions of hollow metal resonant cavities decrease in proportion to the wavelength, the Q-factor drops. Indeed, for a closed metal cavity, as is demonstrated in courses of microwave electrodynamics, the Q-factor is the ratio of the characteristic linear dimension of the cavity, a, to the radiation's depth of penetration of the metal, δ, the skin depth:

$$Q = a/\delta. \tag{7.1}$$

For normal skin effect the skin depth is inversely proportional to the square root of frequency: $\delta \propto \nu^{-1/2}$. By assumption, the linear dimension of a cavity decreases in direct proportion to wavelength: $a \propto \nu^{-1}$. Hence, $Q \propto \nu^{-1/2}$, that is, even on the totally unrealistic assumption that all other conditions remain unchanged the value of the Q-factor drops markedly as we move from the microwave range to the optical.

Thus, we are forced to consider cavities whose dimensions are considerably greater than the wavelength. One additional fact must be taken into consideration, however. In a closed volume whose linear dimensions are considerably greater than the wavelength the number of field oscillators per unit volume and per unit spectral interval coincides with that in free space. This number, equal to

$$n = 8\pi\nu^2/c^3, \tag{7.2}$$

was introduced in Lecture 1 when we discussed the probabilities of spontaneous and stimulated emission (e.g. see Eq. (1.12)). The overall number of field oscillators in volume V and frequency interval $\Delta\nu$ is

$$N = \frac{8\pi\nu^2}{c^3} V\Delta\nu. \tag{7.3}$$

The frequency interval per oscillator (one natural oscillation of the closed volume) is

$$\Delta\nu/N = c^3/8\pi V\nu^2 \propto 1/\nu^2, \tag{7.4}$$

that is, is inversely proportional to ν^2. At the same time, the width of the frequency band per oscillation is determined by the Q-factor of the vibration.

In the case of a closed cavity of large dimensions the value of a in (7.1) is frequency independent ($a \approx V^{1/3}$), and for normal skin effect the Q-factor proves to be proportional to $\nu^{1/2}$. Hence, the width of the resonance curve of the corresponding oscillation, $\Delta\nu_{osc}$, proves to be proportional to $\nu^{1/2}$:

$$\Delta\nu_{\text{osc}} = \nu/Q \propto \nu^{1/2}. \tag{7.5}$$

Comparison of (7.4) with (7.5) shows that as frequency or volume increases the resonance curves of the oscillations of a closed cavity begin to overlap. This means that the cavity is losing its resonant properties.

Hence, in the optical range a cavity with linear dimensions of the order of the wavelength cannot be employed in view of technological difficulties and because of a drastic drop in the value of the Q-factor; a cavity in the form of a metal enclosure with linear dimensions much greater than the wavelength cannot be used because of the high density of the natural oscillations, which leads to loss of resonant properties. What needed are large cavities with a rarefied spectrum of natural oscillations.

The most promising way of purification the spectrum of the eigenoscillations of large-volume cavities while preserving high values of the Q-factor has proved to be the use of open cavities. These have acquired the widest application.

Let us take an open cavity consisting of two flat disks of radius a each with a reflectivity $R \leq 1$ that are separated by a distance l, are parallel to each other and positioned on a single axis perpendicular to it. A simple estimate can be made by assuming that a system of plane waves propagates within the volume between the disks. We used the same approach in Lecture 6 to derive the expression (6.2) for the Q-factor of such a cavity. As we noted the effect of the reflecting surfaces of the disks can be seen as an increase in the length l of the path travelled by the flat waves by a factor of $1/(1 - R)$. This effective increase in path length means an e-fold attenuation of the plane wave after $1/(1 - R)$ reflections.

In addition to a wave that propagates at exactly right angles to the disks' surfaces, other waves that propagate in directions almost perpendicular to these surfaces may be excited in the volume between the disks. If a plane wave propagating at a certain angle to the cavity's axis is able to reflect $1/(1 - R)$ times from the surfaces before it leaves the cavity, the corresponding resonance has a Q-factor approximately half of that for a wave propagating at right angles to the surfaces. Consequently, the angle

$$\theta = 2a(1 - R)/l \tag{7.6}$$

is the critical angle defining the cone limiting the directions of propagation of waves corresponding to oscillations with high-Q values. Hence, out of the overall number of oscillations given by (7.3) only those whose waves propagate within a solid angle $\Omega = \pi\Theta^2$ have high-Q values.

Multiplying both sides of (7.3) by $\Omega/4\pi$ and performing simple transformations, we arrive at an expression for the overall number of such oscillations within a frequency band $\Delta\nu$:

$$N_0 = 32\pi^2 \frac{a^4(1 - R)^2}{\lambda^3 l} \frac{\Delta\nu}{\nu}, \tag{7.7}$$

with

$$N = 32\pi^2 \frac{a^2 l}{\lambda^3} \frac{\Delta\nu}{\nu}, \tag{7.8}$$

and $\lambda = c/\nu$ the wavelength.

A comparison of these formulas suggests a considerable reduction, by a factor of $l^2/a^2(1 - R)^2 \gg 1$, in the number of natural oscillations that land in a unit spectral interval in the case of an open cavity. The reason for such a strong rarefaction of the spectrum of natural oscillations is the absence of lateral sides in the open cavity.

It is convenient to represent formula (7.7) in another form. According to (7.7), the frequency interval per oscillation in the open cavity is

$$\frac{\Delta\nu}{N_0} = \frac{\lambda^3 l\nu}{32\pi^2 a^4(1 - R)^2}. \tag{7.9}$$

On the other hand, the bandwidth for a single oscillation, $\Delta\nu_{\text{osc}}$, is determined by the cavity's Q-factor (see Eq. (6.2)):

$$\Delta\nu_{\text{osc}} = \frac{\nu}{Q} = \frac{\lambda\nu (1 - R)}{2\pi l}. \tag{7.10}$$

The resonance curves for distinct oscillations do not overlap when

$$\frac{\Delta\nu_{\text{osc}}}{\Delta\nu/N_0} = 16\pi N_{\text{F}}^2(1 - R)^3 < 1, \tag{7.11}$$

where we have introduced the following notation for the Fresnel number:

$$N_{\text{F}} = a^2/l\lambda. \tag{7.12}$$

The above reasoning assumes that the geometrical optics approximation is valid and that no diffraction phenomena are present. The geometrical optics criterion is $N_F > 1$. The reader will recall that this criterion was in fact used in Lecture 5 to derive formula (5.14) for the effective spectral density of the input noise of a travelling-wave quantum amplifier. Returning to (7.11), we see that when the mirrors of an open cavity have a fairly high reflectivity, that is, the cavity's Q-factor is fairly high, we can achieve a satisfactory purification of the spectrum of eigenoscillations even for large Fresnel numbers.

Thus, open cavities are convenient resonant systems for optical quantum electronics. The eigenoscillations (types of oscillation) are commonly known as modes. By definition, a mode in a cavity is a field distribution that reproduces itself during multiple propagation of the wave between the cavity's mirrors. Real energy losses lead to the damping of oscillations corresponding to one or another mode, provided that the mode's development is not sustained by emission from the active media.

Let us suppose that the losses of the energy of the radiation propagating in the form of an oscillation mode of the cavity between the cavity's mirrors can be described by a certain equivalent absorption coefficient α:

$$\frac{dI}{dz} = -\alpha I. \qquad (7.13)$$

The magnitude of α is determined by the losses from absorption and scattering of light in the media between the mirrors and from diffraction and reflection. It is convenient to represent this absorption coefficient in the form $\alpha = A/l$, where A is the radiation-energy absorption coefficient per pass of cavity of length l. Introducing the energy density $\varrho = I/c$ and bearing in mind that $dz = c\,dt$, we arrive at an equation

$$\frac{d\varrho}{dt} = -\frac{Ac}{l}\varrho, \qquad (7.14)$$

whose solution

$$\varrho = \varrho_0 \exp\left(-\frac{Ac}{lt}\right) = \varrho_0 \exp\left(-\frac{t}{\tau_{\text{eff}}}\right) \qquad (7.15)$$

attests to the exponential nature of the damping of the cavity's eigenoscillations with the characteristic time constant

$$\tau_{\text{eff}} = l/(Ac)$$
$$= 1/(\alpha c). \tag{7.16}$$

The time τ_{eff}, known as the photon's lifetime in the mode, is related to the mode's Q-factor via a simple relationship. By the definition (6.1),

$$Q = 2\pi \frac{\varrho_0}{\varrho_0[1 - \exp(-T/\tau_{\text{eff}})]} = 2\pi \frac{\tau_{\text{eff}}}{T}$$

$$= \omega\tau_{\text{eff}}, \tag{7.17}$$

where T is the period of the eigenoscillations of the mode considered, with $T \ll \tau_{\text{eff}}$ for modes with sizable Q-factors. A specific lifetime corresponds to each type of loss. Since losses add up, the resulting lifetime of a mode is determined by the obvious formula

$$1/\tau_{\text{eff}} = \sum 1/\tau_{\text{eff}}^{(i)}. \tag{7.18}$$

A similar expression holds true for the Q-factor:

$$1/Q = \sum 1/Q^{(i)}, \tag{7.19}$$

where the superscript i stands for the type of energy loss that determines the corresponding partial Q-factor (lifetime).

In open cavities the problem of diffraction losses is especially important. First of all, this type of loss is responsible for the purification of the spectrum of eigenoscillations when we go over from resonant cavities to open cavities by excluding, while the resonant oscillation mode sets in, oscillations that propagate at noticeable angles to the cavity's axis. It is the presence of diffraction losses that distinguishes open optical cavities from microwave resonant cavities. Of course, this is not the only source of losses, and in many cases even not the primary source. But even in the case of ideal mirrors and an ideal active media the energy losses caused by diffraction at the edges of mirrors with a finite aperture remain unremovable in principle and, hence, an important source of losses.

Within the framework of the geometrical-optics approach to describing open cavities we, obviously, cannot allow for the diffraction losses. Geometrical optics holds for large Fresnel numbers $N_F = a^2/l\lambda$. It is natural to assume then that this parameter determines the size of the energy losses.

A rough estimate can be made in the plane-wave approximation. According to Young, diffraction at the edge of a screen can be interpreted as transverse diffusion of the light field amplitude in the shadow zone. At a distance l from the screen the diffusion zone is $\sqrt{l\lambda}$ large. Hence, a beam of light carrying a near-plane wave reflected, say, from the left mirror of radius a and travelling a distance l to the right mirror of the same radius broadens in its radius by $\sqrt{l\lambda} \ll a$. The radiation that lands inside a ring of area $2\pi a\sqrt{l\lambda}$ escapes from the cavity. The fraction of such escaping radiation is $2\sqrt{l\lambda}/a$, provided that the light field amplitude is distributed evenly over the beam's cross section. By squaring this quantity we arrive at an estimate for the diffraction energy losses per pass:

$$A_{\mathrm{dif}} = 4l\lambda/a^2 = 4/N_{\mathrm{F}}. \tag{7.20}$$

The greater the Fresnel number, the smaller the diffraction losses. The above estimate, which has meaning only for large N_{F}, provides a rough picture of the behavior pattern of diffraction losses as N_{F} varies. In reality the field distribution over the beam's cross section differs considerably in the cavity modes from a uniform one, noticeably diminishing toward the edges. For this reason the diffraction losses prove to be considerably lower than those predicted by formula (7.20).

The problem of diffraction losses in open cavities is closely linked with the possibility of stable modes existing in such cavities. Indeed, diffraction losses hinder the return to the cavity of the total energy of the initial radiation in each pass of radiation between the mirrors and reflection from a mirror. It is, therefore, natural to ask whether after numerous passes the field distribution in the cavity approaches a steady state, which reproduces itself during each subsequent pass, in other words, whether there is an oscillation mode in the open cavity. A related question concerns the number of possible modes that differ in their field configurations and losses.

The answer was given in 1960-61 in the well-known works of A.G. Fox and T. Lee, who developed a graphic picture of natural mode formation in an open cavity by studying the variations in the distribution of the amplitude and phase of the initial plane wave after its multiple subsequent passes through the cavity.

Let a uniform plane wave originate at the left mirror and propagate to the right one. During propagation a fraction of the energy escapes owing to diffraction from the peripheral region of the wave even before the wave has reached the right mirror. Reflection also reduces the intensity of the wave's peripheral region. The reflected wave propagating from right to left loses its energy in a similar manner. As a result of multiple passes the field at the edges of the wavefront becomes ever weaker.

In calculations the arbitrary initial field distribution at the left mirror serves as the source of the field generated at the right mirror as a result of the first pass of the wave. Then the obtained distribution is used in exactly the same manner to calculate the field distribution created at the left mirror as a result of the second pass. The calculations are repeated many times for subsequent passes.

To calculate the electromagnetic field at one of the mirrors in terms of the integral of the field at the other, the scalar formulation of the Huygens principle is employed. This is justified if the mirrors are large compared with the wavelength and if the electromagnetic field is near-transverse and linearly polarized.

Calculations were conducted by computers. It was found that after many reflections (about 300) a steady-state distribution of the field did indeed set in, with an amplitude that falls off at the mirror edge. Diffraction losses were also found to be by several orders of magnitude lower than those predicted by formula (7.20). The diffraction-loss curve obtained from computer calculations within a broad range of Fresnel numbers can be approximated by the formula

$$A_{\text{dif}} = ae^{-bN_{\text{F}}}. \tag{7.21}$$

In the important case considered in Lecture 8, of the so-called confocal cavity, for the fundamental mode we have $a \approx b \approx 10$, which, of course, leads to infinitesimal diffraction losses, especially for large values of N_{F}.

Computer calculations have substantiated the intuitive conclusion that after multiple passes the field distribution near the mirror varies little from reflection to reflection. In the steady state the fields near the mirrors are the same to within a complex-valued factor. Then, expressing the field at one of the mirrors in terms of the field at the other via the Huygens principle in the Fresnel-Kirchhoff form, we arrive at an integral equation for the sought field distribution function at a mirror.

Indeed, the field u_1 in the Fresnel zone at one of the mirrors, generated by the radiation reflected from the other mirror of area A, is given by an integral over A:

$$u_1 = \frac{jk}{4\pi} \int_A u_2 \frac{e^{-jkr}}{r} (1 + \cos\theta) \, ds, \tag{7.22}$$

where u_2 is the field at the aperture of the "emitting" mirror, k a propagation constant, r the distance from a point on the "emitting" mirror to the observation point, and θ the angle that vector $\mathbf{r}$ forms with the normal to the mirror plane. After q passes the field at one of the mirrors is related to the field reflected by the other mirror via formula (7.22) in which u_1 must be replaced by u_{q+1} and u_2 by u_q.

Computer calculations have demonstrated the feasibility of assuming that after multiple passes the field distribution at the mirrors changes little from reflection to reflection and becomes steady-state. The fields at the mirrors coincide to within a complex-valued constant. We can, therefore, write

$$u_q = (1/\gamma)^q v, \tag{7.23}$$

with v the distribution function, which does not vary from reflection to reflection, and γ the complex-valued quantity that characterizes the conditions for radiation propagation between reflections. Substituting (7.23) into (7.22), we arrive at the integral equation

$$v = \gamma \int_A vK \ ds, \tag{7.24}$$

whose kernel is

$$K = \frac{jk}{4\pi} (1 + \cos \theta)e^{-jkr}. \tag{7.25}$$

Its eigenfunctions are the modes (normal oscillations, eigenoscillations, normal types of oscillations, oscillation modes, etc.) of the cavity studied, and ln γ determines the wave's damping and phase shift per pass, being, consequently, the propagation constant of the respective modes.

The analysis carried out by Fox and Lee for cavities of the Fabry-Perot interferometer type in several geometric configurations (rectangular flat mirrors, circular flat mirrors, and confocal spherical and parabolic mirrors) suggests the following important conclusions.

1. Open cavities of the Fabry-Perot interferometer type are characterized by a discrete set of oscillation modes for both flat and concave mirrors.

2. Uniform plane waves are not the normal modes of open cavities.

3. Electromagnetic waves corresponding to the normal modes of a cavity

are almost completely transverse. For this reason we speak of a transverse electromagnetic mode, or TEM.

4. Higher modes always have higher diffraction losses than the fundamental mode.

5. In the fundamental mode the field amplitude rapidly drops off at the mirror edges. Hence, the respective diffraction losses are much lower than those obtained on the assumption that the plane waves are uniform and in reality can be ignored.

The results of Fox and Lee have shown the fruitfulness of analyzing fields and waves in open cavities by solving integral equations that link the fields at the cavity mirrors via the Huygens principle in the Fresnel-Kirchhoff integral framework.

Problems to Lecture 7

7.1. At $R = 1$ formula (7.7) for the overall number of oscillations within a frequency band $\Delta\nu$ yields $N_0 = 0$. For what oscillations is this estimate valid?

7.2. What is the physical meaning of the dimensionless parameter N_F?

7.3. Why is the condition $N_F > 1$ the criterion of applicability of the geometrical-optics approximation?

7.4. Estimate the number of possible oscillation modes by applying formula (7.8) for an Ar^+ laser with the following parameters: $\lambda = 488.0$ nm, $l = 1$ m, $a = 1$ cm, and $\Delta\nu = 3.5$ GHz. By how many times will the number of modes decrease if $R_{out} = 0.95$?

7.5. Assuming no other losses than those associated with extraction of radiation through the output mirror, estimate τ_{eff} and Q-factor of the cavity of the laser of Problem 7.4.

7.6. What must be the minimum reflectivity of the output mirror of a helium-neon laser with $l = 1$ m, $a = 0.5$ cm, and $\lambda = 0.63$ μm so that the resonance curves for separate oscillation modes do not overlap?

LECTURE 8

Gaussian Beams

The confocal cavity. The field distribution. Gaussian beams.
Size of spot. Radiation beam divergence. Wavefront curvature
radius. Gaussian beam transformation by lenses. Cavity mode
matching. Focusing Gaussian beams. The longitudinal and
transverse sizes of the focal region

The Fox-Lee analysis has demonstrated that in open structures of the Fabry-
Perot interferometer type there are characteristic oscillation modes. At
present there are many modifications of open cavities, differing in configuration and mutual position of the mirrors. The simplest and most convenient cavity is the one formed by two spherical reflectors of equal
curvature whose concave surfaces face each other and are separated by a
distance equal to the curvature radius (the radius of either sphere). The
focal length of a spherical mirror is equal to half of the sphere's curvature
radius. Hence, the focal points of the reflectors coincide, and for this reason
the cavity is known as confocal (Figure 8.1). Interest in the confocal cavity
stems from the convenience in aligning it: in contrast to the plane-parallel
Fabry-Perot cavity, a confocal cavity does not require a strictly parallel ad-
justment of the reflectors. The only requirement is that the axis of a confocal cavity intersect each reflector at a point far enough from the reflector's
edge. Otherwise the diffraction losses may be two high.

Let us study the confocal cavity in greater detail. Suppose that the size
of the cavity is large compared to the radiation's wavelength. Then the cavity's modes, the field distribution, and the diffraction losses can be obtained
on the basis of the Huygens-Fresnel principle by solving the appropriate
integral equation. For spherical mirrors with a rectangular or circular aperture this integral equation allows for variable separation with respect to
transverse coordinates and is reduced to one-dimensional integral equations.
If the reflectors of a confocal cavity have a cross section in the form of
a square with side $2a$ small if compared with the distance l between the
mirrors, which is equal to the mirrors' curvature radius R, and the Fresnel
numbers are large, then the eigenfunctions of the Fox-Lee integral equation
can be approximated by products of Hermite polynomials $H_n(x)$ and the
Gaussian function exp $(-x^2/w^2)$.

In a Cartesian system of coordinates with its origin at the cavity's center,
that is, the confocal point, and the z axis coinciding with the cavity's axis

(see Figure 8.1), the transverse field distribution is given by the formula

$$S(x, y) = H_m(x/w) H_n(y/w) \exp [-(x^2 + y^2)/2w^2], \qquad (8.1)$$

where w defines the size of the region in the cross section upon leaving which the in-the-cavity intensity, proportional to S^2, diminishes e-fold. In other words, w is the width of the intensity distribution. The energy of the wave propagating in the direction z through the xOy plane is concentrated primarily in a spot of area πw^2.

The first five Hermite polynomials are

$$H_0(x) = 1, \; H_1(x) = 2x, \; H_2(x) = 4x^2 - 2,$$
$$H_3(x) = 8x^3 - 12x, \; H_4(x) = 16x^4 - 48x^2 + 12. \qquad (8.2)$$

To the eigenfunctions of the Lee-Fox equation that yield the transverse distribution (8.1) there correspond natural frequencies determined by the condition

$$4l/\lambda = 2q + (1 + m + n). \qquad (8.3)$$

Since the longitudinal component of the eigenfunctions is very small, the solutions of the initial equation and, hence, the transverse distributions (8.1) describe TEM$_{mnq}$ modes. Figure 8.2 shows the first three Hermite-Gauss functions for one of the transverse coordinates. These functions were constructed using formulas (8.1) and (8.2) and graphically demonstrate the variation of the transverse field distribution with an increase in the transverse index n.

Resonances in a confocal cavity exist only for integral values of $4l/\lambda$. The mode spectrum of a confocal cavity is degenerate, since increasing $m + n$ by two units and simultaneously decreasing q by one unit leaves

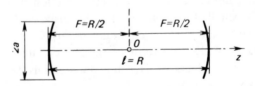

Figure 8.1. The schematic of a confocal cavity. The xOy plane is perpendicular to the z axis, and the confocal point is at the origin O.

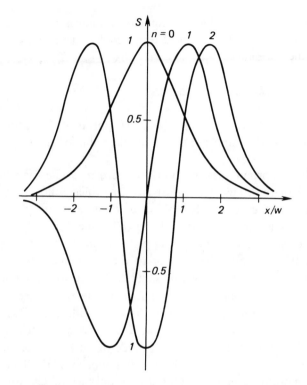

Figure 8.2. The field distribution in the confocal cavity along one of the transverse coordinates for the first three vibrational modes ($n = 0, 1, 2$). The peak values are normalized to unity.

the frequency unchanged. The subscripts m and n in TEM$_{mnq}$ refer to variations of the field in the directions x and y and, generally speaking, assume the values 0, 1, 2, ..., while q is the number of half-waves that fit into the length of the cavity along the z axis.

TEM$_{00q}$ is the fundamental mode, with the respective transverse field distribution determined by the simple Gaussian function $\exp\left[-(x^2 + y^2)/2w^2\right]$. The intensity width varies along the z axis according to the law

$$w^2 = w_0^2 + (z/kw_0)^2, \tag{8.4}$$

where $k = 2\pi/\lambda$, and w_0 stands for the beam radius in the cavity's focal plane, that is, at $z = 0$, which is commonly known as the neck radius of the caustic. The quantity w_0 is determined by the cavity's length and is given by the formula

$$w_0 = \sqrt{l\lambda/4\pi} = \sqrt{l/2k}. \tag{8.5}$$

At the mirror surface, that is, at $z = l/2$, the spot area of the fundamental mode, as shown by (8.4) and (8.5), is twice the value of the cross-sectional area of the caustic neck.

It is highly important that the transverse size of a Gaussian beam, $2w$, is independent of the transverse mirror size $2a$, which follows from the assumption that the Fresnel number $N_F = a^2/l\lambda$ is large and that the ratio a^2/l^2 is low. This led to a solution of the (8.1) type. Since (8.5) allows us to write the Fresnel number in the form $N_F = a^2/4\pi w_0^2$, the requirement that this number be large is equivalent to the requirement that the mode spot area at the mirror, $w^2 = 2w_0^2$, be small compared with the mirror area.

Solution (8.1) was obtained for the field inside the cavity. But when one of the mirrors is semitransparent, as happens with active laser cavities, the outgoing wave is a traveling wave with a transverse distribution (8.1).

Actually, separating the fundamental mode of an active confocal cavity is a way of obtaining a Gaussian beam of monochromatic light. Bearing in mind the importance and the interesting properties of Gaussian beams, let us examine them in greater detail.

Suppose that a plane is the surface of the wavefront of a monochromatic wave with a Gaussian distribution of the amplitude at this flat wavefront,

$$a(x, y) = E_0 \exp \left[-(x^2 + y^2)/2w_0^2\right]. \tag{8.6}$$

According to the Huygens-Fresnel principle, the initial wavefront generates a wave whose field is determined by the Fresnel-Kirchhoff integral.

$$E(x, y) = \int\limits_{-\infty}^{\infty} \int \frac{a(x', y')}{r} \cos (\omega t - kr) \, dx' dy', \tag{8.7}$$

where $r^2 = z^2 + (x - x')^2 + (y - y')^2$, and x, y, and z are the Cartesian coordinates introduced above (Figure 8.3). We replace $1/r$ with $1/z$ and put $r \approx z + [(x - x')^2 + (y - y')^2]/2z$ in $\cos (\omega t - kr)$, that is, we replace

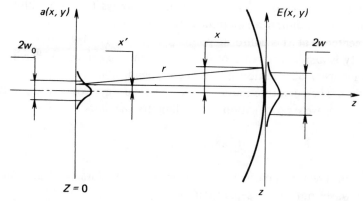

Figure 8.3. Calculating the wavefront generated at a distance z by a plane wave with a Gaussian distribution of the field in the plane $z = 0$.

the hypotenuse with the appropriate leg in the strongly elongated right triangle when dealing with the reduction in the field amplitude and allow for the nonparallelism of z and r when dealing with the phase shift of the wave. Then direct calculation yields

$$E(x, y) = \frac{2\pi}{k} E_0 w_0^2 \exp\left(-\frac{x^2 + y^2}{2w^2}\right) \left[w_0^4 + \left(\frac{z}{k}\right)^2\right]^{-1/2}$$

$$\times \cos\left[\omega t - k\left(z + \frac{x^2 + y^2}{2R}\right) - \alpha\right], \qquad (8.8)$$

where $R = z + (kw_0^2)^2/z$, $\tan \alpha = kw_0^2/z$, and w is given by formula (8.4).[3] In view of all that has been said in this lecture and in Lecture 7, it is not surprising that solution (8.8) coincides completely with the fundamental mode of a confocal cavity, TEM_{00q}.

If we ignore the weak dependence of α on z, the constant-phase surface of a Gaussian beam is determined by the equation

$$z + \frac{x^2 + y^2}{2R} = \text{const.} \qquad (8.9)$$

---

[3] Here E_0 has the meaning of a normalization constant; see Problem 8.11 and its solution.

For $x^2/R^2 \ll 1$ and $y^2/R^2 \ll 1$ (which is always the case within the framework of our initial assumptions) this equation specifies a sphere of radius R centered at the confocal point. Thus, the TEM_{00q} mode of a confocal cavity is a spherical wave propagating from the center and having a Gaussian intensity distribution in a plane perpendicular to the direction of propagation. The curvature radius of the spherical wavefront varies, in the process of the wave's propagation, according to the law

$$R = z + (kw_0^2)^2/z, \tag{8.10}$$

and at great distances from the origin ($z \gg kw_0^2 = l/2$) coincides with the wavefront-cavity separation, $R \approx z$.

This means that in the far zone the wavefront of a Gaussian beam resembles a spherical wave propagating from a point on the beam's axis at its focal neck. At $z = l/2$ the radius R is equal to l, that is, at the mirror surface the wavefront coincides, as expected, with the spherical surface of the mirror. Figure 8.4 shows the envelope of a Gaussian beam in a cavity and the wavefronts.

On the other hand, it is important to note that $R \to \infty$ as $z \to 0$. The symmetry plane of a cavity or, which is the same, the cavity's focal plane is the constant-phase plane. This means that at the focal neck the wave is plane but is spatially limited by an effective size w_0. This size determines the divergence of the TEM_{00q} mode.

The amplitude distribution over the wavefront of a Gaussian beam, (8.8), is axisymmetric and has a width ω given by (8.4). At great distances from the cavity ($z \gg kw_0^2 = l/2$) $w = z/kw_0$, which corresponds to the following angular divergence:

$$\theta = w/z = 1/kw_0. \tag{8.11a}$$

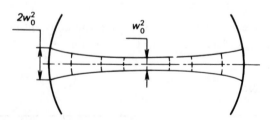

Figure 8.4. The envelope of a Gaussian beam intensity in a confocal cavity and the wavefronts.

As a result of such divergence the main part of the energy of a Gaussian beam is concentrated within a solid angle of

$$\Omega = \pi\theta^2 = \lambda/l. \tag{8.11b}$$

Thus, the divergence of laser light in the fundamental mode is determined by the longitudinal size of the laser cavity rather than by the transverse. The reason is that the smallest effective aperture on which a freely propagating Gaussian beam is diffracted is the focal cross section of the caustic. The diffraction divergence is characterized by the ratio of the wavelength λ to the width of the intensity distribution in the neck region, w_0. In turn, the solution of the self-consistent integral equations of the Fox-Lee type leads to formula (8.5) for w_0.

Actually, (8.8) describes a diffracted wave produced in the self-diffraction of the Gaussian beam. The diffraction pattern described by (8.8) is characterized by a monotonic decrease in intensity as we move away from the axial direction, that is, there are no oscillations in the brightness of the diffraction pattern and the intensity of the wave falls off rapidly at the wings of the distribution. Obviously, the diffraction of a Gaussian beam on any aperture is of the same nature, provided that the aperture size is sufficiently greater than the beam's intensity distribution width w.

It is appropriate to note here that the monotonic nature of the Gaussian-beam diffraction pattern corresponds in the radio-frequency range to lobeless directional diagrams of transmitting-receiving radar dishes or the receiving dishes of microwave radio telescopes. This is achieved by a fall-off in intensity on the peripheral sections of the antenna's flare that forms the directional diagrams.

The absence of intensity oscillations (minor lobes) is explained by the gradual decrease in the field amplitude as we move away from the beam's axis rather than by any specific law of decrease (in our case the Gaussian law).

Formula (8.8) for the Gaussian beam's field was obtained for a certain position of the initial plane, $z = 0$, in which the wavefront is flat and the distribution width is minimal. The calculations can be repeated, however, with the Gaussian distribution in any other plane as the initial, and the result will be the same. Hence, if at some point in space a wave beam is characterized by a spherical wavefront and a Gaussian transverse amplitude distribution, these properties are retained in the entire space. As the wave propagates, only the curvature radius of the wavefront, (8.10), and the width of the amplitude distribution, (8.4), change. A wave of this type is known as a Gaussian wave or a Gaussian beam. The beam width w and the curvature radius R of the phase front completely define a Gaussian beam at a given point on the beam axis. Reversal of the sign of R means the reversal of the curvature of the phase front, that is, the transformation of a diverging beam into a converging one, and vice versa.

For example, an ideal thin lens transforms a diverging Gaussian beam into a converging one, but the beam remains Gaussian. If the transverse dimensions of the lens are so great that we can ignore the diffraction of the beam on the lens, the lens only changes the curvature of the wavefront. We know from geometrical optics that an ideal thin lens deflects all the light rays falling on it parallel to the optical axis in such a manner that they intersect the optical axis at the same distance from the lens, known as the focal length F, which means that after passing the lens a plane wave becomes a spherical wave with a wavefront curvature radius equal to $-F$. Hence, a thin lens changes the curvature of the wavefront of a wave passing through it by a quantity equal to $-1/F$. Then the curvature radius of the wavefront of a Gaussian beam immediately after the wavefront passes the lens is given by the formula

$$1/r = 1/R - 1/F, \qquad (8.12)$$

where R is the radius of the wavefront immediately prior to the wavefront's passage of the lens. For a lens with a sufficiently short focal length ($F < R$) r is negative, that is, the curvature of the wavefront after the lens has a sign opposite to that of the wavefront before the lens, and we obtain a converging Gaussian beam.

In view of the importance of the problem of focusing Gaussian beams in quantum electrons, we will discuss this problem in greater detail.

Suppose that we have an ideal lens with a focal length F. Let a diverging Gaussian beam whose neck region (radius w_0) lies at a distance z from the lens be incident on the lens from the left and let the wavefront's curvature radius just before passage of the lens be R. Then the curvature of the wavefront just after passage of the lens is given by formula (8.12). The beam width at the lens we denote by D. Obviously, D is the same both right and left of the lens (Figure 8.5).

We denote by x the sought distance from the lens to the point on the optical axis of the lens at which the cross-sectional area of the converging lens is minimal and amounts to the sought value v_0^2. Since after passing

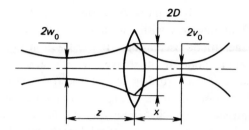

Figure 8.5. Calculating the neck radius of a focused Gaussian beam and the lens-neck separation.

the lens the beam remains Gaussian, it is obvious that Eqs. (8.4) and (8.10) still hold for x, r, v_0, and D. As a result for x and v_0^2 we have the following system of equations:

$$x + k^2 v_0^4/x = r, \quad v_0^2 + x^2/k^2 v_0^2 = D^2, \tag{8.13}$$

which can easily be solved. Simple manipulations yield

$$x = \frac{k^2 D^4}{r^2 + k^2 D^4} \, r, \tag{8.14}$$

$$v_0^2 = \frac{D^2}{r^2 + k^2 D^4} \, r^2. \tag{8.15}$$

Here r is given by formula (8.12), and $D^2 = w_0^2 + z^2/k^2 w_0^2$, in accordance with Eq. (8.4). The above formulas are of a sufficiently general nature and make it possible to study the transformation of one Gaussian beam into another, also Gaussian. The most common problem here is the matching of fields in two different cavities.

Suppose that a passive cavity is used as a Fabry-Perot interferometer, say, for studying the emission spectrum of a laser, that is, the radiation emitted by an active cavity. As we know, the field distribution in the fundamental mode of a confocal cavity constitutes a Gaussian wave with a caustic neck radius determined by the cavity's length l (see Eq. (8.5)) and the wavefront curvature radius at the mirror determined by the mirror's curvature radius. Hence, the modes of these two cavities do not, generally, coincide. When a beam of light corresponding to a mode of one of the cavities enters the other cavity and the mode parameters of the cavities have not been matched, mismatch leads to mode transformation. The fundamental mode of the laser light interacts with the higher modes of the passive cavity and generates higher mode oscillations. If the fraction of the energy pumped by the fundamental mode of the active cavity is sizable, there may be serious errors in the results of studies of the spectral composition of the laser light. What is important here is the possibility of transforming the parameters of a Gaussian beam by employing a lens. Formulas (8.14) and (8.15) make it possible to calculate the focal length and the position of the required lens if we know the neck position and size of the beams in both cavities.

Returning to the important problem of focusing laser light, let us consider the behaviour of solutions (8.14) and (8.15) at great distances from the caustic neck of the initial Gaussian beam, that is, for $z \gg k w_0^2$. As applied to laser light, this corresponds to the case of great distances from the laser cavity, $z \gg l/2$ (see (8.5)). Suppose also that $F \ll z$. Then $D_0^2 \approx z^2/k^2 w_0^2$, $r \approx -F$, and formulas (8.14) and (8.15) yield

$$v_0 = w_0 F/z, \tag{8.16}$$
$$x = -F. \tag{8.17}$$

Thus, at great distances from the caustic neck a lens with a relatively short focal length concentrates the initial radiation of a Gaussian beam in its focal region and increases the intensity by a factor of z^2/F^2.

Formally, from (8.16) it follows that for $z \to \infty$ the spot forming at the focal point of the lens has an infinitesimal radius. However, the spot on the lens becomes infinitely large ($D \to \infty$), which contradicts the initial assumption that the lens does not diaphragm the beam and, hence, does not violate the beam's Gaussian nature. A drastic reduction in the aperture of a Gaussian beam dramatically changes the nature of the diffraction, and the respective Fresnel-Kirchhoff integral can no longer be written in the form (8.8). Also, even if the beam remains Gaussian, diffraction imposes restrictions on the size of D.

The diffraction divergence angle of a Gaussian beam is $\lambda/2\pi w_0$. This means that a Gaussian wave cannot be focused on a spot with a radius smaller than $\lambda/2\pi$. From (8.16) it follows that $F = z v_0/w_0$. Since v_0 cannot be less than $\lambda/2\pi$ and $D = z\lambda/2\pi w_0$, we conclude that formulas (8.14)-(8.16) are valid only if

$$F > D. \tag{8.18}$$

Thus, considerable spatial concentration of the energy of the fundamental mode of laser light is possible if the radiation is focused by a thin lens placed at a great distance from the laser cavity ($z \gg kw_0^2 = l/2$). The focal length of the lens must be less than z but greater than the size D of the spot on the lens. In this case the focusing conditions are written as

$$z \gg l/2, \quad z \gg F > D \qquad (8.19)$$

and, hence, can easily be met.

The longitudinal size of the focal region, where the radiation energy is concentrated most strongly, can be found by applying formula (8.4) to the focused radiation. The intensity decreases by a factor of two as we move away from the point of maximum concentration, $x = -F$, where the distribution width is v_0, by a distance of

$$\Delta x = kv_0^2 = (2\pi/\lambda)v_0^2. \qquad (8.20)$$

In focusing 1-μm laser light into a spot with a radius of 10 μm the energy flux is near-constant in a near-cylindrical region 1200-μm long.

Note that the wavefront is flat at the focal point and near-flat in the entire focal neck region.

Formulas (8.16) and (8.17) have been obtained for great values of z from the general solution (8.14), (8.15). However, in this limiting case they can be obtained directly. If z is great, the wave being focused is close to a plane wave; hence, it is focused at a point called, by definition, the focal point. For large z and F the radius of the spot on the lens is defined by the formula $D = z/kw_0$ when the wave moves from left to right and by the formula $D = F/kv_0$ when the wave moves from right to left. Hence, $v_0 = w_0F/z$.

Let us now consider the opposite particular case. We place the caustic neck of the radiation being focused at the front focal point of the lens, that is, at a distance $z = F$ from the lens plane. Now where will the Gaussian beam be focused and what will be the radius of the new neck? Substituting $z = F$ into (8.4), (8.10), and (8.12) yields the following formulas for the radius of the spot on the lens and the curvature radius of the wavefront immediately after the lens:

$$D^2 = \frac{k^2 w_0^4 + F^2}{k^2 w_0^2}, \quad r = -\frac{k^2 w_0^4 + F^2}{k^2 w_0^4}F. \qquad (8.21)$$

If we now substitute (8.21) into (8.15), we get

$$v_0^2 = F^2/k^2 w_0^2 = D^2 - \omega_0^2, \tag{8.22}$$
$$x = -F. \tag{8.23}$$

For $z = F$ the lens is positioned at the point where the spherical mirror of an equivalent confocal cavity forming the focused Gaussian beam would be. In other words, if $z = F$, then simultaneously $z = l/2$. For $z = l/2$, that is, at the mirror surface, the area of the fundamental-mode spot is twice the cross-sectional area of the cavity's caustic neck (see (8.4) and (8.5)). Hence, $D^2 = 2w_0^2$ and

$$v_0^2 = w_0^2. \tag{8.24}$$

Thus, an ideal thin lens with a focal length F transforms a diverging Gaussian beam into a similar converging beam, provided that the focal neck of the initial beam is placed at the focal point of the lens. In other words, the lens conserves the minimum cross section of a Gaussian beam and shifts this cross section from one of its focal points to the other, provided that at the start the minimum cross section is at the respective focal point.

Obviously, an additional lens, with the same focal length F, placed at a distance of $2F$ from the first transforms the secondary Gaussian beam (in relation to the first lens) in exactly the same manner. Placing in periodic sequence identical lenses separated by a distance of $2F$, we get a confocal lens waveguide that makes it possible to transfer a nondiverging (on the average) light beam over great distances (Figure 8.6). The path of a wave in a confocal lens waveguide constitutes a periodic sequence of alternating identical converging and diverging Gaussian beams. The field distribution between the lenses is identical to that between the mirrors of a confocal

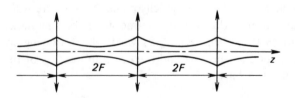

Figure 8.6. A confocal lens waveguide (the lenses are depicted as vertical arrows).

cavity. Naturally, the normal modes of a confocal lens waveguide coincide with the modes of a confocal cavity.

The similarity between cavities and lens waveguides is quite great and is often used to analyze the properties of cavities of different types.

Problems to Lecture 8

8.1. Find the spot radius at the neck of the fundamental mode of an argon laser in a confocal cavity 1-m long; $\lambda = 514.5$ nm.

8.2. For the laser of Problem 8.1, calculate the diameter of the spot on the mirror and at a distance of 1 m from the mirror, and the beam's angular divergence.

8.3. At what distance from the neck is the wavefront curvature radius in a confocal cavity maximal?

8.4. By how many times will the divergence angle of a beam of the fundamental mode of a confocal cavity increase if the output mirror has a flat rear surface and is made of glass with a refractive index n? (*Hint*: Assume the mirror substrate to be a negative lens.)

8.5. Using formula (8.20), evaluate the caustic length in the case of the smallest possible neck of a Gaussian beam.

8.6. Estimate by how many times the intensity of the laser of Problem 8.1 can be increased by using a lens to focus the light on the object.

8.7. Derive a formula for the position of the neck in an arbitrary cavity. (*Hint*: Assume that the mode field is generated by a hypothetical confocal cavity and employ formula (8.10).)

8.8. Derive formulas for the neck size and the dimensions of the spots on the mirrors of a symmetric cavity (not necessarily confocal). (*Hint*: Apply the method used in solving Problem 8.7 and use formula (8.4).)

8.9. What is the radial power distribution in a Gaussian beam?

8.10. Calculate the total power output of a Gaussian beam with intensity I_0 at the beam's axis and a field spot size w_E.

8.11. Suppose that a laser with a confocal cavity lases with a power output P_0 in the fundamental mode. Find the intensity distribution in the far zone and express the result in terms of the total angular divergence of the radiation. (*Hint*: Use formula (8.8) assuming that parameter (8.8) in it is the normalization constant.)

LECTURE 9

Cavity Stability

Stability of lens waveguides. A waveguide with identical lenses. A waveguide with alternating lenses of different focal lengths. The stability condition and the stability diagram. The equivalence of a lens waveguide and an open cavity. Types of stable cavities. Transverse mode selection with a diaphragm. Unstable cavities

Let us use the similarity between lens waveguides and open cavities to consider the important problem of cavity stability. A cavity is stable if the alternating reflection of the radiation in it from the cavity's mirrors leads to such periodic focusing of the radiation that in the geometrical-optics approximation the radiant energy does not escape from the cavity. In an unstable cavity during each pass of the radiation between the cavity mirrors a sizable fraction of the stored energy escapes from the cavity. In other words, a stable cavity is characterized by the presence of a time-independent field distribution that stably recurs in the multiple passes of the radiation between the cavity mirrors and has such low diffraction losses that the lifetime of this distribution, τ_{dif}, obeys the condition $w\tau_{dif} \gg 1$ (see Eq. (7.17)).

A stable lens waveguide corresponds to a stable cavity. A waveguide is stable, in turn, if the light beam propagating in it over large distances does not escape from it. Let us analyze the stability of a waveguide consisting of lenses with the same focal lengths F aligned coaxially in a series and separated by a distance l. We will consider the path of a light ray in a lens waveguide in the paraxial approximation, in which the well-known thin-lens formula is valid:

$$1/a_1 + 1/a_2 = 1/F, \qquad (9.1)$$

where a_1 is the distance from a luminous point placed on the principal axis to the optical center of the lens, and a_2 the distance from the optical center of the lens to the image of the point. This general formula can be represented in a form more convenient for the analysis that follows. Suppose that a ray intersects the lens plane at a distance r from the principal axis. This ray enters the lens at an angle α_1 with respect to the normal to the lens plane and leaves at an angle α_2. In the paraxial approximation, that is, for $r \ll a_1$, (9.1) is equivalent to

$$\alpha_2 - \alpha_1 = -r/F, \tag{9.2}$$

where positive angles are reckoned, as usual, counterclockwise (Figure 9.1).

Let us examine three adjacent lenses of a lens waveguide, $n - 1$, n, and $n + 1$. For the ray considered here, the distance from the optical axis of the nth lens, that is, the distance between the ray and the axis at point n, is r_n. The angle between the optical axis of the waveguide and the ray at the ray's exit from the nth lens is α_n. For two adjacent lenses formula (9.2) assumes the form (Figure 9.2)

$$\alpha_n - \alpha_{n-1} = -r_n/F. \tag{9.3}$$

In turn, the distances from the ray to the axis in the neighboring lenses are related thus:

$$r_{n+1} = r_n + \alpha_n l, \quad r_n = r_{n-1} + \alpha_{n-1} l, \tag{9.4}$$

valid, of course, only in the paraxial approximation. Subtracting the second equation from the first and allowing for (9.3), we arrive at the following recurrence formula:

$$r_{n+1} + (l/F - 2)r_n + r_{n-1} = 0, \tag{9.5}$$

which makes it possible to determine the position of the ray on any lens of the waveguide if we know the ray's position on the first two lenses.

Thus, in the geometrical-optics approximation we have a recurrence relation making it possible to proceed, step by step, from one lens to another

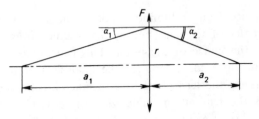

Figure 9.1. Derivation of the lens formula (9.2).

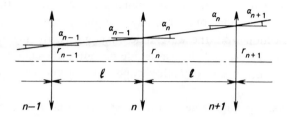

Figure 9.2. Derivation of the stability condition for a lens waveguide.

and determine the path of any ray in a paraxial beam of light propagating in the lens waveguide and, hence, analyze the stability of the waveguide. The similarity with the case of successive passes of radiation between the cavity mirrors in the Fox-Lee method is obvious.

A numerical analysis of the successive steps is not necessary here because (9.5) allows for an analytical solution. We look for the solution to (9.5) in the form

$$r_n = Ae^{jn\theta}, \tag{9.6}$$

with A a constant. Substituting (9.6) into (9.5), using the Euler formula $\exp(j\theta) = \cos\theta + j\sin\theta$, and requiring that (9.5) be valid for the real and imaginary parts of the recurrence relation separately, we find that (9.6) satisfies (9.5) if

$$\cos\theta = 1 - l/2F. \tag{9.7}$$

Expression (9.6) is a particular solution to (9.5). Let us analyze this expression without considering the general solution. The waveguide is stable when r_n oscillates, under variation of n, from $-A$ to $+A$, where A has the meaning of the position of the ray at the entrance to the waveguide.

The necessary and sufficient condition for the existence of nonrising oscillations in r_n is the reality of θ. For real θ the function $\cos\theta$ varies between -1 and $+1$. Hence, the admissible domain of variability of l/F is determined by the following inequalities:

$$-1 \leqslant 1 - l/2F \leqslant 1. \tag{9.8}$$

For values of l/F lying outside of this domain, the cosine function trans-forms into a hyperbolic function, θ becomes complex-valued, the amplitude r_n of the ray deflection from the waveguide's axis grows, the ray exits from the side of the waveguide, and the waveguide becomes unstable. Hence, the inequalities (9.8) constitute the stability condition for the waveguide con-sidered.

Let us now take up a more general case. Suppose that the focal lengths of adjacent lenses be distinct and equal to F_1 and F_2. The lenses are sepa-rated by a distance of l. The values of F_1 and F_2 remain constant along the entire length of the lens waveguide, alternating in such a manner that, say, the even-numbered lenses have the focal length F_1 and the odd-numbered the focal length F_2.

A recurrence formula similar to (9.5) can be obtained by consistently applying the lens formula (9.2) to the diagram in Figure 9.3. The slopes and positions of the rays at the lenses of the waveguide with alternating lenses of different focal lengths are related by the following formulas:

$$\alpha_{2n} - \alpha_{2n-1} = -r_n/F_1, \tag{9.9}$$
$$\alpha_{2n+1} - \alpha_{2n} = -\varrho_{n+1}/F_2. \tag{9.10}$$

The positions of the rays on adjacent lenses in the paraxial approximation are related by the following formulas:

$$r_n = \varrho_n + \alpha_{2n-1}l, \tag{9.11}$$
$$\varrho_{n+1} = r_n + \alpha_{2n}l. \tag{9.12}$$

Subtracting (9.12) from (9.11) and allowing for (9.9), we find that

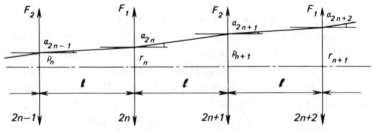

Figure 9.3. Derivation of the stability condition for a lens waveguide with alternating lenses of focal lengths F_1 and F_2.

$$\varrho_{n+1} + \varrho_n = (2 - l/F_1)r_n. \tag{9.13}$$

Similarly,

$$r_{n+1} + r_n = (2 - l/F_2)\varrho_{n+1}. \tag{9.14}$$

If now in (9.14) we replace n with $n - 1$,

$$r_n + r_{n-1} = (2 - l/F_2)\varrho_n, \tag{9.15}$$

combine (9.15) and (9.14), and introduce $\varrho_{n+1} + \varrho_n$ via (9.13), we arrive at a recurrence formula containing only r_{n-1}, r_n, and r_{n+1}:

$$r_{n+1} + [2 - (2 - l/F_1)(2 - l/F_2)]r_n + r_{n-1} = 0. \tag{9.16}$$

Reasoning along the same lines for the positions of the rays at the odd-numbered lenses, ϱ_n, we can arrive at a similar recurrence formula:

$$\varrho_{n+1} + [2 - (2 - l/F_1)(2 - l/F_2)]\varrho_n + \varrho_{n-1} = 0. \tag{9.17}$$

The recurrence formulas (9.16) and (9.17) are similar in appearance to (9.5). This means that we can apply the results of solving (9.5) in the form (9.6) and (9.7). Hence, for a lens waveguide with alternating lenses of focal lengths F_1 and F_2 that are aligned coaxially in series and separated by a distance l, the stability condition is

$$-1 \leqslant \cos \Phi = -\frac{1}{2} [2 - (2 - l/F_1)(2 - l/F_2)] \leqslant 1. \tag{9.18}$$

Simple manipulations transform (9.18) into the following simple condition:

$$1 \geqslant (1 - l/2F_1)(1 - l/2F_2) \geqslant 0. \tag{9.19}$$

For $F_1 = F_2 = F$ the inequalities (9.19) and (9.8) are equivalent, just as $1 \geqslant x^2 \geqslant 0$ is equivalent to $1 \geqslant x \geqslant -1$.

Let us introduce the following notation: $g_1 = 1 - l/2F_1$ and $g_2 = 1 - l/2F_2$. Then the limits of variation of the permissible values of $l/2F$

are determined by the following simple equations:

$$g_1g_2 = 1, \qquad\qquad (9.20)$$
$$g_1g_2 = 0. \qquad\qquad (9.21)$$

This allows for a simple diagrammatic representation of the stability condition for a lens waveguide in the (g_1, g_2) plane.

In Figure 9.4 the hyperbolas $g_1g_2 = 1$ and the coordinate axes corresponding to Eq. (9.21) outline the stability region. To make the region stand out we have hatched it. The permissible values of g_1 and g_2 lie inside the hatched area and on the boundary. The diagram also contains points of special interest. The origin B ($g_1 = 0$, $g_2 = 0$) corresponds to the confocal system with $F_1 = F_2 = F = l/2$. We see that a confocal lens waveguide "lies" on the boundary between the regions of stable and unstable paths. Point C ($g_1 = 1$, $g_2 = 1$) corresponds to the limiting case of an infinite focal length. Finally, point A ($g_1 = -1$, $g_2 = -1$) corresponds to a lens waveguide with identical lenses whose focal length $F = l/4$ is the smallest possible for this type of waveguide.

We now return to open cavities and consider once more the similarity between a lens waveguide and a cavity.

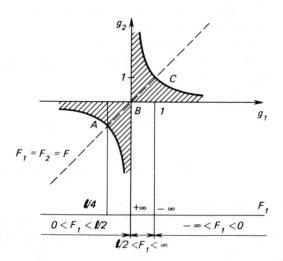

Figure 9.4. A stability diagram. The lines at the bottom depict the ranges of variation of the focal length F_1 corresponding to the variation of parameter $g_1 = 1 - l/2F_1$ from $-\infty$ to $+\infty$.

A typical laser cavity consists of two slightly concave mirrors with a high reflectivity positioned opposite each other. The curvatures may coincide or differ. In the paraxial approximation a concave mirror is equivalent to a flat mirror combined with a planoconvex spherical lens positioned directly in front of the mirror. The cavities depicted in Figures 9.5a and 9.5b are optically equivalent. Then in its reflection from the mirror one of the travelling components of the standing wave of the resonant-mode field crosses the equivalent lens $2F_1$ twice. From the viewpoint of focusing, each traveling wave propagates in the cavity in the same way as it does in the lens waveguide shown in Figure 9.5c. The planoconvex lenses in Figure 9.5b have focal lengths $2F_1$ and $2F_2$. But the wave passes through these lenses twice, changing its direction of propagation directly at the flat surface of a lens. Hence, each lens here acts as two closely placed lenses, whose powers add up. This means that the lenses F_1 and F_2 in Figure 9.5c are equivalent to the lenses $2F_1$ and $2F_2$ in Figure 9.5b, respectively. We can then assume that the cavity is transformed into an equivalent wavelength if we ignore the fact that the wave changes its direction of propagation in the process of reflection and instead assume that the wave travels without reflection and that there is a second lens positioned right after the first and identical to the first. We conclude that a lens waveguide can be interpreted as an open cavity unfolded along the axis. Note that the ideas developed here are based on the geometrical-optics approximation.

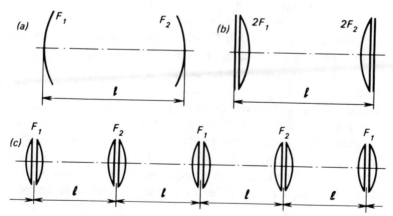

Figure 9.5. The equivalence of a cavity and a lens waveguide.

Lecture 8 showed that the normal propagation modes of a confocal lens waveguide coincide with the normal oscillation modes of a confocal cavity. Generally, in determining the normal modes of a lens waveguide one uses the fact that the field distribution on each lens (or through each lens) is the same to within a phase factor. The next step is, on the basis of the Huygens principle, to write the field in the reference plane of one lens in terms of the Kirchhoff-Fresnel diffraction integral of the field in the reference plane of the preceding lens. The requirement that the field be the same to within a phase factor leads to an eigenvalue problem written in the form of integral equations, which in the case of an open cavity coincide with the Fox-Lee equation. The field distributions obtained from solving these equations can be represented in terms of products of Hermite polynomials by the Gaussian function and coincide with the products in the case of cavities.

Thus, both the geometrical-optics and the wave approach prove the equivalence of open cavities and lens waveguides. Hence, the stability conditions (9.19) and the stability diagram depicted in Figure 9.4 characterize open laser cavities.

Let us consider the stability diagram of cavities in greater detail.

Point B ($g_1 = 0$, $g_2 = 0$) corresponds to the case, discussed many times in this course, of confocal cavities. A cavity of this type "lies" on the boundary separating the stable and unstable regions, but its stability is, essentially, formal. The slightest nonsymmetry in the mirrors easily brings a confocal cavity into an unstable state. Therefore, though the easiest object for analyzing and serving as a model in many theoretical constructions, the confocal cavity is rarely used in practice.

The straight line ABC in Figure 9.4 corresponds to cavities with mirrors of the same curvature radius R (symmetric cavities). Point A ($g_1 = -1$, $g_2 = -1$) corresponds to what is known as a concentric cavity, in which the centers of curvature of the mirrors coincide: $l = 4F = 2R$. Point B corresponds to a confocal cavity. Point C ($g_1 = 1$, $g_2 = 1$) corresponds to a plane-parallel cavity ($F = \infty$). All these cavities, not just only the confocal cavity, lie on the boundary separating the stable region from the unstable. Consequently, when it is desirable to retain the symmetry of the cavity, a quasi-confocal cavity is employed, that is, a cavity in which the mirror separation differs little from that of a confocal cavity:

$$l/2F = 1 \pm \alpha, \quad \alpha \ll 1. \tag{9.22}$$

Even a small α makes the cavity stable without substantially changing the configuration of the field in relation to that in a confocal cavity.

However, the plane-concave cavity (Figure 9.6) has gained the widest acceptance. In such a cavity one mirror is flat ($F_1 = \infty$) and the curvature radius of the other is chosen such that its focal point lands on the flat mirror ($F_2 = l$). The cavity is stable, $g_1 g_2 = 1/2$. The flat mirror in the focal plane divides the confocal cavity in half, replacing the real field in the distant part of the cavity with the image of the field in the remaining part. Hence, half of the field distribution characteristic of a confocal cavity sets in in the plane-concave cavity. The wide application of the plane-concave cavity is explained by the great convenience in constructing the output mirrors as flat, rather than spherical, partially transparent mirrors.

The construction of a laser cavity is closely linked to the mode composition of the laser light. The divergence of radiation discussed in Lecture 8 describes the diffraction divergence of the fundamental-mode Gaussian beam. What leads to a sharp drop in the directivity of the radiation is, obviously, the presence of transverse modes, whose peak-intensity directions differ from each other and from the cavity's axis, which specifies the fundamental-mode radiation direction (see (8.1) and (8.2)). Hence, the importance of transverse mode selection.

What is required in the majority of cases is isolation of the fundamental mode. This mode is characterized by the smallest diffraction losses, which increase sharply with the transverse mode index. But in stable cavities the diffraction losses are so low that the difference between them cannot be used for mode discrimination. Hence, mode selection can be based only on the differences in the field distribution of modes with different transverse indices. Since the fundamental mode has a Gaussian distribution that is symmetric about the cavity's axis and has a minimum width in the transverse plane, the simplest and most reliable way of selecting modes is to

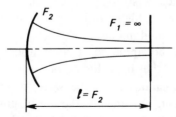

Figure 9.6. A plane-concave cavity.

diaphragm the beam inside the cavity. If the diaphragm aperture is small, the Fresnel number for the cavity, $N_F = a^2/l\lambda$, is determined by it. As the Fresnel number decreases, the difference in the diffraction losses for the fundamental mode and for higher-order modes increases, which makes possible mode selection.

Knowing the calculated dependence of the diffraction losses of the fundamental mode and the mode with the nearest higher-order transverse indices on the Fresnel number, we can find the required diaphragm radius. This, however, introduces losses into the fundamental mode, too. A simple estimate of the transverse size of the diaphragm can be made by assuming that it must be about equal to the transverse size of the field distribution of the mode following the fundamental mode and the diaphragm must be positioned at the point where the mode sizes differ most strongly. Usually both the size of the diaphragm's aperture and the position of the diaphragm are selected experimentally.

An essential drawback of the discussed method of fundamental-mode selection in a stable cavity is the smallness of the transverse dimensions of the mode. This facilitates selection but pushes down the output power since not all of the bulk of the active media is occupied by the electromagnetic field. To increase the output power one must increase the mode volume. A cardinal solution here is transition to unstable cavities. The use of unstable cavities constitutes an effective means for transverse mode selection.

It is clear from what has been said that a cavity is unstable when an arbitrary light ray, reflecting alternately from each of the cavity mirrors, distances itself from the cavity axis infinitely. In other words, optical cavities are unstable when their parameters land in the region of instability in the stability diagram (Figure 9.4). In view of the ray instability in cavities of this type the diffraction losses are high even in the fundamental mode and usually exceed all other losses, and they increase as we go to higher-order modes. Hence, the overall losses strongly depend on the transverse indices, which leads to suppression of higher-order transverse modes and, hence, to isolation of the fundamental mode.

Obviously, unstable cavities can be employed in lasers whose active media has a large gain. Otherwise, the high losses per pass, which are directly related to cavity instability, cannot be compensated for and the self-excitation conditions will not be met. Fortunately, the use of unstable cavities is most desirable in lasers with high stored energy and gain. The point is that in an unstable cavity the volume occupied by the fundamental mode field is great because, in contrast to a stable cavity, no periodic focusing of the field inside an unstable cavity takes place during alternative reflections from the mirrors and the field does not tend to be concentrated near the cavity's axis. It has proved expedient to use the rays that tend to leave the cavity as the useful output radiation of the laser.

A great advantage of unstable cavities is the possibility of controlling the amount of energy leaving the cavity and of attaining an optimal coupling of the cavity and space.

Problems to Lecture 9

9.1. Replace with a geometrical construction the expression obtained as a result of solving Problem 8.7.

9.2. The construction done in the previous problem makes it possible to verify the stability of a cavity. Prove this.

9.3. Where is the neck of a stable cavity with a flat output mirror?

9.4. An experimenter has two concave mirrors of radii $R_1 = 1$ m and $R_2 = 2$ m. What maximum and minimum lengths can a stable cavity with these mirrors have?

9.5. The same question as in the previous problem but the mirror with radius R_1 is convex.

LECTURE 10

Unstable Cavities

The geometrical-optical approach. Gain and radiation losses. Symmetric and telescopic cavities. The equivalent Fresnel number. Longitudinal mode selection. Frequency selection. Spatial selection by thin absorbers. Dispersive cavities. Distributed feedback

We now know that for an unstable cavity either

$$g_1 g_2 < 0 \qquad (10.1)$$

or

$$g_1 g_2 > 1, \qquad (10.2)$$

where $g_1 = 1 - l/2F_1$ and $g_2 = 1 - l/2F_2$. Correspondingly, unstable cavities break down into two classes, negative-region cavities, (10.1), and positive-region cavities, (10.2). On the stability diagram (see Figure 9.4) the positive instability region occupies the first and third quadrants outside the hyperbola $g_1 g_2 = 1$ and the negative instability region the second and fourth quadrants.

Analysis of unstable cavities can be carried out by the sufficiently profound methods of geometrical optics. The thing is that in stable cavities, where as a result of multiple reflection of the wave from the mirrors mode formation obeys the laws of diffraction, lower-order modes are characterized by insignificant diffraction losses. The Gaussian transverse distribution limits the size of the mode spot, and only a minute fraction of the energy diffracts around the edge of a mirror. The fact that the distribution is Gaussian is determined by the focusing property of the mirrors in the stable spherical cavity configuration. In an unstable cavity no such focusing is present; the light is not concentrated near the cavity's axis. And although the losses attributed to radiation curving around the mirror edge can always be classified in the general case as diffraction losses, it is convenient to consider unstable cavities from the viewpoint of geometrical optics, assuming the rays that escape the limits of a mirror to be the source of geometrical losses. One must bear in mind that geometrical losses in unstable cavities are closely related to diffraction losses in stable cavities, and the two should coincide at the stability boundary.

Suppose that the Fresnel numbers of the cavity mirrors are great and diffraction losses can be ignored. Since the radiation does not "gather" near the cavity's axis, it is natural to assume that the mirrors are "filled" with radiation uniformly. This distinguishes unstable cavities from stable cavities. At the same time the radiation's wavefront, specified by boundary conditions imposed by spherical mirrors, is spherical. In this respect unstable cavities are similar to stable.

Thus, we assume that in the geometrical-optics approximation the field distribution in an unstable cavity, that is, its mode, constitutes a linear combination of two spherical waves with a uniform field distribution across the front and emerging from two centers on the cavity's axis. If the path traveled by the light along the entire cavity is to be closed, these centers must be images of each other in the respective mirrors. In other words, rays emerging from one center transform, after reflection from a mirror, into rays emerging from the other center. Depending on the position of the centers of the spherical waves that form the unstable-cavity mode, different types of unstable cavities are realized with respect to the cavity mirrors.

Let us first consider the geometry of an unstable cavity in the general form. Figure 10.1a shows the diagram of an unstable cavity formed by two spherical

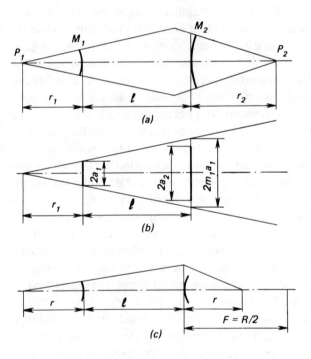

Figure 10.1. An unstable cavity: (a) the general case, (b) calculation of radiation losses per pass, and (c) a symmetric double-end cavity.

mirrors, M_1 and M_2. Let us assume that a wave leaving mirror M_1 is a spherical one centered at point P_1, which may not be the center of curvature or the focal point of this mirror. A fraction of this wave escapes M_2 and the remainder will be reflected by M_2. Suppose that the reflected spherical wave originates at point P_2. We denote the distances from points P_1 and P_2 to mirrors M_1 and M_2, respectively, by r_1 and r_2 and the curvature radii of the mirrors by R_1 and R_2.

As noted earlier, points P_1 and P_2 must be images of each other in the corresponding spherical mirrors. Then, applying the spherical-lens formula in the paraxial approximation,

$$1/a_1 + 1/a_2 = 1/F = 2/R, \tag{10.3}$$

first to the mutually conjugate distances r_1 and $r_2 + l$ and then to the mutually conjugate distances r_2 and $r_1 + l$, we arrive at the following equations:

$$\frac{1}{r_1 + l} - \frac{1}{r_2} = -\frac{2}{R_2}, \quad \frac{1}{r_2 + l} - \frac{1}{r_1} = -\frac{2}{R_1}. \tag{10.4}$$

Simultaneous solution of Eqs. (10.4) enables finding the position of the centers of the spherical waves that form the cavity mode. It is more essential, however, that r_1 and r_2 determine the cavity's radiation losses.

Indeed, as the beam travels from mirror to mirror and back, it increases its cross-sectional area

$$M = m_1 m_2 \tag{10.5}$$

times, where m_1 and m_2 are the magnifications per pass. Figure 10.1a readily demonstrates that

$$m_1 = (r_1 + l)/r_1, \quad m_2 = (r_2 + l)/r_2. \tag{10.6}$$

The cross-sectional area of the beam does not increase without limit, however, as the beam travels more and more times inside the cavity because the mirrors have finite transverse dimensions. A fraction of the radiation passes the mirrors near their edges and thus is not reflected. Since the size of the spherical-wave spot increases M-fold and, by assumption, the intensity distribution at the mirror surfaces is uniform, the intensity decreases M^2-fold in the cavity. Magnification M is independent of the size of the mirrors. Hence, the total radiant flux remaining within the cavity decreases M^2-fold as the beam makes a round trip in the cavity. This means that the round-trip fractional radiation losses to the space surrounding the cavity are

$$A = 1 - 1/M^2 = (M^2 - 1)/M^2 \tag{10.7}$$

and are determined by the values of r_1, r_2, and l, or the cavity's configuration.

Let us discuss this important matter in greater detail. Suppose that a diverging spherical wave with an angular magnification m_1 is incident from the left on a mirror M_2 with a cross sectional area πa_2^2 (Figure 10.1b). At the mirror's cross section the transverse size of this wave (with a uniform intensity distribution across this section) is $2m_1 a_1$, where a_1 is the radius of the left mirror M_1. Mirror M_2 reflects toward M_1 only a fraction of the radiation arriving from the left at the point where M_2 is positioned. This fraction is

$$\Gamma_{21} = a_2^2/m_1^2 a_1^2. \tag{10.8}$$

Similarly, mirror M_1 reflects toward M_2 the fraction of the radiation arriving from the right equal to

$$\Gamma_{12} = a_1^2/m_2^2 a_2^2. \tag{10.9}$$

As a result of two reflections, the fraction of the radiation remaining between the mirrors, that is, in the cavity, is

$$\Gamma = \Gamma_{12}\Gamma_{21} = 1/m_1^2 m_2^2 = 1/M^2. \tag{10.10}$$

This means that, according to (10.7), the round-trip radiation losses in the cavity are

$$A = 1 - \Gamma = (M^2 - 1)/M^2 \tag{10.11}$$

and do not depend on the size of the cavity mirrors. The simple formulas we have derived here illustrate the physical origin of this important fact. The thing is that a decrease in the dimensions of one of the mirrors leads to a proportional decrease in the aperture angle of the waves traveling in both directions. Hence, the relative transverse dimensions (Figure 10.1b) do not change, which also means that the fractional power losses do not change.

Thus, calculation of radiation losses in unstable cavities is carried out by methods of geometrical optics. Such losses are often called geometrical or geometrical-optical. On the whole, the wave approximation yields a close estimate.

Solution of Eqs. (10.4), although quite possible, generally leads to unwieldy formulas for r_1, r_2, and M. Here are two particular solutions. The symmetric double-end cavity has been the most thoroughly studied theoretically (Figure 10.1c). Since in this case $R_1 = R_2 = R$, we have $r_1 = r_2 = r$. Equations (10.4) yield

$$r = \frac{l}{2} \left(\sqrt{1 + 2R/l} - 1 \right) \tag{10.12}$$

and, respectively,

$$M = \frac{l + r}{r} = \frac{\sqrt{1 + 2R/l} + 1}{\sqrt{1 + 2R/l} - 1}. \tag{10.13}$$

The radiation losses and magnification of a symmetric cavity are usually calculated with respect to a single pass. It is to this type of cavity that the wave approximation has been applied most successfully via integral equations of the Fox-Lee type and the geometrical-optics approach was shown to be meaningful.

An interesting variant of unstable cavities is the nonsymmetric confocal cavity, for which $R_1 + R_2 = 2l$. In the last relation the curvature radii of the mirrors are written in the algebraic sense, that is, for a convex mirror the curvature radius is negative. From the practical standpoint the most interesting case is usually the unidirectional beam extraction. Hence, the confocal unstable cavity, which consists of a concave mirror ($R_1 > 0$) and a convex mirror ($R_2 < 0$), has gained the widest acceptance. (Such a cavity is also known as telescopic.)

Figure 10.2a shows the diagram of a telescopic cavity. We analyze it from the standpoint of geometrical optics in Figure 10.2b. Using the notations of Figure 10.2b, we arrive at the following form of Eqs. (10.4):

$$\frac{1}{r_1} + \frac{1}{l + r_2} = \frac{2}{R_1}, \quad \frac{1}{r_2} + \frac{1}{r_1 - l} = \frac{2}{R_2}, \tag{10.14}$$

where r_1, r_2, R_1, and R_2 stand for the absolute values of the respective distances. From (10.14) we can easily obtain a formula that links r_1 and r_2,

$$r_2 = \frac{(r_1 - l)R_2/2}{r_1 - l - R_2/2}, \tag{10.15}$$

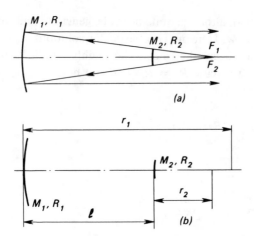

Figure 10.2. A telescopic unstable cavity.

and an equation for r_1:

$$r_1^2 + r_1 \frac{l(l + R_2)}{R_1/2 - R_2/2 - l} - \frac{l(l + R_2)R_1/2}{R_1/2 - R_2/2 - l} = 0. \qquad (10.16)$$

In the confocality limit (as $R_1 - R_2 \to 2l$) we have $r_1 \to \infty$ and $r_2 \to R_2/2$ (i.e. to the focal point). The oscillation mode of the cavity in this case is a combination of a spherical wave and a plane wave, which explains the name "telescopic". The magnification of a telescopic cavity is given, as Figure 10.2 readily shows, by

$$M = (r_2 + l)/r_2 = F_1/F_2 = R_1/R_2, \qquad (10.17)$$

and the coupling with the external space by

$$A = (M^2 - 1)/M^2 = (R_1^2 - R_2^2)/R_1^2. \qquad (10.18)$$

We note also that a telescopic cavity belongs, so to say, to the positive branch of unstable cavities, since a direct calculation shows that for this cavity $g_1 g_2 > 1$.

Though the geometrical-optics approximation has proved to be fairly good for unstable cavities, the wave approximation provides a fuller picture. From this approximation it follows that the phase of the wave solution corresponds to a near-spherical wavefront with a radius almost equal to the one obtained by geometrical means. Oscillation modes, that is, self-regenerating spatial distributions of the field, do exist. But the radial distribution of the field amplitudes in the cavity differs from the geometrical-optical, namely, an annular structure of diffraction origin manifests itself.

In analyzing the diffraction losses it has proved necessary to introduce an equivalent Fresnel number, N_F^{equiv}, such that at half-integral values of it the lowest-order mode, with the lowest losses, is clearly isolated, with the difference between the losses for the ground mode and those for the other modes being fairly high. The connection between the geometrical-optics approach just discussed and the wave approximation lies in the fact that the equivalent Fresnel number N_F^{equiv} is expressed in terms of the geometrical magnification M. In the case of a symmetric double-end cavity,

$$N_F^{equiv} = \frac{M^2 - 1}{2M} N_F, \qquad N_F = \frac{a^2}{l\lambda}. \tag{10.19}$$

For a telescopic cavity,

$$N_F^{equiv} = \frac{M - 1}{2} N_F, \qquad N_F = \frac{a_2^2}{l\lambda}. \tag{10.20}$$

When the values of N_F^{equiv} are half-integral, that is, when the lowest-order mode is well-isolated, the losses at this mode are noticeably smaller (but not considerably) than those predicted by geometrical optics. The difference is most noticeable for values of M slightly larger than unity and for all practical purposes can be ignored for $M \gtrsim 2.5$, which fully corresponds to the picture of a gradual transition from the wave region to the geometrical.

Concluding our discussion of unstable cavities, we note once more their main merits: first, the large volume occupied by the mode and the absence of a Gaussian compression of the field distribution toward the cavity's axis; second, good transverse mode selection due to large geometrical-optical radiation losses. We have not touched on the rigorous wave theory of unstable cavities. But intuitively it is clear that by their very nature geometrical-optical losses are close to diffraction losses, especially near the stability boundary. Hence, as mentioned earlier, the overall losses in unstable cavities depend strongly on the transverse index, which leads to mode selection according to this feature. Finally, from the practical viewpoint a great merit of unstable cavities is the possibility of using in them only reflection optics for both the cavity construction and beam extraction. Hence, metallic mirrors can be used, which is especially important for high-power optics of the IR range.

One of the main disadvantages of unstable cavities is that they can be used only with high-gain active media. In many cases, although not in all, the fact that the cross section of the output beam is of annular shape may be an inconvenience. For a telescopic cavity the inner diameter of the ring is $2a_2$ and the outer $2Ma_2$, where a_2 is the radius of the convex mirror in Figure 10.2a. However, in the far zone or in the focal plane of the lens that focuses the beam the black spot disappears. Existence in the cross section of the beam of diffraction rings following from the wave theory of unstable cavities usually presents no additional difficulties.

To conclude our lectures devoted directly to cavities it is well to note that various matrix and diagrammatic methods, whose description can be found in numerous reference books, have been developed to analyze and calculate optical systems, cavities, lens waveguides, transducers, and Gaussian beam matching devices successfully.

Let us now return to the problem of mode selection. It has been stressed many times in our discussions that the transition to open cavities accompanied by a rapid decrease in wavelength is actually due to the necessity of drastic purification of the oscillation spectrum, which condenses in proportion to the square of the frequency, ν. The above material has demonstrated that in open cavities this purification is achieved by increasing the radiation losses in the unwanted mode and retaining the low level of losses in the selected (useful) modes.

At the same time, in open cavities and especially in stable ones, the natural frequency spectrum is still too rich for various laser applications. Methods aimed at further "cleaning up" the spectrum or, in other words, at improving the mode composition of the radiation have become known as mode-selection methods. All are based on the earlier-mentioned idea of increasing the energy losses in the cavity for the unwanted modes and retaining the high Q-factor of the cavity for the required mode. The difference in the transverse structure of the field is used in transverse mode selection (see Lecture 9). The longitudinal modes have the same transverse structure but different numbers of half-waves fit into the space between the cavity mirrors. Hence, longitudinal modes differ in frequency and the position of the nodes of the standing wave along the cavity's axis.

The most general method of longitudinal mode selection uses the difference in frequency for these modes and, hence, requires placing narrow-band dispersive elements into the laser cavity. Among such elements are Fabry-Perot etalons, prisms and diffraction gratings, and mirrors with frequency dependent reflectivities.

The simplest approach to longitudinal mode selection is to use the frequency dependence of the gain of the active laser medium. The mode separation for longitudinal modes is (see Eqs. (6.14)-(6.17))

$$\Delta \nu_q = c/2l. \tag{10.21}$$

If the mode separation exceeds the amplification linewidth,

$$\Delta \nu_q > \Delta \nu_{line}, \tag{10.22}$$

and the central frequencies of a mode and the amplification line are close,

$$\nu_q \approx \nu_{line}, \tag{10.23}$$

single-mode lasing (in the sense of a longitudinal mode) and, hence, single-frequency lasing are generated. The lasing frequency in this case, in accordance with (6.33), is determined by the tuning of the mode frequency to the line frequency and their Q-factors ratio. This method of longitudinal mode selection can be successful in the case of gas lasers with sufficiently narrow amplification lines. An example is the carbon-dioxide low-pressure laser with an amplification linewidth of 60 MHz and a cavity length of 1 m ($\Delta \nu_q$ = 150 MHz). However, in the majority of cases the active-media amplification lines are much broader and this method leads to unacceptably short cavities.

The dispersive mirror method has gained wide acceptance. Let us consider the diagram in Figure 10.3. For $l_2 \ll l_1$ the two right mirrors in this triple-mirror cavity can be considered a single mirror with a reflectivity R_2 that is frequency dependent. To estimate the R_2 vs. λ dependence we can employ Eq. (6.7) for the transmission coefficient of a regenerated Fabry-Perot etalon.

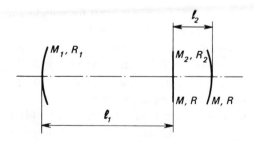

Figure 10.3. A cavity with a dispersive mirror R_2.

At $K = 1$ in the plane-wave approximation the transmission coefficient of two parallel mirrors each with a reflectivity R and separated by a distance l_2 is, in accordance with (6.7),

$$D = \frac{(1 - R)^2}{1 - 2R \cos (4\pi l_2/\lambda) + R^2}. \tag{10.24}$$

Then the reflectivity of the equivalent mirror, R_2, is given by the formula

$$R_2 = 1 - D = 2R \frac{1 - \cos (4\pi l_2/\lambda)}{1 - 2R \cos (4\pi l_2/\lambda) + R^2}. \tag{10.25}$$

We see that R_2 vanishes at $\cos (4\pi l_2/\lambda) = 1$, and at such frequencies the Q-factor of the cavity formed by mirrors R_1 and R_2 is nil, which leads to selection of the appropriate longitudinal modes. The diagram in Figure 10.3 and Eq. (10.25) lay no claims to a quantitative description of the process of longitudinal mode selection by the introduction of an additional dispersive element into the laser cavity. They are given only to illustrate the ideas of the method.

For $l_2 \ll l_1$ the nature of the frequency dependence of the dispersive mirror R_2 is such that a cavity constructed according to the diagram in Figure 10.3 is more suitable for suppressing separate modes than for selecting a single mode.

Practically speaking, there exist many modifications of the selecting mirror. In constructing the type of cavity now being discussed one must (a) take onto account the necessity of matching the transverse field distribution, that is, of mode matching, in all the resonant volumes of the multimirror cavity obtained in this manner and (b) allow for the effect that these partial cavities have on each other.

The most convenient compound selective mirrors for selecting a single longitudinal mode have proved to be those whose configuration differs considerably from the linear. Asymmetric configurations containing Michelson and Fox-Smith cavities and the symmetric T-shaped cavity have gained wide acceptance. The coupling of the main cavity to the selective mirror is maintained in such cases by inclusion of a beamsplitting plate.

Actually, in longitudinal mode selection one can use the difference in the longitudinal field distribution for different modes. Each longitudinal mode constitutes a standing wave, and the distance between the standing-wave nodes differs from mode to mode.

Let us consider the mode with the longitudinal index

$$q = 2l/\lambda. \tag{10.26}$$

The distance between the nodes of the qth and $(q + 1)$st modes closest to the mirror is

$$\Delta_1 = \frac{\lambda_q}{2} - \frac{\lambda_{q+1}}{2} = \frac{l}{q^2} = \frac{\lambda^2}{4l} \tag{10.27}$$

and very small. As we move away from the mirror the distance between the nodes grows and for the node of the qth mode with number $q/2$ is

$$\Delta_2 = \left(\frac{\lambda_q}{2} - \frac{\lambda_{q+1}}{2}\right)\frac{q}{2} = \frac{l}{2q} = \frac{\lambda}{4}, \tag{10.28}$$

which means that a node of the qth mode coincided with an antinode of the $(q + 1)$st mode. This happens at a distance

$$L_2 = \frac{\lambda_q}{2}\frac{q}{2} = \frac{l}{2} \tag{10.29}$$

from the mirror, that is, at the middle of the cavity. But at the same point there are an antinode of the $(q + 2)$nd mode, a node of the $(q + 3)$rd mode, an antinode of the $(q + 4)$th mode, etc. Hence, there is no way in which such spacing of the nodes and antinodes of longitudinal modes can be used for effective mode selection.

Obviously, the total number of modes among which selection must be carried out is

$$m = \Delta\nu_{\text{line}}/\Delta\nu_q. \tag{10.30}$$

Let us assume that we are looking for a way to select the qth mode and that the mode's central frequency ν_q coincides with the central frequency of the amplification line, ν_{line}. Then the number of farthest longitudinal modes which it is desirable to get rid of is $q \pm m/2$. In the vicinity of the first node of the qth mode the distance between the nodes is still very small:

$$\Delta_3 = \frac{\lambda_q}{2} - \frac{\lambda_{q \pm m/2}}{2} = \pm\frac{m}{2}\frac{l}{q^2} = \pm\frac{m\lambda^2}{2\,4l}. \tag{10.31}$$

The number N of the node in the qth mode that is the first to coincide with the antinode of mode $q + m/2$ or $q - m/2$ is determined by the equation

$$\frac{m}{2} \frac{1}{q^2} N = \frac{\lambda}{4} \tag{10.32}$$

and amounts to

$$N = \frac{\lambda}{2} \frac{q^2}{ml} = \frac{q}{m}. \tag{10.33}$$

This will happen at a distance

$$L = N \frac{\lambda_q}{2} = \frac{l}{m} = \frac{\Delta \nu_q}{\Delta \nu_{\text{line}}} l \tag{10.34}$$

from the mirror. At this point the distance between the nodes of the closest modes q and $q + 1$ is, in accordance with Eqs. (10.27), (10.33), (10.30), and (10.26),

$$\Delta = \Delta_1 N = \frac{\Delta \nu_q}{\Delta \nu_{\text{line}}} \frac{\lambda}{2}, \tag{10.35}$$

which for many gas lasers constitutes a noticeable fraction of the wavelength.

What has just been said suggests the following method of longitudinal mode selection.

Let us place a thin, partially transparent absorber at the Nth node of the sought for mode q. If the thickness of the absorber is considerably less than the wavelength, the presence of the absorber at the node, where the electric field strength of the qth mode is zero, does not cause any energy losses in the qth mode, but all the other modes at this point on the cavity's axis have a nonzero electric field strength and, therefore, will experience energy losses, which leads to their discrimination (Figure 10.4). This method of spatial longitudinal mode selection by thin absorbers has found application in gas lasers, primarily in the visible range. Thin, partially transparent absorbers are manufactured by sputtering metallic films 1-10 nm thick on optically polished transparent substrates. But because of the low optical durability of such films their application is restricted to medium-power lasers.

Introducing narrow-band dispersive elements into the laser cavity makes it possible not only to select longitudinal modes but, even when the amplifica-

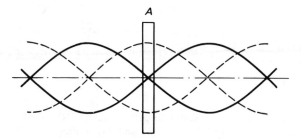

Figure 10.4. Transverse mode discrimination by a thin absorber. The dashed curves represent the standing wave of an unwanted mode, and the solid curves the standing wave of an isolated mode (A is the semitransparent absorbing film).

tion line is sufficiently broad and the tuning frequency of the dispersive element controllable, to transform the laser to the tunable type. Cavities with dispersive elements (prisms, diffraction gratings, and Fabry-Perot etalons), intended for tunable lasers, have become known as dispersive cavities. The simplest dispersive cavity is formed by introducing a prism into an ordinary cavity (Figure 10.5).

The common approach to reducing the losses inflicted by the introduction of a prism into the cavity is to ensure that the radiation is incident on the prism at the Brewster angle. Usually the prism's dispersion is not sufficient to attain a high degree of monochromaticity and fine frequency tuning. Hence the wide acceptance that more complicated dispersive cavities have attained (Figure 10.6). In such cavities the Fabry-Perot etalon enables selecting a single longitudinal mode and the grating suppresses lasing at unwanted transmission maxima of the etalon. The telescope serves to broaden the beam originating in the active media, which is necessary for optimal operation of both etalon and grating.

Figure 10.5. A prism dispersive cavity.

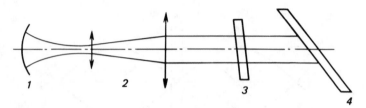

Figure 10.6. A compound dispersive cavity: (1) mirror, (2) telescope, (3) Fabry-Perot etalon, and (4) grating.

Dispersive cavities similar to the one shown in Figure 10.6 have made it
possible to generate radiation that is single-mode in the longitudinal and transverse indices and whose frequency can be tuned within a wide range.

Cavities based on the phenomenon of distributed feedback, or distributed
feedback cavities, have gained fairly wide acceptance in tunable lasers and
in others too and, equally so, in integrated optics and in building miniature
lasers. These lasers are sometimes called distributed feedback lasers. Distributed feedback is maintained by ensuring spatial modulation of the optical
parameters of the active media in the direction of propagation of the emitted
radiation with a period approximately equal to one-half of the wavelength
of this radiation in the medium. The underlying mechanism of this type of
feedback is the Bragg scattering on, theoretically speaking, the grating grooves
due to the spatial modulation of χ' and χ'' of the media. Ignoring the mode
structure of the field in such a distributed feedback cavity, we can exemplify
the qualitative picture of field distribution in it in the following manner.

A wave propagating, say, from left to right, experiences consecutive reflections from the grooves of the grating as it passes through the media and
decays. But the process also generates diffracted waves propagating in the opposite direction, from right to left. The reason that in our representation the
initial wave decays is that the energy is pumped into the counterrunning wave,
which increases in amplitude under in-phase addition. The net field in the
distributed feedback cavity is the superposition of these two waves.

Distributed feedback is spectrally selective by its very nature. The tuning
wavelength of a distributed feedback cavity is given by the Bragg condition
for backward scattering and determined by the grating constant. Various
methods have been developed for manufacturing both three-dimensional gratings and two-dimensional ones on the surface of a thin layer (a planar
waveguide) of the active medium. These methods can be both static and
dynamic.

In the radio-frequency range, Yagi antennas and multidipole delay systems of traveling wave masers are equivalent to the distributed feedback cavities of the type considered.

Problems to Lecture 10

10.1. Draw the ray diagrams when the light does two and three complete passes of the cavity in Figure 10.2.

10.2. Using the result of the previous problem, determine the minimum size of the concave mirror if the tube diameter in a copper-vapor laser is 2 cm.

10.3. Calculate the parameters of an unstable telescopic cavity for a copper-vapor laser if the tube diameter is 2 cm, the cavity length is set at 1 m, and after two passes only one-sixteenth of the initial power remains in the cavity.

10.4. Determine how well the higher-order modes are suppressed in the cavity of Problem 10.3.

10.5. For a carbon-dioxide laser ($\lambda = 10.6$ μm) calculate the parameters of an unstable telescopic cavity whose length is set at 3 m, N_F^{equiv} at 5.5, and the coupling with the external space, A, at 0.2.

10.6. Determine the tube diameter of the laser of Problem 10.5.

LECTURE 11
Mode Locking

Lasing in several longitudinal modes. The irregular nature of the lasing spectrum. Mode pulling. Mode locking. Pulse length and repetition period in mode locking. Active and passive mode locking. Self-locking. Q-switching. Lamb dip

In the previous lectures we established the possibility of lasing in the optical range and considered the properties of the open cavities required to maintain lasing. The mode structure of the field in cavities determines the mode composition of the laser radiation. The presence of many modes in the laser radiation leads to a multitude of interesting effects.

If the transverse modes are suppressed but generation of several longitudinal modes is possible and the inhomogeneous broadening $\Delta\nu_{\text{line}}$ of the laser transition exceeds the mode separation $\Delta\nu_q$, that is,

$$\Delta\nu_{\text{line}} > \Delta\nu_q, \tag{11.1}$$

lasing sets in at several frequencies separated approximately by the interval

$$\Delta\nu_q = c/2l, \tag{11.2}$$

known as the intermode beat frequency. Lasing occurs near the frequency corresponding to the central frequency of a lasing line, ν_{line}, which is approximately equal to the frequency ν_q of the qth longitudinal mode:

$$\nu_{\text{line}} \approx \nu_q = qc/2l, \quad q = 2l/\lambda \gg 1. \tag{11.3}$$

The number of laser modes is determined by the $\Delta\nu_{\text{line}}/\Delta\nu_q$ ratio. It is assumed, of course, that conditions for self-excitation are met for the entire inhomogeneously broadened line, $\Delta\nu_{\text{line}}$. Since inhomogeneous broadening corresponds to emission (absorption) of radiation in different spectral ranges by different particles, the conditions of self-excitation in multimode lasing are met independently for several (approximately $\Delta\nu_{\text{line}}/\Delta\nu_q$) independent oscillators placed, however, into a common cavity. Hence, the overall oscillation spectrum corresponds to a chaotic sum of emissions of several oscillators and has an irregular nature. The laser-output resulting field can be written as follows:

$$E(t) = \sum_{q_0 - N/2}^{q_0 + N/2} A_q \sin \left[2\pi(\nu_{line} + q\Delta\nu_q)t + \varphi_q \right], \tag{11.4}$$

where $N \approx \Delta\nu_{line}/\Delta\nu_q$, and $q_0 = 2l/\lambda_0$ is the number of the longitudinal mode corresponding to the line's center.

The phase of each of the N independent oscillators, φ_q, is in no regular way linked to the phase of any of the other oscillators, which leads to a chaotic spectral distribution. One must bear in mind, however, that the monochromaticity of the radiation remains fairly high in this case, too, since the oscillating frequencies of individual spectral components do not fall outside the limits of $\Delta\nu_{line}$. Let us consider the inhomogeneous broadening caused by the Doppler effect. Within the optical range, in accordance with Eq. (2.30), $\Delta\nu_{line} \approx$ 1000-1500 MHz. For gas lasers with a characteristic length $l = 1$ m we have $\Delta\nu_q = 150$ MHz. Hence, N reaches a value ranging from 5 to 10. It is this number of independent oscillators that works simultaneously if nothing is done either to enforce mode locking or suppress emission from all the oscillators except the specified.

Thus, the multimode character of the radiation emitted by a cavity leads, for a sufficiently broad amplification line, to the existence of several practically equidistant oscillating frequencies corresponding to several independent oscillators. The second important effect is that the oscillating frequencies are not exactly equidistant, that is, are not separated by an exact distance of $c/2l$; rather, they are shifted slightly toward the center. When Eq. (6.33) for the oscillating frequency was derived, we discussed the phase conditions of self-excitation, that is, the phase balance condition in self-excitation. The resonance amplification (absorption) line introduces its dispersive properties into the cavity. When the line is so broad that it covers several modes, this fact manifests itself in the following manner.

For modes whose frequencies are lower than the central one, the anomalous (resonance) material dispersion lowers the refractive index of the medium, the optical path shortens, and, hence, the resonance frequencies of the respective modes grow, that is, move toward the center of the line. On the other hand, for modes whose frequencies are higher than the central, the anomalous material dispersion elevates the refractive index, the optical path increases, and the resonance frequencies of the respective modes lower, that is, again move toward the center of the line. In other words, for a fixed value of q and in the presence of an additional variation in the refractive index, δn, the integral-number-of-half-waves condition takes on the form $l(n + \delta n) = q(\lambda + \delta\lambda)/2$, which leads to an appropriate change in the mode's frequency: $\delta\nu/\nu = -\delta\lambda/\lambda$.

The effect is determined by the anomalous dispersion curve when population inversion is present, and increases as we move away from the line's center. Since in the event of active medium excitation the n vs. ν dependence may be extremely complex, especially for inhomogeneously broadened lines, such effects as mode frequency splitting, mode pulling toward the line's center, and mode pushing from the line's center may be observed; these effects usually lead to disruption of the equidistance property of the oscillating frequencies of the separate longitudinal modes. The effects are mild; for instance, for a helium-neon laser the corresponding frequency shifts amount to 20-200 kHz. There are, however, situations in which mode pulling must be taken into account.

We see that in multimode lasing the output radiation varies with time in an irregular pattern, so that not only are the phases of the lasing modes distributed at random but the lasing processes in them do not occur simultaneously and have different amplitudes. But if the modes are forced to oscillate with approximately equal amplitudes and with phases rigidly locked in a certain manner, an interesting phenomenon emerges.

From Fourier analysis we know that a periodic train of identical pulses repeating with a fixed period T (Figure 11.1) can be represented by a series of discrete harmonic oscillations:

$$F(t) = \frac{F_0}{2} + \sum_{m=1}^{\infty}\left[F_m \cos\left(2\pi\,\frac{mt}{T}\right) + F_m' \sin\left(2\pi\,\frac{mt}{T}\right)\right]. \quad (11.5)$$

The spectral expansion (11.5) corresponds to the equidistant spectrum (adjacent to the zero frequency) with a frequency interval between neighboring com-

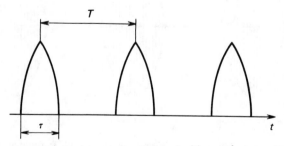

Figure 11.1. A periodic train of video pulses.

ponents equal to the pulse repetition frequency. As is well known, the overall width of the spectral expansion of $F(t)$ is inversely proportional to the length τ of one of the pulses in the train. In (11.5) the spectrum width is determined by the value of m at which the spectral amplitudes F_m and F'_m are not yet very small.

Function $F(t)$ describes a train of so-called video pulses. These pulses with a repetition interval (period) T and a length $\tau \ll T$ may serve as an envelope of a high-frequency process whose carrier frequency ν is much higher than $1/\tau$ (Figure 11.2). Such a high-frequency pulse-periodic process can be written, say, in the form

$$E(t) = AF(t) \cos (2\pi\nu t) + BF(t) \sin (2\pi\nu t), \qquad (11.6)$$

with A and B arbitrary constants. From Eqs. (11.6) and (11.5) it follows that the frequency spectrum of the process described by $E(t)$ is an equidistant train of spectral components with an interval between adjacent components (equal to the repetition frequency $1/T$), the train being grouped around frequency ν. The number of components (i.e. the overall width of the spectrum) is inversely proportional to the length τ of one pulse.

Thus, a periodic train of short pulses of high-frequency oscillations corresponds to the sum of many spectrally equidistant monochromatic oscillations with rigidly locked phases.

The longitudinal mode spectrum of an open cavity closely resembles an

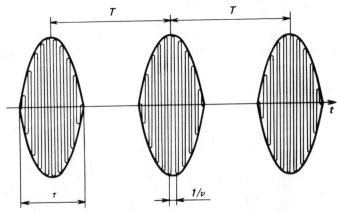

Figure 11.2. The pulse-train envelope in a high-frequency process.

equidistant spectrum, as we have just seen. For this reason lasers with many longitudinal modes in the amplification line of the active substance can emit radiation in the form of trains of short pulses. To achieve this we must ensure that (a) a large number of modes take part in the process, (b) the frequency spectrum of the lasing modes is strictly equidistant, and (c) the mode phases are rigidly locked. The lasing is then said to be mode-locked.

Such a situation may be achieved by periodic modulation of the parameters of the laser cavity. This must ensure that the laser radiation is modulated and, hence, that its spectrum contains additional components detuned from the carrier frequency by intervals that are integral multiples of the modulation frequency. If the modulation frequency is equal to the intermode beat frequency (the mode interval) (11.2), the side frequencies in the spectrum of each mode will coincide with the frequencies of adjacent modes and act as driving forces on each other. Since each of the lasing modes, in itself, is an independent self-oscillating system, a driving force whose frequency is close to the natural frequency of the system ensures forced locking. Since the locking of self-oscillating systems is characterized by a finite locking band, a small diversity of the mode spectrum from that of an equidistant spectrum, say, owing to the pulling effect, plays no role. Locking forces the modes to form an equidistant spectrum. The total number of locked modes is determined by such factors as the amplification linewidth, the losses in the cavity, the cavity's dispersive properties, the nonequidistant nature of the cavity's mode, and the depth and type of modulation.

In the simplest case of amplitude modulation we can picture the development of mode locking in a manner shown in Figure 11.3. Here we see the splitting of each mode into components and their subsequent locking.

Rigid mode locking requires maintaining a constant phase difference between the locked vibrations.

Suppose that the phase difference between the neighboring modes is $\Delta\varphi_q$, the frequency difference is $\Delta\omega_q$, and the central mode frequency ω_q is approximately ω_{line} and the central mode phase is zero. In all the modes the amplitudes of the vibrations are the same and equal to E_0. Then the total field of $2n + 1$ rigidly locked modes is given by the sum

$$E(t) = \sum_{k=-n}^{k=n} E_0 \exp\{j[(\omega_{\text{line}} + k\Delta\omega_q)t + k\Delta\varphi_q]\}. \tag{11.7}$$

Summation in (11.7) can easily be carried out if we employ the identity

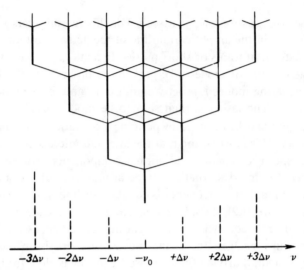

Figure 11.3. Development of mode locking under amplitude modulation with an intermode beat frequency $\Delta \nu$.

$$\sum_{-n}^{n} e^{jk\alpha} = 1 + 2 \sum_{1}^{n} \cos k\alpha \quad \text{and the known sum}$$

$$\sum_{0}^{n} \cos (k\alpha) = \frac{\cos \dfrac{n\alpha}{2} \sin \left(\dfrac{n+1}{2} \alpha \right)}{\sin \dfrac{\alpha}{2}}.$$

As a result of simple manipulations we find that

$$E(t) = E_0 \frac{\sin \left[\dfrac{2n+1}{2} (\Delta \omega_q t + \Delta \varphi_q) \right]}{\sin \dfrac{\Delta \omega_q t + \Delta \varphi_q}{2}} \exp (j \omega_{\text{line}} t). \qquad (11.8)$$

Hence, the total field generated by the interference of $2n + 1 = N$ locked vibrations is an oscillation periodically modulated in amplitude with a carrier frequency ω_{line} equal to the central mode frequency (the frequency at the amplification-line center) and with the envelope

$$A(t) = E_0 \frac{\sin\left[\dfrac{N}{2}(\Delta\omega_q t + \Delta\varphi_q)\right]}{\sin\dfrac{\Delta\omega_q t + \Delta\varphi_q}{2}}. \tag{11.9}$$

Figure 11.4 shows the intensity envelope $A^2(t)$ for the case in which seven modes ($N = 7$) are locked.

Several important conclusions can be drawn from (11.9). First, the peak values of the envelope are NE_0, which leads, as is usually the case with interference phenomena, to an N-fold increase of the peak intensity as compared with the case of the incoherent summation of N independent oscillations. The pulse repetition period is determined by the moments when the denominator in (11.9) vanishes and amounts to

$$T = 2\pi/\Delta\omega_q = 1/\Delta\nu_q = 2l/c, \tag{11.10}$$

that is, coincides with the intermode beat period or, in other words, the time that it takes the beam to make a complete trip around the cavity. This means that the train of pulses emerging in mode locking can be thought of as a single pulse travelling in the cavity between the mirrors in the direct and reverse directions alternately.

The length of a single pulse in a train or, as is commonly said, of a single spike in a train of mode-locked pulses (defined as the time interval in the

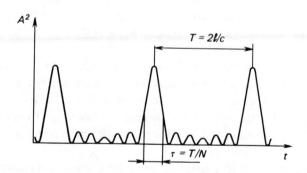

Figure 11.4. The intensity envelope in the case where seven modes are locked.

course of which practically all the energy corresponding to the beam making a complete trip around the cavity is emitted) is given by the period in which the numerator in (11.9) vanishes, and amounts to

$$\tau = 1/N\Delta\nu_q = T/N. \tag{11.11}$$

Since the number of locked modes cannot exceed the total number of modes that fit into the amplification linewidth $\Delta\nu_{\text{line}}$ ($N \leqslant \Delta\nu_{\text{line}}/\Delta\nu_q$), the minimum pulse length is restricted by the inverse of the amplification linewidth: $\tau \geqslant 1/\Delta\nu_{\text{line}}$. This is a direct corollary of the Fourier theorem. The characteristic lengths of such ultrashort laser spikes are 1 ns for typical gas lasers and 1 ps for solid lasers.

Equations (11.10) and (11.11) form the basis for describing mode locking. There are numerous methods of mode locking. The essence of each lies in the periodic modulation of the cavity parameters so as to obtain spectral components in the radiation that are shifted in frequency by the mode interval $\Delta\nu_q$. Mode locking in the case of external forced modulation of the cavity parameters is commonly called active mode locking. Various types of amplitude and phase modulation, cavity length modulation, and modulation of the active-media gain find application in active mode locking. When mode locking is achieved by using saturable absorbers (bleachable filters) placed into the laser cavity, it is said to be passive.

Passive mode locking, whose theory is quite complicated, can be examined qualitatively in two equivalent ways. Suppose that an absorber with two energy levels and a resonance frequency equal to the laser frequency is placed inside the laser cavity. Suppose also that the absorption line is homogeneously broadened and characterized by a cross section σ and a relaxation time τ_{rel}. If the initial absorption of the filter introduced into the cavity is not too great, lasing is achieved. The laser field causes the absorption of the filter to become saturated. When several modes are generated, the filter's saturated absorption is determined by the total radiation field and, in accordance with Eqs. (3.29) and (3.30), which give a nonlinear (quadratic) dependence of the absorption on the field strength, contains terms that vary with the intermode beat frequency, provided that the beat period exceeds the relaxation time ($T > \tau_{\text{rel}}$). Hence, the filter's transmission is modulated at the intermode beat frequency, which results in mode locking.

The second approach to passive mode locking is based on interpreting a train of pulses with a repetition period T as a single short (in comparison to T) pulse that travels many times back and forth between the cavity mirrors.

Suppose that the bleachable filter is placed near one of the mirrors. A pulse randomly generated and not yet clipped reaches the absorbing media and clears it owing to absorption of the energy from the leading edge of the pulse. The thus clipped pulse passes through the cleared absorber, is reflected by a mirror, and propagates in the opposite direction. If $T > \tau_{rel}$, in the second pass of the absorber the pulse again interacts with the already unsaturated filter, becomes still shorter, and so on. One must also bear in mind that the pulse becomes ever stronger in each pass between clipping acts in the absorber. This clipping process continues in an ideal situation as long as τ remains longer than $1/\Delta\nu$.

Thus, a bleachable filter operates in mode locking as an amplitude modulator that automatically tunes the modulation frequency to the intermode beat frequency.

Note that the random character of the onset of the passive mode locking process can easily be discovered when the method is realized experimentally.

If passive mode locking is realized not by the introduction of saturable absorbers or some other kind of nonlinear media into the laser cavity but by the nonlinearity of the laser's active medium, the term used to describe it is "self-mode-locking". Essentially, the difference between "self-mode-locking" and "passive mode-locking" is purely semantic.

To conclude this topic, we note the possibility of transverse-mode locking, which leads to periodic spatial scanning within the solid angle corresponding to the directions of emission of the locked transverse modes. The scanning takes place with the intermode beat frequency, and the scanning beam has an angular dimension corresponding to the width of a single mode.

A technique closely related to the method of transverse-mode locking is Q-switching which leads to the generation of so-called giant pulses. Essentially, differing from mode locking and being close to it only in some technical details, the Q-switching method emerged as a way of controlling the behavior in time of the pulsed lasing of solid lasers, which will be discussed later in detail. In a nutshell the situation here is as follows. Suppose that the properties of the laser's active media are such that pumping at a rate Λ is achieved. Then, ideally, the lasing power in the cw mode, P_{cw}, is $h\nu\Lambda$. The energy W emitted during the lifetime of the upper laser level, τ_{lft}, and corresponding to this power is $h\nu\Lambda\tau_{lft} = P_{cw}\tau_{lft}$.

Now let us assume that pumping is maintained but the conditions for lasing are not right because the cavity mirrors are blocked in some way, say, by an absorbing filter. Then the pump energy grows on the upper laser level but is not emitted, that is, the level is not depleted radiation-wise. However, because τ_{lft} is finite, the level cannot store an amount of energy greater than W. Suppose then that almost instantly the condition for lasing becomes right, say, the absorbers blocking the mirrors are removed. This means that the Q-factor of the cavity is suddenly switched on (hence the name Q-switching). If by this time the conditions for self-excitation are well met due to energy storage, all the energy is emitted in a single giant pulse. If the pulse length is τ_{pulse}, the peak power can be estimated at

$$P_{\text{pulse}} = W/2\tau_{\text{pulse}} = P_{\text{cw}} \tau_{\text{lft}}/2\tau_{\text{pulse}}. \qquad (11.12)$$

During τ_{pulse} the upper level is depleted completely. Figure 11.5 shows the sequence of events taking place as a result of the Q-factor being switched on at time t_0.

Q-switching becomes useful when the lifetime of the upper laser level, τ_{lft}, is great, since it is then that giant pulses are generated.

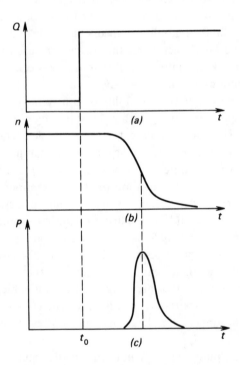

Figure 11.5. Q-switching (a) at time t_0 leads to changes in population-inversion density (b) and power output (c).

Before we start a more thorough, though still approximate, analysis of Q-switching, a remark is in order, namely, that the terms "Q-switching," "Q-switched pulse," and "Q-switched mode" appeared at the time when Q-switching was maintained through periodic modulation of the Q-factor, say, by rotating one of the cavity mirrors.

Let us now imagine a laser operating in the Q-switched mode, assuming that the Q-factor is switched on instantaneously and ignoring the relaxation of population-inversion densities and the pumping during the time that the giant pulse operates.

Let the lifetime of a photon in the cavity with a nonzero Q-factor be τ_{ph}. The reader will recall that, according to (7.17), this lifetime is related to the cavity's Q-factor by a simple formula: $Q = \omega \tau_{ph}$. Since $1/\tau_{ph}$ is the rate at which photons leave the cavity, the rate equation for the number of photons, Φ, can be written in the form

$$\frac{d\Phi}{dt} = \Phi\left(\alpha c - \frac{1}{\tau_{ph}}\right),$$ (11.13)

where c is the speed of light, and α the active-media gain. Introducing the notation $\alpha_{thr} = 1/c\tau_{ph}$ for the gain's threshold value and recalling that gain is proportional to the population-inversion density n, we can write Eq. (11.13) in the form

$$\frac{d\Phi}{dt} = \left(\frac{n}{n_{thr}} - 1\right)\frac{\Phi}{\tau_{ph}}.$$ (11.14)

Here $n/n_{thr} = \alpha/\alpha_{thr}$, $\Phi n/n_{thr}\tau_{ph}$ is the rate at which the number of photons in the cavity increases, and Φ/τ_{ph} the rate at which the number decreases. The increase prevails over the decrease if $n > n_{thr}$. Since each act of emission of a single photon lowers the population-inversion density by two units, the rate of population-inversion density decrease is proportional to the rate of photon-number increase, with the proportionality factor being equal to 2:

$$\frac{dn}{dt} = -2\frac{n\Phi}{n_{thr}\tau_{ph}}.$$ (11.15)

Simultaneous solution of the nonlinear equations (11.14) and (11.15) can easily be carried out numerically, but some qualitative conclusions can be drawn on the basis of available analytical solutions. Dividing (11.14) by (11.15), we arrive at the equation

$$\frac{d\Phi}{dn} = \frac{1}{2}\left(\frac{n_{\text{thr}}}{n} - 1\right), \tag{11.16}$$

whose solution is known:

$$\Phi = \Phi_0 + \frac{1}{2}\left[n_{\text{thr}} \ln \frac{n}{n_0} - (n - n_0)\right], \tag{11.17}$$

where Φ_0 and n_0 are the number of photons and the population-inversion density at the initial moment, t_0, that is, the moment when the Q-factor was switched on. It is natural to assume that $\Phi_0 \ll \Phi$. Then

$$\Phi = \frac{1}{2}\left[n_{\text{thr}} \ln \frac{n}{n_{\text{thr}}} - (n - n_0)\right]. \tag{11.18}$$

Now, we don't know the solution to (11.14), that is, $\Phi(t)$. It is obvious, however, that for $t \gg \tau_{\text{ph}}$ the number of photons, Φ, tends to zero. The population-inversion density does not vanish in the process; rather, it reaches a finite value n_{fin}, which can be found at $\Phi = 0$ from (11.18) by solving the transcendental equation

$$\frac{n_{\text{fin}}}{n_0} = \exp \frac{n_{\text{fin}} - n_0}{n_{\text{thr}}}. \tag{11.19}$$

This equation can be used, for one thing, to determine the usefully utilized fraction of the energy stored during population inversion:

$$\eta = \frac{n_0 - n_{\text{fin}}}{n_0} = 1 - \frac{n_{\text{fin}}}{n_0} = 1 - \exp \frac{n_{\text{fin}} - n_0}{n_{\text{thr}}}. \tag{11.20}$$

Equation (11.20) justifies the intuitively clear conclusion that when the initial population-inversion density exceeds the threshold value considerably ($n_0/n_{\text{thr}} \to \infty$), all the energy stored during population inversion is used for lasing: $\eta \to 1$.

The laser power is linked to the number Φ of photons in the laser cavity and the photon lifetime τ_{ph} by the following obvious relation:

$$P = \Phi h\nu/\tau_{\text{ph}}. \tag{11.21}$$

From (11.21) and (11.18) it follows that

$$P = \frac{h\nu}{2\tau_{\mathrm{ph}}} \left[n_{\mathrm{thr}} \ln \frac{n}{n_0} - (n - n_0) \right], \tag{11.22}$$

where the population-inversion density n is time-dependent: $n = n(t)$. If we knew $n(t)$, Eq. (11.22) would give us the shape of the radiation pulse. Numerical solutions of Eqs. (11.14) and (11.15) show that for $n_0/n_{\mathrm{thr}} \gg 1$ the leading edge of the pulse is shorter than the photon lifetime in the cavity, τ_{ph}, while the trailing-edge length is of the order of τ_{ph}.

The peak power carried by a giant pulse (the Q-switched pulse) can be found via (11.22). The condition $dP/dn = 0$ implies that the maximum radiation power emitted in giant-pulse lasing is attained at $n = n_{\mathrm{thr}}$ (see Eq. (11.22)). Here

$$P_{\max} = \frac{h\nu}{2\tau_{\mathrm{ph}}} \left[n_{\mathrm{thr}} \ln \frac{n_{\mathrm{thr}}}{n_0} - (n_{\mathrm{thr}} - n_0) \right], \tag{11.23}$$

which for $n_0/n_{\mathrm{thr}} \gg 1$ yields

$$P_{\max} \approx n_0 \frac{h\nu}{2\tau_{\mathrm{ph}}}. \tag{11.24}$$

Since the initial population-inversion density n_0 has the meaning of the product of the upper-laser-level pumping rate by the upper level's lifetime, $n_0 = \Lambda\tau_{\mathrm{lft}}$, we see that the estimate (11.12) actually coincides with (11.24) obtained in a somewhat more rigorous fashion, assuming, of course, that the length of the giant pulse is approximately equal to the photon lifetime in the cavity.

The Q-switched mode is realized by introducing mechanooptical and optoelectronic gates or gates based on bleachable filters. In the latter case the resonance-absorption cross section of the filter substance must be many times greater than the resonance-gain cross section of the active medium: $\sigma_{\mathrm{fil}} \gg \sigma_{\mathrm{gain}}$. Otherwise, practically all the stored energy will be used for a steady-state bleaching of the filter. Besides, to insure the rapid switch-on of the Q-factor the filter's relaxation time must be made much shorter than the population-inversion lifetime: $\tau_{\mathrm{fil}} \ll \tau_{\mathrm{lft}}$ (see Lecture 3).

To conclude this lecture we will briefly discuss an interesting effect caused by Doppler broadening of the amplification line, an effect that manifests itself most strikingly in single-mode gas lasers. As we know, radiation only interacts with particles that are in resonance with the electromagnetic wave, that is, particles lying within a homogeneously broadened spectral line. Hence, in the event of Doppler, or inhomogeneous, broadening the fraction of particles interacting with the field is determined by the homogeneous linewidth to the Doppler linewidth ratio (see Lecture 2). In other words, if we have a travelling plane wave $E \cos (\omega t - \mathbf{k} \cdot \mathbf{r})$, the only particles that interact with the wave are those having a resonance frequency $\omega = \omega_0 + \mathbf{k} \cdot \mathbf{v}$ and falling into a spectral interval with a homogeneous linewidth $\Delta\omega$:

$$| \omega_0 - \omega + \mathbf{k} \cdot \mathbf{v} | \leq \Delta\omega/2. \tag{11.25}$$

Here $\mathbf{k}$ is the wave vector, and $\mathbf{v}$ the particle velocity.

The preferable excitation of particles moving with a certain velocity leads to a noticeable shift in their velocity distribution when the field is strong. The velocity distribution of the particles on the lower level acquires a dip and that of the particles on the upper level a respective peak. These features emerge when the particles move with velocities determined by the field's frequency, in accordance with the meaning of the Doppler effect, as follows:

$$v = \frac{\omega - \omega_0}{\omega_0} c. \tag{11.26}$$

The depth of the dip and the height of the peak are determined by the saturation of the corresponding transition, and the width of both dip and peak is equal to the homogeneous linewidth with allowance for broadening by the strong field.

In many gas lasers the homogeneous amplification linewidth is much smaller than the Doppler linewidth. An intense light wave causing gain saturation, which, as we know, is the nonlinear effect that determines the operation of a laser, "burns out" a dip at the field frequency in the Doppler contour of the amplification line (Bennett's dip). The standing light wave inside the laser cavity is a combination of two waves of the same frequency travelling in opposite directions. Each burns out its own dip. If one of the travelling waves burns out a dip for particles with speeds specified by (11.26), the wave of the same frequency travelling in the opposite direction acts on particles with velocities

$$v = -\frac{\omega - \omega_0}{\omega_0} c. \tag{11.27}$$

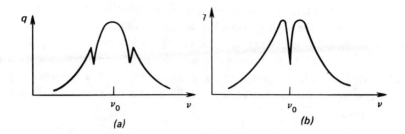

Figure 11.6. Dips "burnt out" in the contour of a Doppler-broadened amplification line (a) $\nu \neq \nu_0$, and (b) $\nu = \nu_0$.

Thus, two dips, positioned symmetrically in relation to the center, appear in the velocity distribution of the particles, and two respective dips appear in the Doppler contour (Figure 11.6). When the laser is tuned to the center of the Doppler contour, the two dips merge. This leads to a sharp drop in the laser output power at the center of the Doppler amplification line. The phenomenon became known as the Lamb dip and is used to stabilize the laser frequency.

Problems to Lecture 11

11.1. Can mode locking be achieved in the case of a truly homogeneous line contour?

11.2. Sketch the approximate shape of the amplification spectrum for a weak signal of the active medium of a mode-locked laser.

11.3. Assuming that $\eta \approx 1$ (see Eq. (11.20)), estimate the length of the giant pulse.

11.4. Figure 11.6 depicts the contours of a Doppler-broadened amplification line as laser radiation "sees" it. What will be the appearance of these contours for a weak signal propagating perpendicular to the cavity's axis?

11.5. Estimate the pulse length for a mode-locked helium-neon laser emitting radiation with a Doppler linewidth of 1500 MHz and having a cavity 1-m long.

11.6. Formula (11.24) implies that the power output tends to infinity as τ_{ph} decreases. Explain the reason.

11.7. Assuming that a giant pulse is 10-ns long, estimate the Q-factor (in the high-Q mode) of a visible laser and the minimum amplification linewidth necessary for generating such a pulse.

11.8. By how much will the repetition frequency of a passively mode-locked laser change if the plate of a saturable absorber is placed not close to the mirror but in the middle of the cavity, or $l/3$ distant from the mirror?

LECTURE 12

The History of Quantum Electronics. Basic Formulas

The origins of quantum electronics. The quantum nature of light, stimulated emission, and bosons. Einstein and Dirac. The first maser. Radio engineering and optics. Microwave spectroscopy. Townes, Prokhorov, and Basov. The three-level method. The proposal of the open cavity. The first lasers. The place of quantum electronics in optics. The basic formulas of quantum electronics

What has been said in previous lectures shows that the main problems of quantum electronics come down to creating an active medium and ensuring the necessary feedback in the appropriate cavity. The active medium and the cavity are, in principle, the most important parts of the majority of devices of quantum electronics, because essentially it is they that form a laser oscillator. The most general and essential requirement here is the creation of conditions for population inversion. Hence, what is important in creating lasers and masers is to establish population inversion. In what follows we discuss this aspect as applied to laser systems, and in view of the fast progress of quantum electronics we will restrict our discussion to the most characteristic methods.

To conclude our exposition of the basics of quantum electronics it is expedient to examine briefly the history of the ideas underlying quantum electronics.

Three fundamental ideas form the basis of quantum electronics. First, the energy of electromagnetic radiation consists of discrete portions called light quanta or photons. This fact manifests itself primarily in the interaction of radiation with matter, when photons are absorbed or emitted. Second, the emission of photons at sufficiently high intensities is determined by stimulated emission. Here the photons belonging to the stimulating and stimulated fluxes are identical, and the emission rate is proportional to the intensity. Third, and last, the quanta of electromagnetic radiation obey the Bose-Einstein statistics. For this reason the number of photons per field oscillator is unlimited. When a single field oscillator (one mode) is occupied by a large number of indistinguishable photons, a classical coherent electromagnetic wave forms.

When Einstein put forward these ideas, he laid the physical basis of quantum electronics. In 1905, fifty years before quantum electronics came into being, Einstein, analyzing statistically the fluctuations of the energy of equilibrium radiation, arrived at the hypothesis of light quanta, or photons, and applied it immediately to the photoelectric effect. Explaining the photoelectric threshold made it possible to establish the quantum nature of electromagnetic radiation. In 1916 Einstein derived Planck's formula in accordance with Bohr's already well-known postulates. This played an important role in comprehending the nature of equilibrium emission by quantum systems, that is, systems possessing discrete energy levels. For us, however, this conclusion is important primarily because it introduced the concept of stimulated emission of radiation, postulated the existence of such emission, and determined the properties of the emitted radiation on the basis of very general thermodynamic considerations. Briefly, Einstein's conclusions can be formulated as follows.

Stimulated emission of radiation is defined as an effect whose probability of a single act is proportional to the energy flux of the radiation that exerts on the particle emitting radiation. The frequency of the emitted radiation is exactly equal to that of the stimulating radiation, and the spatial directivity of the stimulated radiation coincides with that of the stimulating. To within the degeneracy multiplicity of the respective levels, the stimulated emission and absorption processes are equally probable. Somewhat later Einstein brought the polarization of the radiation into the picture, and again on the basis of thermodynamic considerations came very close to formulating the idea of complete identity of the photons of stimulated and stimulating radiation.

Thus, the quantum nature of electromagnetic radiation and the quantization of energy levels of microparticles suggests that there must be a process essential for the emission of identical photons.

By 1924 the Indian physicist Satyendra Nath Bose (1894-1974) and Einstein, furthering the method of deriving Planck's formula by consistently applying the light quanta hypothesis, had created a generalized thermodynamic description of a system of particles with symmetric wave functions, the so-called Bose-Einstein statistics. The fundamental property of the identical particles that obey this statistics is their indistinguishability. The quanta of electromagnetic radiation, photons, are among the particles that obey this statistics, and the generic name for such particles is bosons. Hence, as we have stressed many times in our lectures, the stimulated radiation quanta with identical frequencies, polarizations, directivities, and phases are indistinguishable from each other. The state of the entire radiation field is determined by the number of quanta (photons) per field oscillator. This number may be infinitely great. The fact that photons are bosons makes it possible in quantum electronics to pass from corpuscular representation to wave representation, for which the principle of linear combination of oscillations, including coherent oscillations, is characteristic.

A consistent quantum theory of the emission and absorption of light was formulated by Dirac in 1927. Dirac substantiated Einstein's statistical laws of radiation and, calculating the emission probability, found the link between Einstein's phenomenological coefficients and the characteristics of the emitting atom. A most important result of Dirac's quantum theory of light is that Dirac consistently proved the existence of the stimulated radiation postulated by Einstein and its coherence, which Einstein intuitively predicted.

Hence, by 1927 all the fundamental physical prerequisites for quantum electronics had been created. But it was only toward the end of 1954 and beginning of 1955 that the theoretical bases of quantum electronics were formulated and the first device built. This was the molecular oscillator or ammonia-beam laser.

Quantum electronics emerged when a quantum system with population inversion was placed inside a resonant cavity. This happened at the end of 1954 simultaneously at the Oscillations Laboratory of the Lebedev Physics Institute (Moscow, USSR) headed by Alexander M. Prokhorov and at the Radiation Laboratory of Columbia University (New York, USA) by a group of scientists led by Charles H. Townes.

The big gap between creation of the prerequisites for the appearance of quantum electronics and its actual appearance is proof that the physical idea of stimulated emission could not be used immediately and directly for building sources of radiation of a new type.

It is also no accident that quantum electronics emerged in the radio-frequency range. In a molecular oscillator at $\lambda = 1.25$ cm stimulated emission of radiation was observed for the first time directly in pure form, with the oscillator operating in the self-oscillating mode with positive feedback.

The fact is that to observe stimulated emission of radiation one must have excited atoms and, in addition, the probability of stimulated emission must be higher than that of spontaneous emission. In experimental conditions of classical optical spectroscopy these requirements could not be met.

Although the possibility of building a laser had existed for some 50 years, lasers appeared later than masers, because the emission of radiation by classical sources in the optical range is not coherent, stimulated emission is suppressed by spontaneous, and, which is also important, scientists working in optics lacked a whole range of methods and concepts well-developed in radio-frequency electronics.

By 1940 the fact that a system of excited atoms is capable of amplifying light was firmly ingrained in the minds of scientists working in optics, notably the Soviet scientist Valentin A. Fabrikant. But no one at the time suggested building an optical oscillator on this basis. In optics such a suggestion was impossible, notwithstanding the fact that Einstein and Dirac had had in mind optics when they formulated their ideas of stimulated emission.

In the first half of the 20th century, radio engineering and optics developed along different lines. Quantum ideas prevailed in optics and wave ideas in radio engineering, and the unity of radio engineering, and optics was always stressed from the standpoint of wave theory. The wave concepts taken from optics enriched radio engineering, and vice versa.

The unity of radio engineering and optics stemming from the quantum nature of electromagnetic radiation did not manifest itself for a long time. This situation continued until radiospectroscopy appeared as a section of spectroscopy studying the spectra of molecules, atoms, and ions in the microwave range. Vigorous development of radiospectroscopy began after World War II owing to the intense development of centimeter technology that met the requirements of radiolocation. At the time reliable sources of monochromatic microwave radiation were constructed, waveguide methods of channeling microwave energy and high-Q resonant microwave cavities were developed, and sensitive receivers were built. All this gave radiospectroscopy a head start over ordinary spectroscopy.

Radiospectroscopy is not solely a region into which spectroscopy extended quantitatively at the low-frequency end. The basic feature that set radiospectroscopy apart from optical spectroscopy at the time was the monochromaticity of radiation in the former. In addition, spontaneous emission in the radio-frequency range is much weaker (by a factor of ν^3) than the stimulated and the excited levels are heavily populated, a fact that at once introduced the concept of stimulated emission. As a result radiospectroscopy paved the way for studies in quantum electronics.

In the USSR work in radiospectroscopy was started by Prokhorov at the Oscillations Laboratory of the Lebedev Physics Institute.

Along with purely spectroscopic problems, studies were also carried out in the use of microwave spectra for establishing standards of frequency and time. It was along this avenue that quantum electronics emerged.

Indeed, the precision of operation of a frequency standard based on the stability of the frequency of a spectral absorption line depends on the linewidth. The narrower the line, the higher the precision with which the frequency being stabilized is tuned to the rated value of the absorption-line frequency. For molecular gases at low pressures, Doppler broadening is a characteristic feature in the radio-frequency range. Hence, the narrowest lines are observed in the case of molecular beams with a small velocity spread, but the absorption-line strength for molecular beams is low since in a well-collimated beam there are few particles and the difference between the numbers of excited and nonexcited particles is small.

At this stage the idea emerged that changing the ratio of the nonexcited particles to the excited could considerably increase the sensitivity. If molecules could be sorted out and a beam formed of none but excited molecules, absorption would change sign and become emission. A molecular beam then becomes media with negative losses. Specialists in the theory of oscillations knew that if such a beam could be combined with a resonance circuit, a generator of monochromatic oscillations would be created. By this time within the theory of oscillations there had developed the theory of self-oscillating systems with positive feedback, and resonant cavities, which are extremely well-suited for operating beams, had become developed in the microwave range.

The most important step was to combine all the possibilities, and this was done when a beam of molecules with population inversion was sent through a resonant cavity. The first masers were built along these lines by Nikolai G. Basov and Alexander M. Prokhorov in the Soviet Union and by James P. Gordon, Townes, and Herbert J. Zeiger in the United States. In 1964 the Nobel Prize in physics was awarded to Townes, Prokhorov, and Basov for "work in quantum electronics leading to construction of instruments based on maser-laser principles."

The next important step was taken in 1955, when Basov and Prokhorov suggested an active method for creating population inversion by radiation pumping, the so-called three-level pumping method, which later found wide application and will be discussed at length in subsequent lectures. In 1956, Nicolaas Bloembergen (USA) suggested employing the three-level method to build maser amplifiers using paramagnetic crystals, which considerably broadened the scope of quantum electronics.

The achievements of quantum electronics in the radio-frequency range suggested moving into the shorter-wave region, which was a natural problem for radio engineers and radio physicists. However, one difficulty that became ever more serious on this route was the problem of cavities, devices essential for the operation of generators of monochromatic radiation. Here a significant step was taken by Prokhorov (1958), who suggested using what became known as open cavities. As we know, an open cavity is a Fabry-Perot interferometer, but it was the radio-engineering approach that allowed Prokhorov to use this interferometer as a cavity.

The first achievement of quantum electronics in the optical range was the construction at the end of 1960 of optical generators of monochromatic oscillations, the first lasers, a ruby laser by Theodore H. Maiman and a laser on a mixture of gaseous neon and helium by Ali Javan. These events concluded the prehistory of quantum electronics and marked the beginning of its remarkable progress.

Today quantum electronics encompasses a huge field of R&D. But as an offshoot of microwave electronics it brought the greatest changes to optics. Although the maser and the laser operate on the same principle, there is a very big difference between them. In the radio-frequency range the building of masers meant the appearance of devices new in principles of operation and yet possessing properties common to classical electronics. It goes without saying that masers drastically improved the parameters of radio-frequency devices. The sensitivity of amplifiers increased by a factor of 100 to 1000, and the stability of the oscillator's frequency increased 1000- to 10000-fold. The significance of these achievements of quantum electronics is great, but actually, they amount to a quantitative increase in already known qualities, since even before quantum electronics there existed coherent radio-frequency amplifiers and oscillators of monochromatic radiation in electronics.

Optics, however, is a different matter. By their very nature all sources of light in optics, in contrast to ordinary electronics, are quantum. Hence, the use of the term "optical quantum generator" has proved to be highly unsuccessful. The very idea of photons emerged during analysis of the properties of optical emissions. But before quantum electronics came into being, all quantum sources of light emitted nonmonochromatic incoherent oscillations. Optics had no coherent amplifiers or monochromatic generators of electromagnetic oscillations. Only lasers, in contrast to ordinary sources of light, emit light waves of high spatial directivity, spectral monochromaticity, and temporal coherence. The emergence of lasers, therefore, gave optics an entirely new opportunity to concentrate radiant energy in space, time, and a spectral interval. This was what lifted optics to a new level, characterized by applications that are traditionally nonoptical. These new fields of application came into being only because in modern optics the effect of stimulated emission of light, central to quantum electronics, is used to generate light by methods of quantum electronics.

We stress once more that quantum electronics determines the new possibilities of optics but does not overhaul its fundamental concepts. More than that, quantum electronics, is based on the fundamental concepts of 20th-century optics, and its emergence, first steps, and further development strengthened these concepts and broadened the field of application.

Quantum electronics is essentially not a science with a closed field of investigation. Rather, it is a method for constructing devices of intense action on matter by monochromatic radiation or devices for transferring large volumes of data via monochromatic radiation. This is why quantum electronics is not restricted to lasers. At the same time the designing of lasers and their improvement and the conquest of new wavelength ranges constitute the base of quantum electronics.

In the process of forming laser light a fraction of the energy is lost, with the result that the efficiency of lasers is low. The properties of laser light, which are a direct consequence of the properties of stimulated radiation, predetermine the possibility of applications that more than compensate for energy losses.

The range of laser applications is already incredibly broad, from submillimeter to UV waves, in the cw and pulsed modes. The variety of lasers is enormous, with, of course, a corresponding variety of active media and physical phenomena used for excitation of the media.

Any detailed description or even simple enumeration of all the known lasers is out of the question within the scope of a course of lectures on the basics of quantum electronics. In subsequent lectures we will consider the most typical lasers and most characteristic methods of creating the conditions for population inversion. We must stress once more that a common feature of all these methods is the need to expend so as to create in the laser's active substance a thermodynamically nonequilibrium state in which stimulated emission prevails over absorption.

* * *

To conclude Part One of this book we offer a list of basic formulas of laser physics.

Probability of stimulated transitions per unit time:

$$w_{12}^{\text{stim}} = B_{12}\varrho_\nu, \tag{1.2}$$

$$w_{21}^{\text{stim}} = B_{21}\varrho_\nu. \tag{1.3}$$

Probability of spontaneous emission per unit time:

$$w_{21}^{\text{spon}} = A_{21}. \tag{1.6}$$

The relations between Einstein's coefficients:

$$g_1 B_{12} = g_2 B_{21}. \tag{1.10}$$

$$A_{21} = \frac{8\pi v^2}{c^3} hv B_{21}. \tag{1.11}$$

Natural linewidth:

$$\Delta v_0 = A_{21}/2\pi. \tag{2.2}$$

Collision-broadened linewidth:

$$\Delta v_{col} = 1/2\pi \tau_{col}. \tag{2.31}$$

Lorentzian lineshape:

$$q(v) = \frac{1}{2\pi} \frac{\Delta v_{line}}{(v - v_0)^2 + \Delta v_{line}^2/4}. \tag{2.7}$$

Stimulated-transition rate for monochromatic radiation with radiative energy density ϱ and homogeneous line broadening:

$$W^{stim} = 2B_{21}\varrho/\pi\Delta v_{line}. \tag{2.20}$$

Gaussian lineshape of a Doppler broadened line:

$$q(v) = \frac{1}{\Delta v_T \sqrt{\pi}} \exp\left[-\left(\frac{v - v_0}{\Delta v_T}\right)^2\right]. \tag{2.26}$$

Doppler linewidth:

$$\Delta v_D = \Delta v_T 2\sqrt{\ln 2} = 2v_0 \sqrt{\frac{2kT}{mc^2} \ln 2}. \tag{2.28}$$

Population inversion condition:

$$n_2/g_2 > n_1/g_1. \tag{3.2}$$

Saturation condition:

$$n_1/g_1 - n_2/g_2 = 0. \tag{3.15}$$

Amplification factor, or gain, (absorption coefficient):

$$\alpha = \left(\frac{n_1}{g_1} - \frac{n_2}{g_2} \right) \frac{g_1 2 B_{12} h \nu}{c \pi \Delta \nu_{\text{line}}}, \tag{3.5}$$

$$\alpha = n \sigma. \tag{3.13}$$

Amplification (absorption) cross sections:

$$\sigma = \frac{g_2}{g_1} \frac{\lambda^2}{2\pi} \frac{1}{2\pi \Delta \nu_{\text{line}} \tau_0} = h \nu \frac{2 B_{12}}{c \pi \Delta \nu_{\text{line}}}. \tag{3.14}$$

Saturation intensity in cw mode:

$$I_{\text{sat}} = h\nu / 2\sigma\tau. \tag{3.28}$$

Saturation gain in the event of homogeneous broadening and cw mode:

$$z = \frac{z_0}{1 + (g_1 + g_2) I / 2 g_2 I_{\text{sat}}}. \tag{3.30}$$

Saturation energy density in pulsed mode:

$$F_{\text{sat}} = I_{\text{sat}} \tau = h\nu / 2\sigma. \tag{3.43}$$

Strong gain saturation in the event of homogeneous broadening and pulsed mode:

$$z = z_0 \exp \left(- \frac{g_1 + g_2}{2 g_2} \frac{F_{\text{pulse}}}{F_{\text{sat}}} \right). \tag{3.45}$$

Einstein's coefficient B_{12} versus the matrix element $\langle \mu \rangle$ of the interaction energy:

$$B_{12} = \frac{8\pi}{3} \frac{\langle \mu \rangle^2}{h^2}. \tag{4.25}$$

Rabi' frequency:

$$\Omega_{\text{R}} = \left\langle \frac{\mu E}{h} \right\rangle. \tag{4.49}$$

Bandwidth of a travelling-wave amplifier:

$$\Delta \nu = \Delta \nu_{\text{line}} (\ln 2)^{1/2} (\ln G_0 + \ln L - \ln 2)^{-1/2}. \tag{5.5}$$

Minimum single-mode spectral density of input noise of a travelling-wave amplifier:

$$P_{\text{in}}^{\text{eff}} = h\nu.$$ (5.16)

Maximum output intensity of a travelling-wave amplifier in cw mode:

$$I_{\text{max}} = \frac{\alpha_0}{\beta} I_{\text{sat}}.$$ (5.26)

Maximum output fluence of a travelling-wave amplifier in pulsed mode:

$$F_{\text{max}} = \frac{\alpha_0}{\beta} F_{\text{sat}}.$$ (5.30)

Q-factor of an open cavity:

$$Q = \frac{2\pi l}{\lambda} \frac{1}{1 - R}.$$ (6.2)

Power gain of a pass resonance amplifier:

$$G = \frac{(1 - R)^2 K}{1 - 2RK \cos(4\pi l/\lambda) + R^2 K^2}.$$ (6.7)

Self-excitation condition:

$$RK = 1.$$ (6.9)

Single-mode oscillating frequency:

$$\omega = \frac{\omega_{\text{cav}} \Delta\omega_{\text{line}}/\ln K + \omega_{\text{line}} \Delta\omega_{\text{cav}}}{\Delta\omega_{\text{line}}/\ln K + \Delta\omega_{\text{cav}}}.$$ (6.33)

The highest possible output intensity of a cw oscillator:

$$I_{\text{out}}^{\text{max}} = \frac{\alpha_0}{\beta} I_{\text{sat}}.$$ (6.44)

Maximum output intensity of a medium-length oscillator:

$$I_{\text{out}}^{\text{max}} = \alpha_0 l (1 - \sqrt{\beta/\alpha_0})^2 I_{\text{sat}}.$$ (6.46)

Optimal transmittance of the output mirror:

$$T_{op} = 2\alpha_0 l(\sqrt{\beta/\alpha_0} - \beta/\alpha_0). \tag{6.47}$$

Maximum power output of a unit volume of active medium in cw mode with the rate Λ $[s^{-1}cm^{-3}]$ at which population inversion is created:

$$P_1 = \Lambda h\nu. \tag{6.57}$$

Maximum energy emitted in pulsed mode by a unit volume of active medium with population inversion involving N particles per cubic centimeter:

$$E_{em} = Nh\nu/2. \tag{6.58}$$

Fresnel number:

$$N_F = a^2/l\lambda. \tag{7.12}$$

Q-factor of a mode versus the photon's lifetime in the mode:

$$Q = \omega\tau. \tag{7.17}$$

Confocality condition:

$$2l = R_1 + R_2.$$

The natural frequencies of a confocal cavity for the TEM_{mnq} mode:

$$4l/\lambda = 2q + (1 + m + n). \tag{8.3}$$

The transverse field distribution in the fundamental mode TEM_{00q} of a confocal cavity:

$$S(x, y) = \exp[-(x^2 + y^2)/2w^2]. \tag{8.1}$$

Intensity distribution width in the fundamental mode TEM_{00q}:

$$w^2 = w_0^2 + (z/kw_0)^2. \tag{8.4}$$

Neck radius:

$$w_0 = \sqrt{l/2k}. \tag{8.5}$$

Radius of spot on mirror:

$$w = w_0\sqrt{2}.$$

Curvature radius of the wavefront of a Gaussian beam:

$$R = z + (k\omega_0^2)^2/z. \tag{8.10}$$

Angular divergence of a Gaussian beam:

$$\theta = w/z = 1/(kw_0), \quad \Omega = \lambda/l. \tag{8.11}$$

Gaussian-beam focusing:

$$v_0 = w_0F/z, \tag{8.16}$$

$$x = -F. \tag{8.17}$$

Conditions for focusing a Gaussian beam:

$$z \gg l/2, \quad z \gg F > D. \tag{8.19}$$

The boundaries of the stability region of an open cavity formed by two spherical mirrors ($g_1 = 1 - l/R_1$, $g_2 = 1 - l/R_2$):

$$g_1g_2 = 1, \tag{9.20}$$

$$g_1g_2 = 0. \tag{9.21}$$

Distance between the mirrors in an unstable telescopic cavity (confocality condition):

$$2l = R_1 - R_2.$$

The magnification of an unstable telescopic cavity:

$$M = R_1/R_2. \tag{10.17}$$

Radiation losses per pass in a telescopic cavity:

$$A = (R_1^2 - R_2^2)/R_1^2. \tag{10.18}$$

Intermode beat frequency for longitudinal modes:

$$\Delta \nu_q = c/2l. \tag{10.21}$$

Envelope of the light wave field for N locked longitudinal modes of equal amplitude E_0:

$$A(t) = E_0 \frac{\sin\left[\frac{N}{2}(\Delta\omega_q t + \Delta\varphi_q)\right]}{\sin\dfrac{\Delta\omega_q t + \Delta\varphi_q}{2}}. \tag{11.9}$$

Repetition period of locked-mode pulses:

$$T = 1/\Delta \nu_q = 2l/c. \tag{11.10}$$

Length of a single pulse for N locked modes:

$$\tau = T/N. \tag{11.11}$$

Maximum radiation power emitted in giant-pulse lasing with $n_0 \gg n_{\mathrm{ph}}$ and instantaneous Q-switching:

$$P_{\max} \approx n_0 \frac{h\nu}{2\tau_{\mathrm{ph}}}. \tag{11.24}$$

PART TWO

Lasers

LECTURE 13

Gas Lasers. The Helium-Neon Laser

Features of a gaseous active medium. The basic excitation methods. Electric discharge, gasdynamics, chemical excitation, photodissociation, and optical pumping. Resonant excitation-energy transfer in collisions. The helium-neon laser. The level diagram. Excitation-energy transfer. Competition of emission levels at $\lambda = 3.39\,\mu m$ and $\lambda = 0.63\,\mu m$. Discharge parameters and laser parameters

To study the methods of creating conditions for population inversion we will use as examples the most interesting lasers.

We start with gaseous, or gas, lasers. The fact that the active media of such lasers is a gas leads to a number of remarkable consequences. First, only gaseous media can be transparent in a broad spectral interval from the vacuum UV range to the far IR, essentially microwave, range. As a result gas lasers operate over a broad spectrum of wavelengths corresponding to variations in frequency by a factor greater than 1000.

Next, as compared to solids and liquids, gases have considerably lower densities and higher homogeneity. Therefore, a light beam propagating in a gas is less distorted and scattered than it would be in a solid or liquid. This makes it easier to reach the diffraction limit of the radiation-beam divergence.

When the gas pressure is low, Doppler broadening of spectral lines is a characteristic feature, but the size of it is small compared with the luminescence linewidth in condensed media. This makes it easier to achieve a high monochromaticity for the beam emitted by gas lasers. As a result the characteristic features of laser radiation, high monochromaticity and directivity, manifest themselves most clearly in the radiation of gas lasers.

The particles of the gas interact with each other in the process of gas-kinetic collisions. But this interaction is quite weak and has practically no effect on the position of the energy levels of the particles, manifesting itself only in the broadening of the corresponding spectral lines. At low pressures the collision broadening is small and does not exceed the Doppler. At the same time an increase in pressure leads to a growth in the collision width (see Lecture 2), and we acquire the opportunity, inherent only in gas lasers, to control the amplification linewidth of the active medium.

For the self-excitation conditions to be met, as is known, the amplification in the active medium per pass in the laser cavity must exceed losses. The fact that in gases there are no nonresonance energy losses directly in the active media facilitates these conditions being met. Technically it is difficult to manufacture mirrors with losses considerably lower than 1%. Therefore, the amplification per pass must exceed 1%. The relative ease with which this condition is met in gases, say, by increasing the length of the active media, explains the great diversity of gas lasers within a broad spectral range. At the same time the low density of gases makes it difficult to obtain the high concentration of excited particles characteristic of solids. Hence, the power output of gas lasers is considerably lower than that of lasers operating on condensed media.

The specific features of gases also manifest themselves in a multitude of physical processes that find application in creating the conditions for population inversion. These processes include excitation in collisions in an electric discharge, excitation in gasdynamic processes, chemical excitation, photodissociation, optical pumping (primarily by laser light), and electron-beam excitation.

In the great majority of gas lasers population inversion is created in an electric discharge. Such gas lasers are known as gas-discharge. The gas-discharge method of creating active media is the most general method of obtaining population inversion in gas lasers, since the electrons from the discharge easily excite the gas particles, transferring them to higher energy levels in the process of inelastic collisions. The commonly observed emission from a gas discharge (fluorescent lamps) is caused by spontaneous downward transition from these levels. If the rates of decay of the excited states are favorable for the buildup of particles on an upper energy level and depletion of a lower level, there is population inversion between the two levels. By easily exciting the gas in a broad energy interval, the electrons from the gas-discharge area create conditions for population inversion on the energy levels of neutral atoms, molecules, and ions.

The gas-discharge method can be applied to excite both cw and pulsed lasers. Pulsed excitation is used primarily when the conditions are unfavorable for extablishing population inversion in the cw mode for upper and lower levels and when it is desirable to obtain a high power output unachievable in the cw mode.

Electric discharge in a gas may be either self-maintained or semi-self-maintained. In the latter case the conductivity of the gas is sustained by an external ionization agent and excitation takes place independently of the conditions necessary for breakdown in the gas at an optimum electric field strength in the discharge gap. In gaseous media ionized independently by an external agent the electric field and the current generated by the field determine the excitation energy (power input) introduced into the discharge.

A characteristic feature of gases is the possibility of creating flows of gas in which the thermodynamic parameters of the gas vary drastically. For instance, if a strongly heated gas suddenly expands, say, in supersonic flow through a nozzle, its temperature drops drastically. This new and much lower temperature has a corresponding new equilibrium population distribution over the energy levels of the gas particles. When the temperature is suddenly lowered, this equilibrium is violated for some time. If the relaxation to the new thermodynamic equilibrium for the lower level proceeds faster than for the upper, the gasdynamic flow is accompanied by population inversion existing downstream in a certain extended region. The size of this region is determined by the rate of the gasdynamic flow and the relaxation time of the population inversion in the gas.

This briefly is the gasdynamic method of creating population inversion in which the thermal energy of the heated gas is directly transformed into the energy of monochromatic electromagnetic radiation. An important feature of this method is the possibility of organizing gasdynamic flows of large masses of the active media and thus obtaining a high power output (see Eq. (6.57)).

In chemical excitation, conditions for population inversion are created as a result of chemical reaction in which the products are excited atoms, molecules, or radicals. The gaseous medium is suitable for chemical excitation in that the reagents mix easily and quickly and are easily transported. In gasphase chemical reactions the nonequilibrium distribution of chemical energy among the reaction products manifests itself most markedly and is conserved the longest. Chemical lasers are interesting in that they transform chemical energy directly into the energy of electromagnetic radiation. Bringing chain reactions into the picture leads to a drop in the share of energy spent on initiating the reactions that ensure population inversion. As a result consumption of electric energy by a chemical laser may be very low, which is also a big advantage of the chemical method of creating population inversion. Note also that removal of the reaction products, that is, laser operation in a gas flow, can ensure that the chemical laser will operate in the cw mode. A combination of the chemical and gasdynamic methods of excitation can also be employed.

Closely linked to chemical lasers are those in which population inversion is maintained through photodissociation reactions. As a rule these are fast reactions initiated by a strong pulse of light or an explosion. As a result of dissociation excited atoms or radicals appear. The explosive nature of the reaction predetermines the pulsed mode of operation of such lasers. Since with the proper initiation the volume of the initial gas in which photodissociation occurs at a single time may be large, the pulsed power output and radiation energy involved in this method of creating conditions for population inversion can reach considerable values.

In the case of gaseous active media such a general method of creating conditions for population inversion as optical pumping also acquires specific features. In view of the low density of the gases their resonance absorption lines are narrow. For this reason optical pumping can be effective only if the pumping source is sufficiently monochromatic. Laser sources are usually employed for this. The specific nature of gases in optical pumping manifests itself also in the fact that because of their low density the depth of penetration of the pumping radiation into the gas may be great and heat production in radiation absorption low. As a rule resonant optical pumping of gaseous media has practically no effect on their optical homogeneity.

In electron-beam excitation of gaseous media the gas is ionized by high-energy electrons (0.3-3 MeV). The energy of the fast electrons in the primary beam (the number of such electrons is fairly low) is transformed in a cascade manner into the energy of a large number of low-energy electrons. These, with energies ranging from several electronvolts to several dozen, excite the upper electron levels. Since the mean free path of high-energy electrons in gases is fairly long, the electron-beam method is very convenient for creating large-volume active media at high pressures in gases of any composition.

Electron-beam excitation is a flexible and also a powerful method that virtually can always be applied. Another great advantage of this method is that it can be combined with other methods used to create the active media of gas lasers.

Before we examine in detail how all these methods of creating population inversion are realized in the most interesting gas laser systems, it is well to note two circumstances of a general nature.

First, creating population inversion in gaseous media is greatly facilitated by the fairly low rate of relaxation processes in gases. The corresponding rate constants are usually well-known or can be obtained experimentally quite easily. In the short-wave range and for well-allowed transitions the process that obstructs creating and maintaining conditions for population inversion is the spontaneous decay of the upper level (see Lecture 2). Radiative lifetimes of atoms, molecules, and ions are also either well-known or fairly well-known. Their values, which are known for free particles, are valid for gases, too.

Second, a characteristic feature of gases is the transfer of excitation energy from particles of one type to particles of another via inelastic collisions. Such energy transfer is the more effective the closer the energy levels of the colliding particles are. The point is that the difference that always exists between the energy of the states whose populations are being exchanged in a collision leads to a situation in which excitation transfer is accompanied by release (or absorption) of a certain amount of kinetic energy E_{kin}:

$$N^* + n \overset{K}{\rightleftarrows} N + n^* + E_{kin}, \tag{13.1}$$

where N is the number density of donors of excitation energy and n the acceptor number density, the asterisk indicates that the corresponding particle is excited, and the "K" above the arrows in Eq. (13.1) denotes the respective reaction rate constant. The necessary kinetic energy can be taken from the thermal-energy reservoir associated with the translational motion of the gas particles (or released to the reservoir). For this process to be effective the energy taken from, or released to, the reservoir in a single collision must not be higher than the mean energy of thermal motion of a single particle, kT. In other words, the energy gap must be small:

$$|E_{N^*} - E_{n^*}| \ll kT. \tag{13.2}$$

In this case the so-called resonant (quasi-resonant) excitation-energy transfer takes place.

In general terms the energy transfer process (13.1) is described by a rate equation of the following type:

$$\frac{dn^*}{dt} = -\frac{n^*}{\tau} + K(N^*n - n^*N), \tag{13.3}$$

where τ is an effective relaxation time, and the excitation-energy transfer rate constant is defined, as usual, as

$$K = \langle \sigma v \rangle. \tag{13.4}$$

Here v is the speed of the colliding particles, and the cross section σ of the transfer process tends to the gas-kinetic cross section $\sigma_{gas-kin}$ ($\sigma \to \sigma_{gas-kin}$) when condition (13.2) is met. On the right-hand side of Eq. (13.3) we have allowed for the inverse process, $N + n^* \to N^* + n$. Assuming that for N, N^*, n, and

n^* the law of conservation of the number of particles is satisfied,

$$n + n^* = n_0, \quad N + N^* = N_0, \tag{13.5}$$

we see that from (13.3) it follows that in steady-state conditions,

$$n^* = \frac{KN^*}{1/\tau + KN_0} \, n_0. \tag{13.6}$$

When

$$KN_0 \gg 1/\tau, \tag{13.7}$$

the excitation of acceptors reaches the maximum value for a given excitation of donors.

Thus, the collisional transfer of excitation energy from particles of one type to particles of another, characteristic of gaseous media, is effective only if condition (13.2) is met. The process plays a considerable role in creating active laser media based on particles of type n by exciting particles of type N provided that conditon (13.7) is met.

Excitation-energy transfer considerably broadens the possibility of creating gas lasers by dividing in the active media the functions of excitation-energy storage from those of subsequent energy release in the form of radiation at a desired wavelength. The process occurs in two stages. The first to be excited in one way or another are the particles of an auxiliary gas, the carrier of the excess energy acting as the excitation-energy donor. Then in inelastic collisions the energy is transferred from the carrier gas to the particles of the active gas, the excitation-energy acceptor, thus populating the upper laser level of these particles. The upper energy level of the auxiliary gas must have a large proper lifetime so as to store energy successfully. Schematically the process described is shown in Figure 13.1.

The method studied here has found wide application, since practically in all excitation methods (electric discharge, gasdynamic, chemical, etc.) it often proves more expedient to supply excitation energy not directly to the particles from which emission is desirable but rather to those that easily absorb this energy, do not emit it, but readily pass on the excitation to the necessary particles.

Now let us examine some of the gas lasers. We start with atomic gaseous systems, one striking example of which is the helium-neon laser. It is well-known that historically this laser was the first to come. The initial calculations and suggestions referred to gas lasers primarily because of the already noted

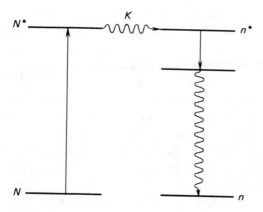

Figure 13.1. Excitation-energy transfer according to the scheme $N^* + n \xrightarrow{K} N + n^*$. The straight arrow pointing up corresponds to the excitation of N-type particles, the straight arrow pointing down corresponds to emission of radiation by n^*-type particles, and the wavy arrow pointing down corresponds to the relaxation of the lower laser level of n-type particles. The absence of proper relaxation of N^*-type particles is clearly seen.

high degree of understanding of the level diagrams and the conditions of excitation in a gaseous medium. Yet the first laser to be built was a ruby laser because by that time single-crystal rubies had been thoroughly studied in EPR spectroscopy and were being widely used in microwave quantum electronics for creating paramagnetic quantum amplifiers (paramagnetic masers). Soon, at the end of the same year 1960, Ali Javan, William R. Bennett, Jr., and Donald R. Herriott built a helium-neon laser operating at $\lambda = 1.15 \, \mu$m. Gas lasers attracted the greatest attention after the helium-neon laser was found to lase on the red line $\lambda = 632.8$ nm in practically the same conditions as at $\lambda = 1.15 \, \mu$m. This primarily stimulated interest in laser applications. The laser beam had become a practical tool.

Technical improvements led to a situation in which the helium-neon laser ceased to be a miracle of laboratory techniques and experimental art and transformed into a reliable device. The laser is now well-known, warrants its reputation, and merits attention.

The active media of a helium-neon laser consists of neutral neon atoms. Excitation is carried out by electric discharge. A simplified and at the same time in a way generalized level diagram for neon is shown on the right-hand side of Figure 13.2. Levels E_3, E_4, and E_5 are excited in an electric discharge in electron collisions. Levels E_4 and E_5 are metastable and level E_3 is short-lived compared to E_4 and E_5. Population inversion would seem, therefore, to set in easily on levels E_4 and E_5 with respect to E_3. However, the metastable level E_2 interferes with this. In the spectra of many atoms, including those of inert gases, there is such a long-lived metastable level. Populated in the course of collisions with electrons, this level prevents level E_3 from becoming depleted, which hinders population inversion.

In the cw mode it is difficult to create population inversion in pure neon. This difficulty, which is common in many cases, is avoided by introducing into the discharge area an auxiliary gas, the excitation-energy donor. Helium is just this gas. The energies of the first two excited metastable helium levels F_2 and F_3 (Figure 13.2) are almost the same as those of the levels E_4 and E_5 of neon. Hence, the conditions necessary for resonant excitation transfer along the following scheme are realized fairly well:

$$\text{He*} + \text{Ne} \rightarrow \text{Ne*} + \text{He} + E_{\text{kin}}. \tag{13.8}$$

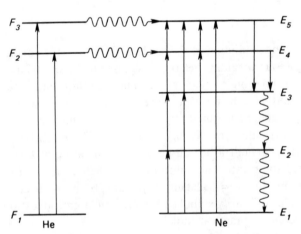

Figure 13.2. Excitation of helium and neon in an electric discharge (the notation for the arrows is the same as in Figure 13.1). The possibility of cascade population of neon energy levels is illustrated.

For pressures of helium and neon that allow for condition (13.7) we can ensure that either E_4 or E_5 or both are populated much more densely than in the case of pure neon and create population inversion on these levels with respect to E_3.

Depletion of the lower laser levels takes place in collision processes, including collisions with the walls of the gas-discharge tube.

Note that the method of transferring energy from a gas not directly involved in the operation of the laser but easily excited to a gas that does not store excitation energy but can easily emit energy in the form of radiation, was first realized in the helium-neon laser and found wide application in the quantum electronics of gas lasers.

Let us now consider in greater detail the level diagram for neutral helium and neon atoms (Figure 13.3).

The lowest of the excited helium states 2^3S_1 and 2^1S_0 correspond to energy levels at 19.82 and 20.61 eV, respectively. Optical transitions from these states to the ground state $^1S_0$ are forbidden in the LS-coupling approximation, which holds true for helium. 2^3S_1 and 2^1S_0 are metastable states with a lifetime of

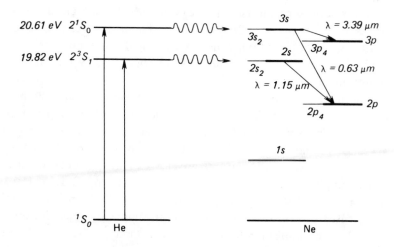

Figure 13.3. The diagram for the lower excited energy levels of helium and neon. The straight arrows pointing up correspond to helium excitation, the wavy arrows from left to right correspond to excitation-energy transfer from helium to neon, and the slanted straight arrows correspond to emission of radiation by neon atoms. The relaxation channels for the lower laser levels of neon are not depicted.

approximately 1 ms. Hence, they easily store energy obtained as a result of excitation by electron impact.

For neon the intermediate jl-coupling approximation is valid. In Figure 13.3 the states belonging to a single configuration are denoted by a heavy line with the active sublevel isolated. The Paschen scheme, which is the most widely used in the literature, is employed to identify the various levels. The levels $3s_2$ and $2s_2$ are close to the metastable helium levels 2^1S_0 and 2^3S_1, with a gap of approximately $300 \, \text{cm}^{-1}$. (At room temperature kT is approximately $210 \, \text{cm}^{-1}$.) The $1s$ state has a long lifetime in view of resonance radiation trapping owing to radiation coupling with the ground state.

In neon the s-states have longer lifetimes than the p-states. This generally makes it possible to secure population inversion on the $2s \to 2p$ and $3s \to 3p$ transitions. One must bear in mind, however, that the $1s$ state of neon becomes well populated in the discharge and at not very high discharge currents the lower laser levels may become populated in a stepwise (cascade) manner in the event of transitions from $1s$ to $2p$ and $3p$.

Introduction into the discharge area of a relatively large amount of helium, which opens a powerful population channel for states $2s$ and $3s$ that is external with respect to neon, lifts all restrictions on the possibility of creating conditions for population inversion in the cw mode. Chronologically the first lasing was achieved on the $2s \to 2p$ transitions. The main power output corresponds to the $2s_2 \to 2p_4$ transition, with $\lambda = 1.15 \, \mu\text{m}$. Then population inversion was realized on the $3s \to 3p$ transition ($3s_2 \to 3p_4$, $\lambda = 3.39 \, \mu\text{m}$) and the $3s \to 2p$ transition ($3s_2 \to 2p_4$, $\lambda = 0.63 \, \mu\text{m}$).

All three forms of lasing occur in approximately the same discharge conditions and are characterized by like dependence of the lasing power on the discharge parameters. Especially important here is the competition of lasing at $\lambda = 3.39 \, \mu\text{m}$ and at $\lambda = 0.63 \, \mu\text{m}$, with the two transitions having a common upper level. Hence, lasing at one wavelength reduces the lasing at the other. What complicates matters is the considerable difference in gain. The $3s_2 \to 3p_4$ transition ($\lambda = 3.39 \, \mu\text{m}$) corresponds to a gain of 20 dB/m, and, therefore, lasing is easily achieved on this transition in simple, e.g. metallic, mirrors. The $3s_2 \to 2p_4$ transition ($\lambda = 0.63 \, \mu\text{m}$) is much more tricky. It corresponds to a small gain of 5-6%/m, which, all other things being equal, can in no way compete with the giant gain of 20 dB/m. For this reason, to be able to lase in the optical range, the helium-neon laser is fitted with multilayer dielectric interference mirrors with a high reflectivity only at the required wavelength. The $2s_2 \to 2p_4$ transition ($\lambda = 1.15 \, \mu\text{m}$) corresponds to a gain of 10-20 %/m, and lasing is achieved by using dielectric mirrors.

The helium-neon laser is a gas-discharge laser. The excitation of helium (and neon) atoms takes place in a weak-current glow discharge. Generally, cw lasers that use neutral atoms or molecules to create the active medium operate on the weakly ionized plasma of the glow-discharge positive column. The current density in the glow discharge amounts to 100-200 mA/cm^2. The strength of the longitudinal electric field is such that the number of ions and electrons emerging in a unit interval of the discharge gap compensates for losses of charged particles in their diffusion to the walls of the gas-discharge tube. In these conditions the positive column is stationary and homogeneous. The electron temperature is determined by the value of the product of the gas pressure p by the inner diameter D of the tube. At small values of pD the electron temperature is high and at great values it is low. The fact that pD is constant determines the scaling conditions on discharges. When the electron number density is constant, the conditions for and parameters of discharges do not change if pD remains constant. In the weakly ionized plasma of the positive column the electron number density is proportional to the current density.

The optimum values of pD for the helium-neon laser and also the partial composition of the gaseous mixture differ somewhat in different parts of the lasing spectrum.

In the 0.63-μm region the strongest line of the $3s \rightarrow 2p$ series, the $3s_2 \rightarrow 2p_4$ line ($\lambda = 0.632\,82\,\mu$m), corresponds to an optimum value of pD of approximately 3.5-4.0 torr $\cdot$ mm. The helium-to-neon partial pressure ratio at which condition (13.7) is met most favorably for this lasing region is 5-to-1. The excitation energy transfers primarily from state $He(2^1S_0)$ to state $Ne(3s)$, as shown in Figure 13.3.

In the 1.15-μm region the strongest line of the $2s \rightarrow 2p$ series is the $2s_2 \rightarrow 2p_4$ line ($\lambda = 1.152\,28\,\mu$m). The state 2^3S_1 of helium supplies the excitation energy (see Figure 13.3). The optimum helium-to-neon partial pressure ratio is 10-to-1, and the corresponding value of pD is approximately 10-12 torrmm.

In the 3.39-μm region the strongest line of the $3s \rightarrow 3p$ series is the $3s_2 \rightarrow 3p_4$ line ($\lambda = 3.3913\,\mu$m), and the upper laser line, as noted earlier, coincides with the upper level of the red lasing line at 0.63 μm. For this reason the optimum discharge conditions prove to be the same as in the 0.63-μm region.

In the most widespread cases, where one and the same sealed-off gas-discharge tube is used in a helium-neon laser with interchangeable mirrors for operation in different wavelength ranges, some compromise is usually chosen in the values in a fairly wide range of parameters: the diameter of the gas discharge tube varies from 5 to 10 mm, the partial pressure ratio from 5 to 15, the total pressure from 1 to 2 torr, and the current from 25 to 50 mA.

That an optimum diameter can be chosen arises from the competition of two factors. First, all other things being equal, if the cross-sectional area of the laser's active medium increases, so does the output power. Second, a decrease in the capillary diameter of the gas-discharge tube drives the gain up in proportion to $1/D$. The latter occurs both because of an increase in the probability of decay of the $1s$ metastable state of neon at the capillary walls and because of an increase in the amount of excited helium (and, in this way, of neon), which means that gain also increases at a constant value of pD, that is, provided that the scaling condition for glow discharges is met as the gas-discharge tube diameter varies.

An optimum discharge current density can be chosen because cascade processes of the type

$$e + Ne(1s) \rightarrow Ne(2p) + e \qquad (13.9)$$

occur at high currents and drive down the population inversion (see Figures 13.2 and 13.3). Processes of this type also prove important if the neon pressure grows, which in turn brings about an optimum value of pressure.

Several dozen milliwatts in the 0.63-μm and 1.15-μm ranges and hundreds in the 3.39-μm range may be considered the characteristic values of the power output of helium-neon lasers. The service life of lasers in the absence of manufacturing errors is restricted only by discharge processes and amounts to years. As time passes, the initial composition of the gas in the discharge region breaks down. The sorption of atoms at the walls and electrodes leads to "hardening" due to the getter effect, the pressure drops, and the helium-to-neon partial pressure ratio changes.

Now let us consider the design of cavities for helium-neon lasers. The high short-duration stability, the simplicity, and the reliability of the design are achieved by placing the cavity mirrors inside the discharge tube. But this arrangement leads to fairly rapid deterioration of the mirrors in the discharge region. Consequently, the construction most widely applied is one in which the gas-discharge tube with windows positioned at the Brewster angle with respect to the optical axis is placed inside the cavity. This arrangement has many advantages: the aligning of the cavity mirrors is simplified, the service life of the gas discharge tube and the mirrors grows and their replacement becomes simpler, and it proves possible to control the cavity, employ a dispersive cavity, and select modes.

An important aspect of quantum electronics is the linewidth of the lasing transition (see Lecture 2). For gas lasers the natural, collision, and Doppler broadenings are important. For a helium-neon laser formula (2.8), where τ_{01} is interpreted as τ_p (the natural lifetime of the p-state of neon) and τ_{02} as τ_s (the natural lifetime of the s-state of neon), gives the value of the natural linewidth $\Delta\nu_0$ at approximately 20 MHz. The collision-broadened linewidth (see Eqs. (2.31) and (2.32)) is determined by the gas pressure. For neon atoms, assuming that the cross section of the corresponding collision process is equal to the gas-kinetic cross section, $\Delta\nu_{col} \approx 1$ MHz at pressures of about 1 torr. The Doppler linewidth (see Eqs. (2.28) and (2.30)) is determined, for one thing, by the radiation's wavelength. For the 0.63-μm line these equations yield $\Delta\nu_D = 1500$ MHz at 400 K, which agrees well with the experimental data. This suggests that the basic mechanism that broadens the emission lines in the helium-neon laser is the Doppler effect. The broadening associated with this effect is moderate, which allows lasing on a single longitudinal mode in such a line, that is, single-frequency lasing at a small and yet physically quite realizable cavity length $l \lesssim 15$ cm (see Eq. (10-21)).

The helium-neon is the most representative of the gas lasers. Its radiation clearly exhibits all the characteristic features of this type of laser, for one thing, the Lamb dip, which was discussed in Lecture 11. The width of this dip is close to that of one of those homogeneously broadened lines that as a group form the inhomogeneously Doppler broadened line. In the case of a helium-neon laser the natural linewidth $\Delta\nu_0$ is such a homogeneous width. Since $\Delta\nu_0 \ll \Delta\nu_D$, the position of the Lamb width (see Figure 11.6) indicates very precisely the center of the lasing-transition line. The curve in Figure 11.6 for the Lamb dip, is obtained in experiments by gradually adjusting the cavity length of a single-mode laser. Hence, the position of the lowest point in the dip can be used, with appropriate feedback controlling the cavity length, to stabilize the lasing frequency. Frequency stability and reproducibility achieved in this way amount to one part in a thousand million. Note, however, that a higher stability (10^{-12}-10^{-13}) is achieved when the dip is "burnt out" in the absorption line of the resonant gas rather than in the amplification line of the active medium. For the lasing line $3s_2 \rightarrow 3p_4$ ($\lambda = 3.39\ \mu$m) the resonant gas is methane.

There is a broad spectrum of gas lasers operating on neutral atoms, including atoms of noble gases. Today there are many types of commercially produced helium-neon lasers.

Problems to Lecture 13

13.1. What is the width of the Lamb dip in a helium-neon laser at $\lambda = 0.63 \, \mu m$?

13.2. Is it accidental that the discharge in the tube of a helium-neon laser glows with a light just as red as the laser emits?

13.3. Using Figure 13.3, estimate the upper bound on the efficiency of helium-neon lasers for each of the three lasing lines.

13.4. What happens if helium-neon lasers are made in the form of intracavity tubes with Brewster windows?

13.5. How can the absorption line of methane ($\lambda = 3.39 \, \mu m$) be used to stabilize the lasing frequency of a helium-neon laser?

13.6. What must be done to make the dip in the absorption line of methane as narrow as possible?

LECTURE 14

Ion Lasers. Metallic Vapor Lasers

The argon laser. Level diagram. Two-step excitation. Dependence on discharge current density. The population inversion condition. Gas transfer in a discharge. Laser parameters. The helium-cadmium laser. Penning ionization and excitation. Level diagram. Cataphoresis. Laser parameters. Gas-discharge laser efficiency. Self-terminating transitions. Efficiency, energy, and power output of self-terminating lasers. The copper-vapor laser, level diagram, and laser parameters

The helium-neon laser, considered in Lecture 13, is an outstanding example of neutral-atom lasers. Prominent among noble-gas ion lasers is the argon laser, which has the highest continuous power output in the optical range (up to several hundred watts). A combination of high output parameters with a favorable pattern of operating energy levels ensuring that lasing occurs in the blue-green range of the spectrum, where the radiation receivers are most sensitive, has determined the employment of this laser in many important fields of science and technology. These range from nonlinear optics, studies in light scattering, biological and medical research, and plasma diagnostics to technological treatment such as resistor trimming and microcircuit scribing. An important application of the argon laser is to pump dye lasers, which will be discussed in a subsequent lecture.

Moving on to the topic of our lecture, we note first that lasers whose operation depends on transitions between excited levels of noble-gas ions are characterized by the extremely high current densities required to reach the lasing threshold owing to the need to maintain a sufficiently high degree of ionization in the gas. The point is that in an ion gas laser the upper laser level becomes populated as a result of two successive collisions of an atom with discharge electrons. The first ionizes the atom and the second excites the ion. Hence, the creation of population inversion is a two-step process, and the effectiveness of each step is proportional to the discharge current. Hence, the effectiveness of excitation on the whole is at least proportional to the square of the discharge current, which requires a high current density to achieve a noticeable population inversion. For argon lasers the characteristic values of current density are several hundred amperes per square centimeter, and for high intensities these values may reach several thousand amperes per square centimeter.

Noble-gas ion lasers require high energy expenditures because efficiencies are small and, hence, parasitic loss of energy is high, which leads to thermal decay, erosion, and other undesirable effects. On the other hand, these lasers have the advantage that noble gases are relatively cheap for production, are easily obtainable in pure form, do not react with cathodes, getters, walls, etc., are not toxic, and do not require heating to obtain a desirable density. The spectrum of such gases is well-studied.

A simplified level diagram of an argon ion laser is given in Figure 14.1. Since all the lines of a cw argon laser belong to transitions between the $4p$ and $4s$ configurations, all levels of each configuration is represented in Figure 14.1 by a single line. As mentioned earlier, excitation of the upper laser levels of Ar^+ occurs stepwise:

$$Ar + e \rightarrow Ar^+ + 2e, \quad Ar^+ + e \rightarrow (Ar^+)^* + e. \tag{14.1}$$

In other words, excitation of the levels belonging to the $4p$ configuration occurs from the ground state $3p^5$ of Ar^+. As Figure 14.1 shows, direct excitation

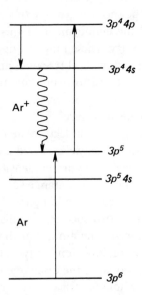

Figure 14.1. Lasing level diagram for argon. The straight arrows pointing up correspond to ionization and excitation, the straight arrow pointing down corresponds to emission of radiation, and the wavy arrow pointing down corresponds to radiation depletion of the lower lasing level.

from the ground state $3p^6$ of Ar atoms requires higher energies. Besides, mechanism (14.1) is corroborated by the dependence of spontaneous emission from the laser levels of Ar^+ on the current density. Unquestionably, other mechanisms are possible, for instance, radiative cascade transitions form the higher energy levels of Ar^+ or excitation transfer in collisions with suitable (energywise) metastable levels of Ar^+. However, process (14.1) not only proves to be the simplest but also gives, in general outline, a correct picture of excitation of the higher laser levels belonging to the $3p^4 4p$ configuration of Ar^+. Let us denote the particle number density on these levels by N, the ion number density in the ground state $3p^5$ by N_i, and the electron number density by N_e. On the whole, the discharge plasma is electrically neutral, that is, $N_e \approx N_i$. Then the pumping rate for the upper laser levels in process (14.1) is

$$\Lambda = dN/dt \propto N_e N_i \approx N_e^2. \tag{14.2}$$

In a steady-state discharge the electron number density is proportional to the discharge current density: $N_e \propto J$. Hence, $\Lambda \propto J^2$. The pumping rate in steady-state conditions determines the maximum power output per unit volume (see Eq. (6.57)). A more thorough examination developed on the basis of these simple ideas leads to the scaling law for argon lasers in the form

$$P/V = 10^{-5} J^2, \tag{14.3}$$

where P/V [W/cm^3] is the total volume density of the power output of an Ar^+-laser in all modes and on all lines of the blue-green region of the spectrum in the cw mode, and J [A/cm^2] the discharge current density. Formula (14.3) agrees fairly well with the experimental data.

Thus, we can assume that in the active medium of an argon laser the collisions of neutral atoms with electrons result first in the excitation of neutral atoms and ions in the ground state. Then the ions in the ground state are excited in collisions with electrons. But in studying any scheme of population inversion formation we must always consider not only pumping to the higher level but also depletion of the lower level.

It has been established that for Ar^+ the lower laser levels belonging to the $3p^4 4s$ configuration are depleted because of vacuum UV radiation at a wavelength around 72 nm (see Figure 14.1). The radiative (spontaneous) lifetime of the lower levels (10^{-9} s) is much shorter than that of the upper levels (10^{-8} s). Such a relaxation time ratio ensures satisfaction of the population inversion condition (3.2). Indeed, the transport equations of the (13.3) or

(13.6), (13.7) type can be written in our case, assuming that stimulated transitions are negligible, in the form

$$dn_2/dt = K_2 N_e n - n_2/\tau_2, \qquad (14.4)$$

$$dn_1/dt = K_1 N_e n - n_1/\tau_1, \qquad (14.5)$$

$$n + n_1 + n_2 = N, \qquad (14.6)$$

where (see Figure 14.1) n_2 is the particle number density in the $3p^4 4p$ configuration (the upper laser level), n_1 the particle number density in the $3p^4 4s$ configuration (the lower laser level), n the particle number density in the $3p^5$ configuration (the ground state of the ion), τ_2 and τ_1 the corresponding lifetimes, N the overall ion number density, N_e the electron number density, and K_1 and K_2 the excitation rate constants for states $4s$ and $4p$, respectively. In steady-state conditions corresponding to the cw mode $dn_2/dt = dn_1/dt = 0$. Then the steady-state population inversion, which determines the small-signal gain, amounts to

$$n_2 - n_1 = \frac{(K_2\tau_2 - K_1\tau_1)NN_e}{1 + (K_2\tau_2 + K_1\tau_1)N_e}. \qquad (14.7)$$

Obviously, the population inversion condition amounts here to $K_2\tau_2$ being greater than $K_1\tau_1$ or, if the rates of excitation by electron impact of the upper and lower laser levels do not differ too strongly,

$$\tau_2 > \tau_1, \qquad (14.8)$$

which is quite general.

The following particular cases are of interest here. For $K\tau N_e \gg 1$,

$$n_2 - n_1 = \frac{K_2\tau_2 - K_1\tau_1}{K_2\tau_2 + K_1\tau_1} N, \qquad (14.9)$$

which in view of the fact that $N \propto N_e$ contradicts the experimental data reflected in Eq. (14.3). In the opposite particular case ($K\tau N_e \ll 1$),

$$n_2 - n_1 = (K_2\tau_2 - K_1\tau_1)NN_e \propto J^2, \qquad (14.10)$$

which agrees well with the experimental data.

Population inversion has been obtained in more than ten lines. The strongest lines are the green at 514.5 nm and the blue at 488.0 nm. The level diagram depicted in Figure 14.1 gives a clear idea of the possible efficiency of argon lasers. According to the diagram, the efficiency is close to 7%, while actually this figure is much lower.

The capillary discharge used to initiate lasing in argon lasers occupies in its properties an intermediate position between a glow discharge and a highly ionized arc, being closer to an arc. The discharge occurs at a low pressure, whose optimum value, from the standpoint of lasing, amounts to 0.25-0.5 torr. As has been repeatedly emphasized, high current densities must be used to sustain a high degree of ionization and a high electron temperature.

An important feature of argon lasers and, for that matter, of all noble-gas ion lasers, is the rapid increase in the power output with the discharge current (see Eqs. (14.3) and (14.10)). The reason is that the saturation of the mechanism for the creation of population inversion (14.7) may occur at discharge current densities much higher than attainable values.

High current densities lead to excessive thermal loads on the walls of the laser tube, which also undergo intensive ion bombardment. The technical problems involved in constructing gas-discharge tubes and the electrodes of argon lasers are serious. Measurements of the Doppler linewidth, found to be approximately 3500 MHz, suggest an ion temperature of about 3000 K, which makes the erosion effect of these ions on the materials of the discharge tube and electrodes a serious problem.

Heavy thermal loads necessitate effective cooling of the laser by running water. This imposes strict requirements on the thermal conductivity of the discharge tube walls. Beryllium ceramics (BeO) has proved to be the best material. A sequence of metal and dielectric washers is an alternative construction.

An interesting feature of argon lasers is the transfer of Ar^+ ions in the gas-discharge channel from the anode to the cathode due to the high current density, which results in longitudinal pressure gradients emerging in the channel and cessation of the discharge. To avert this effect the gas-discharge tube carries a bypass channel that ensures reverse circulation of the gas. To ensure that no discharge occurs in the bypass channel its length is greater and its diameter is smaller than those of the laser tube.

Continuous argon lasers with total power output ranging from 5 to 15 W on all lines have gained the widest application. To select the required lines dispersive cavities are employed, although the competition of the various lasing lines is small or practically absent, in contrast to the situation with helium-neon lasers. Several hundred watts is the highest power output that a continuous argon laser can produce.

Another aspect of argon laser radiation worth noting is that in the same discharge conditions formation of Ar^{++} takes place, and the radiation of these ions lands in the UV range. The strongest lasing lines in the cw mode are those with $\lambda = 363.8$ nm and $\lambda = 351.1$ nm. Several watts is the power output on these lines.

An important place among ion lasers is occupied by metallic vapor lasers. Here the helium-cadmium laser takes the center of the stage. Operation of this laser is based on the collisional transfer of excitation energy from a metastable helium atom in the 2^3S_1 state to a cadmium atom, with resulting ionization of the cadmium atom and subsequent excitation of the resulting ion. This process, known as Penning ionization, proceeds according to the following scheme:

$$He^* + Cd \xrightarrow{K_P} He + (Cd^+)^* + e + E_{kin}. \tag{14.11}$$

The Penning ionization rate constant $K_P = \langle \sigma_P v \rangle$ is determined by the cross section σ_P, which in the case at hand ($\sigma_P = 6.5 \times 10^{-15}$ cm^2) exceeds the gas-kinetic cross section.

A process of the

$$A^* + B \to A + (B^+)^* + e \tag{14.12}$$

type is possible only if the excitation energy of atom A^* is greater than the ionization energy and subsequent excitation of ion B^+. The process is most effective when the excited state A^* is metastable.

In general terms process (14.11) resembles the process discussed earlier in connection with the helium-neon laser (see (13.8)). In the case of the helium-cadmium laser an exact resonance of the excited states He^* and $(Cd^+)^*$ is not required since the excess energy is carried off by the electron, which is characteristic of Penning ionization.

The importance of the storage of excitation energy by the metastable states of the helium atom and of the transfer of this energy to cadmium is reflected in the name of the laser, helium-cadmium. The shorter name often used is "cadmium laser."

The level diagram of a helium-cadmium laser is quite simple (Figure 14.2) and corresponds to a situation in which a single electron is outside a closed shell. The metastable states excited in the discharge, 2^1S_0 and 2^3S_1, can excite the states of ion Cd^+, namely, $^2D_{3/2}$, $^2D_{5/2}$, $^2P_{3/2}$, and $^2P_{1/2}$. Although resonance plays no role in the Penning process, excitation is transferred most effectively in a process with the lowest energy gap, that is, from 2^3S_1 to

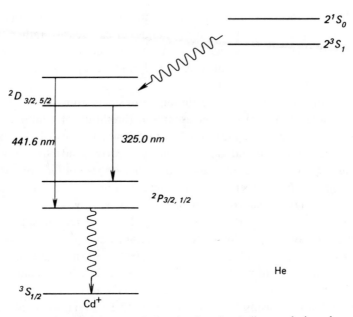

Figure 14.2. Diagram of lasing levels of a helium-cadmium laser. The wavy arrow pointing down corresponds to radiation depletion of the lower lasing level, and the slanted wave arrow corresponds to excitation-energy transfer from helium to cadmium.

$^2D_{3/2,5/2}$. However, population inversion emerges and is maintained continuously, in accordance with condition (14.8), thanks to the more rapid decay of the lower (P) levels than of the upper (D) levels. The radiative lifetime of the D-states amounts to approximately 10^{-7} s and of the P-states to 10^{-9} s. No radiation trapping occurs in the transition from the lower laser level to the ground ion state because of the low concentration of ions Cd^+.

Thus, the helium-cadmium laser is similar to the helium-neon laser in the mechanism of exciting the upper laser levels and to the argon laser in the mechanism of depleting the lower levels. Comparison of the level diagrams for these three lasers (see Figures 13.3, 14.1, and 14.2) makes it possible to continue a comparison of the continuous gas lasers considered. The lasing lines of the helium-cadmium laser at $\lambda = 325.0$ nm and $\lambda = 441.6$ nm correspond to transitions originating at different levels. For this reason the radiation of the helium-cadmium laser contains no competition between lasing lines at $\lambda = 3.39\,\mu$m and $\lambda = 0.63\,\mu$m, so characteristic of the radiation of the helium-neon laser. Penning ionization and excitation of cadmium ions is a single-step process. Hence, the rate of pumping of the cadmium-laser active medium is proportional to the discharge current density rather than to the square of this density, which is the case with the argon laser. This leads to a considerably lower current density and much less power dissipated per unit length of the discharge tube.

The interest to the helium-cadmium laser stems from the fact that this laser is a reliable source, with a low excitation threshold, of continuous coherent light in the blue (441.6 nm) and UV (325.0 nm) regions of the spectrum. A power output of several dozen milliwatts corresponds to a power supply producing several hundred watts, and the laser does not require water cooling, which greatly facilitates its use in physical experiments. The fact that the radiation of this laser is short-wave is directly connected with the fields of its use. Among these are laser photochemistry and research in various types of molecular scattering of light such as Raman, Rayleigh, and Brillouin, whose intensity is proportional to $1/\lambda^4$. Another feature of the helium-cadmium laser is the high monochromaticity of the radiation. The low excitation threshold results in a situation in which the active medium is not strongly overheated and the Doppler lasing linewidth amounts to 1-1.5 GHz. The lasing spectrum for this reason exhibits an isotopic splitting of the cadmium line. If only the isotope ^{114}Cd is involved, the lasing line is extremely narrow. Hence, it is easy to realize the single-frequency and single-mode modes in the helium-cadmium laser.

Many laser transitions have been discovered in metallic vapors, but in many cases the difficulties associated with maintaining a homogeneous distribution of the vapors of the chosen metals over the discharge area make the problem of building such lasers difficult. In the case of cadmium the problem has been resolved by employing the cataphoresis of the cadmium ions Cd^+ in the gas-discharge tube of the laser. Cataphoresis is an electrochemical process in which the cadmium ions migrate in the DC gas discharge under the influence of an external electric field. In a two-phase system cataphoresis results in a preferable flow of the component with the lower ionization potential to the cathode. In large-diameter tubes, back diffusion compensates for cataphoresis and prevents the formation of large concentration gradients. But in the small-diameter tubes commonly used in lasers diffusion is weaker than cataphoresis. This may result in high concentration gradients, which means that the mixture will be suitable for lasing only in a small fraction of the discharge length.

This problem has been solved by constant additional feeding of cadmium to the discharge near the anode end of the tube. Cataphoresis is used here to channel cadmium ions through the entire system from the anode to the cathode at a controlled velocity. Cadmium is extracted from the gas phase via condensation on the cold walls of the expanded section of the laser tube directly before the cathode (Figure 14.3). The discharge sustains the temperature of the capillary at a level high enough to protect the capillary walls from cadmium condensation without heaters or any additional thermal insulation. Helium pressure in the system is fairly high to prevent cadmium diffusion toward the hermetically sealed cell windows. This method is widely used in gas lasers to prevent condensation of high-temperature vapors on the surface of the optical elements of the laser tube.

Usually the following parameters correspond to the rough sketch of the device depicted in Figure 14.3. The discharge tube is 1-1.5 m long and 2-2.5 mm in diameter, and the helium pressure inside it is several torr. First the glow discharge is ignited in helium, then the vessel containing the cadmium is heated to 230-250 °C, and the cadmium flux reaches $(1-1.5) \times 10^{-3}$ g/h at a discharge current of 100 mA, with the cadmium vapor pressure being $(3-4) \times 10^{-3}$ torr. At a voltage of 4.5 kV and a current of 0.1 A the energy release per unit length of the gas-discharge tube is about 3 W/cm, which makes it possible to employ simple air cooling and ordinary glass tubes. The power output corresponding to all these parameters amounts to 100-200 mW at $\lambda = 325.0$ nm, with the optimum transmittance of the output mirror ranging from 5 to 7%.

The design shown in Figure 14.3 makes it possible, after the cadmium in the initial vessel has been completely vaporized, to interchange the position of the cathode and the anode and begin vaporizing the cadmium condensed in the precooled vessel.

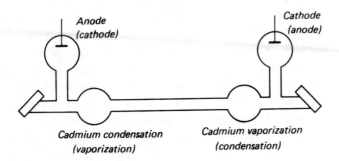

Anode
(cathode)

Cathode
(anode)

Cadmium condensation
(vaporization)

Cadmium vaporization
(condensation)

Figure 14.3. The schematic of the gas-discharge tube of a helium-cadmium laser.

We also note that the migration of the ions of the active medium manifests itself differently in the argon laser and in the helium-cadmium laser.

Let us now consider the efficiency of gas-discharge lasers. This can be represented in the form

$$\eta = \eta_{\text{pump}} h\nu/E_{\text{up}}, \tag{14.13}$$

where η_{pump} is the pumping efficiency, more precisely, in the case of gas-discharge lasers, the fraction of the pumping energy introduced into the discharge area that is used to excite the upper laser level. We denote the energy of this level by E_{up}.

As we know, the population inversion conditions in the cw mode require the rapid decay of the lower laser level (e.g. see condition (14.8)). If using the predominant collisional decay of the lower level is out of the question, spontaneous decay is the only possibility. This means that the lower level must be elevated. Then the ratio $h\nu/E_{\text{up}}$, which has the meaning of the highest possible laser-transition efficiency, is small. Indeed, continuous gas-discharge lasers operating on atoms or ions do use elevated levels ($E_{\text{up}} \approx$ 10-20 eV). Usually $h\nu/E_{\text{up}} < 0.1$. Besides, the elevation of the operating levels has a bad effect on the effectiveness of level excitation. The main fraction of the energy in a typical discharge of an atomic gas is spent on ionization and excitation of low-lying levels. As a result the efficiency of such lasers does not exceed 10^{-3}-10^{-4}, a fact observed for the discussed lasers.

A remark is in order here. Involving low-lying vibrational levels of molecules in the lasing process leads to high values of η. But this will be discussed later in the lectures devoted to molecular gas lasers.

A radical solution of the efficiency problem for atomic systems is possible, however. It lies in replacing the cw mode with an essentially pulsed mode of lasing. The point is that in a gas discharge in an atomic system the main share of the discharge energy in most cases goes into excitation of the first resonance level of the atom. This level has the highest excitation cross section in collisions with electrons. Methods of calculating the cross sections of excitation of various states in multielectron atoms by electron impact are little developed. But it is known that the levels belonging to the best-allowed electrodipole transitions to the ground state have the highest excitation cross sections. This is why the first resonance level is excited most easily and is, therefore, well-suited for the role of the upper level in a laser transition. Then in an atomic system (neutral atoms or free ions) only a metastable level, which usually lies below the first resonance level, can serve as the lower level in a transition. In view of what has been said, a forbidden transition is excited by electrons to a lesser degree than an allowed. Hence, excitation brings about population inversion that exists at least for a short time after excitation has ceased. If population inversion is fairly high and the self-excitation threshold of the laser is considerably exceeded, lasing sets in, and the radiation "pushes" the particles from the upper level to the lower, whereupon population inversion and, hence, lasing finally cease. After this population inversion is not restored at once, since the lower laser level remains populated practically for its entire lifetime, which is fairly long.

Hence, conditions (14.7) or (14.8) for steady-state population inversion are not met, lasing has a pulsed nature, and the repetition frequency cannot exceed the inverse of the lifetime of the lower level.

In view of what has been said, pulsed lasing involving transitions that end at metastable levels is called self-terminating, or lasing on self-terminating transitions.

We can analyze the operation of self-terminating lasers by using equations of the (14.4)-(14.6) type and adding terms that allow for the dumping of population inversion by radiation. Allowing for stimulated transitions, that is, for terms like $(n_2 - n_1)W_{21}$ and $(n_1 - n_2)W_{12}$, in Eqs. (14.4) and (14.5) makes these equations nonlinear because the stimulated-transition probability is determined by the radiation field density which, in turn, is determined by the number of emitting particles. To a great extent the situation becomes similar to the one encountered when the Q-factor is suddenly switched on in the Q-switching method (see Lecture 11).

Indeed, let us consider a simplified example. In the absence of radiation Eqs. (14.4)-(14.6) describe the kinetic of population of the levels of a self-terminating laser, provided that

$$K_1 N_e \ll K_2 N_e, \quad \tau_1 \gg \tau_2. \tag{14.14}$$

If for the sake of simplicity we assume that $\tau_1 = \infty$ and $K_1 N_e = 0$, then $dn_1/dt = 0$ in the absence of radiation. Hence, $n_1 = \text{const}$. Since in the optical range we can assume that $n_1(0) = 0$, the constant is zero. Then the measure of population inversion, $n_2 - n_1$, coincides with the number of particles on the upper level,

$$n_2 = \frac{K_2 N_e}{K_2 N_e + 1/\tau_2} N\{1 - \exp[-(K_2 N_e + 1/\tau_2)t]\}. \tag{14.15}$$

Suppose that the pumping is very strong ($K_2 N_e \gg 1/\tau_2$). Then in an extremely short time interval $1/K_2N_e$, much shorter than the fairly short lifetime τ_2 of the resonance level ($\tau_2 \gg 1/K_2N_e$), a rapid switch-on of deep population inversion will take place according to the law

$$n_2 = N(1 - e^{-K_2N_et}).$$ (14.16)

If this switch-on occurs in a stable cavity with a certain Q-factor, further evolution of a radiated pulse in the cavity proceeds in the same way as it would in the event of a steady-state population inversion n_2 under a rapid switch-on of the Q-factor.

The switching-on of population inversion is known as gain-switching, a mode that requires for its study the solution of nonlinear equations of the (11.14) and (11.15) type. All the conclusions arrived at in Lecture 11 in relation to Q-switching are valid for gain switching as well.

In deep population inversion the energy emitted during the pulse length is

$$W_{\text{pulse}} = n_2 h\nu/2.$$ (14.17)

The peak power is determined by the characteristic lifetime of a photon in the cavity, $\tau_{\text{ph}} = Q/\omega$ (see Eq. (7.17)), through the following formula:

$$P_{\text{pulse}} = n_2 h\nu/2\tau_{\text{ph}}.$$ (14.18)

The buildup of intensity in a lasing pulse occurs at a rate greater than $1/\tau_{\text{ph}}$ and the decay of intensity at a rate of the order of $1/\tau_{\text{ph}}$. An accordance with Eqs. (6.2) and (7.17), τ_{ph} can be estimated via the formula

$$\tau_{\text{ph}} = \frac{l/c}{1 - R},$$ (14.19)

which for large gains (hence, low R's) and for $l = 10$-100 cm yields $\tau_{\text{ph}} = 0.3$-3 ns as a lower bound.

Now let us return to the problem of gas-discharge laser efficiency. In the pulsed mode formula (14.13) is invalid. As we know, after population inversion has been switched on, stimulated emission proceeds until the population numbers even out. A fraction of the population of the upper level is not used. In the absence of degeneracy this fraction is roughly one-half, the value used in the above estimates. If, however, degeneracy is present, this fraction depends on the ratio of the upper-to-lower statistical weights of the levels, g_{up} and g_{low}, and amounts to $g_{\text{up}}/(g_{\text{low}} + g_{\text{up}})$. As a result the efficiency of a self-terminating laser can be written as

$$\eta = \eta_{\text{pump}} \frac{h\nu}{E_{\text{up}}} \frac{g_{\text{low}}}{g_{\text{low}} + g_{\text{up}}} = \eta_{\text{pump}}\eta_{\text{lim}}, \tag{14.20}$$

with

$$\eta_{\text{lim}} = \frac{g_{\text{low}}}{g_{\text{low}} + g_{\text{up}}} \frac{h\nu}{E_{\text{up}}} \tag{14.21}$$

the limit transition efficiency, reached when the entire energy is spent on excitation of the upper energy level, E_{up}. Actually, η_{lim} is the lasing characteristic of a transition. The factor $g_{\text{low}}/(g_{\text{low}} + g_{\text{up}})$ usually varies between 1/3 and 2/3. The ratio $h\nu/E_{\text{up}}$ for atoms with a not very high lower level may range from 0.5 to 0.7. Hence, there is hope of finding transitions for which $\eta_{\text{lim}} = 0.3\text{-}0.5$. Estimating η_{low} is more complicated, however. Experimental data suggest that it is possible to reach values of η_{low} ranging from 0.3 to 0.5, provided that the energy spent on plasma formation is taken into account. As a result the total efficiency could reach 0.1 or even 0.2, which would be a great achievement, indeed.

Let us now estimate the potentials of self-terminating lasers have energy-wise, so to say. The excitation cross section of a resonance level, $\sigma = 10^{-16}\,\text{cm}^2$, corresponds, for an electron number density $N_e = 10^{16}\,\text{cm}^{-3}$ and a relative electron velocity $v = 10^8\,\text{cm/s}$, to an excitation rate $K_2 N_e = 10^8\,\text{s}^{-1}$. Hence, it takes approximately 10 ns for a population inversion $n_2 = N$ to set in. For a particle number density of $10^{16}\,\text{cm}^{-3}$, which corresponds to a partial pressure of several tenths of a torr, the population-inversion number density reaches $10^{16}\,\text{cm}^{-3}$. This means that in the middle of the visible range of the spectrum approximately two joules of energy per liter of active medium are emitted during a single pulse, in accordance with Eq. (14.17). For $1 - R = 0.3$ and a cavity approximately one meter long, the lasing pulse length, according to (14.19), must be about 10 ns and the peak power, according to (14.18), about 200 MW/l.

The above estimates indicate the expediency of designing the type of lasers under discussion.

There are many metals whose η_{lim} are fairly high. The green line of thallium, $\lambda = 535.0$ nm, corresponds to the greatest value, $\eta_{\text{lim}} = 0.47$. Lead, gold, and copper have somewhat lower values of η_{lim}. Today the copper vapor laser is the best in efficiency and in the mean power output in the pulse-periodic mode.

Figure 14.4 shows the level diagram of a copper atom. The two close-lying

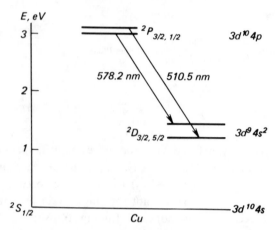

Figure 14.4. Diagram of lasing levels of a copper vapor laser. The corresponding electronic configurations are given in the right column.

levels, $^2P_{1/2}$ and $^2P_{3/2}$, are effectively excited by electron impact, but the excitation cross sections are not known precisely. Estimates yield 9.7×10^{-16} cm^2 for $^2P_{3/2}$ and 4.5×10^{-16} cm^2 for $^2P_{1/2}$, which is higher than for all similar metals. Lasing is observed on transitions from these levels to the metastable levels $^2D_{3/2}$ and $^2D_{5/2}$. The fact that there are only two lasing lines is due to competition.

The lifetimes of the upper levels are quite long thanks to radiation trapping and amount in real conditions to approximately 800 and 400 ns, which greatly lowers the requirements put to the pulsed power source. Lasing has been obtained at 1500 °C (the copper vapor pressure was 0.4 torr, and the particle number density 2×10^{15} cm^{-3}). The equilibrium Boltzmann population of the $^2D_{5/2}$ level (11 203 cm^{-1}) at this temperature was found to be about 5×10^{11} cm^{-3}, and the lasing power output on the green line (510.5 nm) was much higher than that on the yellow line (578.2 nm). The lasing pulse length amounted to 5-10 ns and the peak power output to 200 kW. The average power output achieved at a repetition frequency of 20 kHz was 40 to 50 W, with an efficiency of 0.01. The data correspond to a temperature of 1600-1700 °C, a discharge-tube length of 80-100 cm, and a tube diameter of 15-25 mm.

It has been proved experimentally that the repetition frequency in the discharge can be raised to 100 kHz, which corresponds to a lifetime of 10 μs for the lower level.

Efficiency can be improved if the length of the discharge current pulse is coordinated with the creation of population inversion and its period of existence. Another promising approach is to shift from longitudinal to transverse discharge. The high-temperature mode of operation of the discharge laser tube poses a serious problem. Alumina and beryllia ceramics (based on Al_2O_3 and BeO) are employed for the tube, and to prevent the hot active gas from landing on the windows of the laser cells and the walls in the cold section and ensure that the discharge channel extends from the cold electrodes to the hot gas, neon or helium at a pressure of several torr is used as a buffer gas.

Vaporization of metallic copper can be replaced by dissociating volatile copper-containing molecules or exploding copper wire or by some other means. The choice of method for vaporizing refractory metals requires special study in relation to the concrete problem.

At present a copper vapor laser with a high lasing repetition frequency is the best source of laser radiation in the green region of the spectrum.

Problems to Lecture 14

14.1. Using Eq. (14.3), suggest a simple technical solution that increases the power output, protects the tube walls from erosion, and raises the pumping efficiency.

14.2. What does the condition $K_T N_e \ll 1$ used in deriving Eq. (14.10) mean?

14.3. Will there be competition between the longitudinal modes in an argon laser with the following parameters: homogeneous linewidth 460-800 MHz (Stark broadening) and cavity length 1.2 m?

14.4. What is the width of the Lamb dip in the Ar^+-laser?

14.5. Estimate the drift velocity of ions in the Ar^+-laser if thorough measurements of the gain profile reveal that the profile consists of two Doppler contours separated by a distance $\Delta \nu \approx 500$ MHz.

14.6. Estimate the ion and electron number densities in the active medium of the Ar^+-laser for a current density $J \approx 10^3$ A/cm^2. What fraction of the Ar-atoms is ionized?

14.7. What is the width of the Lamb dip in the helium-cadmium laser?

14.8. Estimate the Doppler linewidth of the radiation emitted by the copper vapor laser ($m_{Cu} = 64$ a.u.).

For the time being we restrict our discussion to molecular lasers of IR spectral range. The radiation of such lasers corresponds to vibrational tra (more precisely, to vibrational-rotational spectra). The vibrational e. levels are obtained through quantization of vibrational energy, and in the first approximation the vibrations are assumed harmonic.

A diatomic molecule, that is, a system with one vibrational degree of freedom, is considered a linear harmonic oscillator, while a polyatomic molecule is considered an oscillating system having many degrees of freedom and performing small vibrations. The number of vibrational degrees of freedom is $3N - 5$ for a molecule with a linear equilibrium configuration and $3N - 6$ for a nonlinear molecule, with N the number of atoms in the molecule.

A diatomic molecule constitutes the simplest case. For small harmonic vibrations, when the restoring force is of an elastic nature, that is, is proportional to the deflection from the equilibrium position, the potential curve has the shape of a parabola (Figure 15.1). Quantization in a parabolic potential well yields, as is known, equidistant energy levels

$$E_{vib} = h\nu_0(V + 1/2), \tag{15.3}$$

where V is the vibrational quantum number, with $V = 0, 1, 2, \ldots$. At $V = 0$ Eq. (15.3) gives the energy of zero-point vibrations, $h\nu_0/2$. For dipole transitions the following strict selection rule holds true:

$$\Delta V = V' - V'' = \pm 1, \tag{15.4}$$

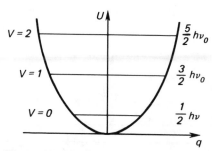

Figure 15.1. Parabolic potential curve of harmonic vibrations. The energy U of such vibrations is $kq^2/2$, where k is the bond's force constant, and q the deflection from the equilibrium position.

which fixes the energy $h\nu$ of an absorbed or emitted photon at $h\nu_0$, as shown in Figure 15.1. The value of ν_0 coincides with the classical frequency of small elastic vibrations:

$$\nu_0 = \sqrt{k/M} , \qquad (15.5)$$

where k is the force constant of a bond in a molecule, and M is the molecule's reduced mass.

A real potential curve resembles a parabola only near the minimum, that is, near the equilibrium position of the nuclei (Figure 15.2). When the deflection from the equilibrium is great, that is, at high excitation, the potential curve differs considerably from a parabola, which explains what is known as the anharmonicity of the vibrations. Because of anharmonicity the vibrational energy levels gradually become closer as the vibrational quantum number V grows, which means that the frequencies of the $V = 1 \rightarrow V = 0$, $V = 2 \rightarrow V = 1$, $V = 3 \rightarrow V = 2$, etc. transitions cease to be equal. In addition to the fundamental frequency $1 \leftrightarrow 0$ ($\Delta V = \pm 1$), harmonics appear in the emission (absorption) spectrum: the first harmonics $2 \leftrightarrow 0$ ($\Delta V = \pm 2$), the second harmonic $3 \leftrightarrow 0$ ($\Delta V = \pm 3$), etc. The intensity of these harmonics decreases rapidly as ΔV grows, however. Hence, in the presence of anharmonicity, Eq (15.3) and the selection rule (15.4) cease to be valid.

In the first approximation, replacing (15.3) with the binomial

$$E_{\text{vib}} = h\nu_0(V + 1/2) - \chi h\nu_0(V + 1/2)^2, \qquad (15.6)$$

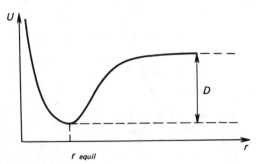

Figure 15.2. A representation of the potential curve for a diatomic molecule: D is the dissociation energy, and r_{equil} the equilibrium distance between the nuclei.

where the anharmonicity constant χ is much less than unity, results in a fairly good description of the real situation. From the standpoint of physics, anharmonicity appears because of dissociation of the molecule. As the vibration amplitude increases, which is the classical analog of an increase in the vibrational quantum number V, the force constant in the molecule drops in value, the molecule becomes loose and falls apart, and the steepness of the right branch of the potential curve in Figure 15.2 decreases (the horizontal section of this branch corresponds to the infinite motion of atoms with respect to each other, that is, dissociation).

The anharmonicity constant χ can easily be linked to the dissociation energy D. From (15.6) it follows that the photon energy $h\nu$ corresponding to a transition with $\Delta V = 1$ can be obtained in the form

$$h\nu = h\nu_0[1 - 2\chi(V + 1)]. \tag{15.7}$$

As the dissociation threshold is approached, the separation between the vibrational levels decreases and the photon energy tends to zero. This is accompanied by the vibrational quantum number V tending to

$$V_{max} = 1/2\chi - 1 \approx 1/2\chi. \tag{15.8}$$

Substituting this value into the expression for the vibration energy of an anharmonic oscillator, (15.6), we get the maximum possible value of the vibrational energy of this oscillator:

$$E_{vib}^{max} \approx h\nu_0(V_{max} - \chi V_{max}^2) = h\nu_0/4\chi. \tag{15.9}$$

By definition, the maximum vibrational energy of an anharmonic oscillator is equal to the dissociation energy of the corresponding molecule: $E_{vib}^{max} = D$ (see Figure 15.2). Hence,

$$\chi = h\nu_0/4D. \tag{15.10}$$

Usually the precision with which formula (15.10) provides an estimate for D using the values of χ obtained spectroscopically via (15.7) varies between 10 and 30%, but it must be stressed that formula (15.10) reflects the essence of the phenomenon correctly. The higher the dissociation energy, that is, the greater the durability of the molecule, the more exact is (15.10) in describing the molecule in the harmonic approximation, at least on the lower levels.

But, as we recall, there is also rotational energy. Since $E_{rot} \ll E_{vib}$, rotation leads to the splitting of the vibrational level into rotational sublevels. Actually, the IR spectrum of a molecule always contains a vibrational band owing to the presence of rotational energy levels. These are found by quantization of the rotational energy if the molecule is approximated by a rigid body with distinct principal moments of inertia: $I_x \neq I_y \neq I_z$.

For a diatomic molecule, $I_x = I_y = M\varrho^2$ and $I_z = 0$ (z is the molecule's axis), with M the reduced mass, and ϱ the distance between the nuclei in the molecule. Quantization yields

$$E_{rot} = BJ(J + 1), \tag{15.11}$$

where $J = 0$, 1, 2, 3, ... is the rotational quantum number, and $B = \hbar^2/2I = \hbar^2/2M\varrho^2$ the rotational constant. The degeneracy multiplicity of a rational level is $2J + 1$. What has just been said is also true for linear poly-atomic molecules with $I = \Sigma m_i r_i^2$.

The rotational structure of a vibrational band is determined by the variation in rotational energy brought on by a vibrational transition,

$$E'_{rot} - E''_{rot} = B'J'(J' + 1) - B''J''(J'' + 1). \tag{15.12}$$

For dipole transitions the following strict selection rule holds true:

$$\Delta J = J' - J'' = 0, \pm1, \tag{15.13}$$

which yields what is known as the P-, Q-, and R-branches. By definition, $\Delta J = -1$ in a P-branch, $\Delta J = 0$ in a Q-branch, and $\Delta J = \pm1$ in an R-branch. For linear molecules there is an additional selection rule that prohibits a transition with $\Delta J = 0$, with the result that there is no Q-branch.

In the harmonic approximation and with allowance made for the selection rules (15.3) and (15.4), the transition frequencies in the P- and R-branch are given, in accordance with (15.3) and (15.11), by the following simple formulas:

$$h\nu = h\nu_0 - 2BJ \tag{15.14}$$

for the P-branch, and

$$h\nu = h\nu_0 + 2BJ \tag{15.15}$$

for the R-branch, where J is the rotational quantum number of the initial sublevel (in absorption). Formulas (15.4) and (15.5) clearly show that the separation of adjacent levels involved in vibrational-rotational transitions in the

P- and R-branch of a vibrational band is determined by the rotational constant and is equal to $2B$.

We have arrived at formulas (15.14) and (15.15) on the assumption that B does not vary during vibrational excitation. But as the vibration amplitude becomes greater, the molecule becomes less rigid and its effective size grows. For this reason the rotational constant B usually decreases somewhat as the vibrational quantum number V grows. In the P- and R-branch this decrease is usually barely noticeable, but it manifests itself in the Q-branch, in which the frequencies of vibrational-rotational transitions are given, in accordance with (15.3), (15.4), (15.11), and (15.13), by the formula

$$h\nu = h\nu_0 + J(J + 1)(B'' - B'). \qquad (15.16)$$

The fact that the difference in the values of the rotational constants, $B'' - B'$, of two adjacent (in V) vibrational states, V' and $V'' = V' + 1$, is small results in the Q-branch being much narrower than the P- and R-branch. As noted earlier, the degeneracy multiplicity of rotational states is $2J + 1$. This statistical weight must be taken into account in determining the equilibrium population of a rotational sublevel:

$$n_J \propto (2J + 1) \exp\left[-BJ(J + 1)/kT\right]. \qquad (15.17)$$

The competition of the pre-exponential factor with the exponential function leads to the nonmonotonic way in which the states with different values of J are populated at a fixed temperature of the gas. Finding the derivative of (15.17) and nullifying it allows us easily to establish that the most populated is the rotational state with the rotational quantum number

$$J_{\max} = \sqrt{kT/2B} - 1/2 \approx \sqrt{kT/2B} . \qquad (15.18)$$

Figure 15.3 shows the transitions in the P-, Q-, and R-branch of the vibrational band $V' \rightarrow V' - 1$, and the lower part of Figure 15.3 gives the spectrum of these branches, which illustrates what has been said.

As a rule, even when the equilibrium population distribution over the vibrational levels is strongly disrupted, the equilibrium distribution (15.17) sets in very fast in the system of rotational sublevels. The reason is that the rotational splitting is much smaller than kT. Hence, in the process of gas-kinetic collisions the molecules easily exchange their rotational energy. In other words, the rotational relaxation time does not exceed the time interval between gas-kinetic collisions.

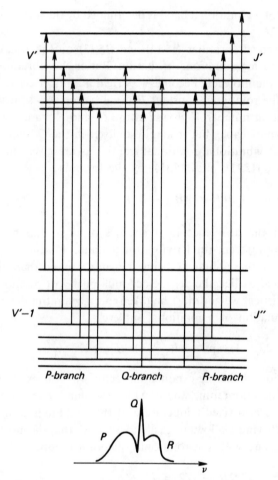

Figure 15.3. *P-*, *Q-*, and *R*-transition pattern in the $V' \rightarrow V' - 1$ vibrational band.

In the first approximation a polyatomic molecule is considered a set of independent harmonic oscillators, that is, in the small-vibrations approximation the vibrational energy is given by the formula

$$E_{vib} = \sum_{i=1}^{r} h\nu_i(V_i + 1/2), \tag{15.19}$$

where r is the number of vibrational degrees of freedom and V_i the vibrational quantum number of the ith oscillator, and the frequency of each of these oscillators, ν_i, is called the frequency of normal vibrations of the oscillator.

The idea of normal vibrations is important for interpreting molecular spectra and establishing the structure of molecules. As is known from the theory of vibrations, normal vibrations are harmonic vibrations that could exist in linear oscillating systems with many degrees of freedom if there were no dissipation of energy. In each normal vibration all points of the system oscillate with the same frequency. The number of normal vibrations is equal to the number of vibrational degrees of freedom of the system. All normal vibrations are independent in the sense that in principle only one of the possible normal vibrations can be excited. The set of frequencies of the normal vibrations is determined by the set of natural resonance frequencies of the oscillating system. The similarity to open-cavity modes is obvious. Actually, the concepts of normal vibrations in molecules and of modes in open cavities are identical. The term "mode" is being used ever more frequently in describing molecular vibrations.

In addition to the possibility of all atoms vibrating inside a molecule simultaneously, some parts of the molecule may vibrate while others may not, for instance, the length of certain bonds and/or the angles between the bonds may oscillate. Knowing the shape of normal vibrations (normal modes) in a molecule makes it possible to characterize the motion of separate parts of a molecule relative to each other and to discriminate between the normal vibrations by their localization. For the sake of clarity all normal vibrations of molecules are broken down into two classes, stretching vibrations (in which the lengths of bonds oscillate) and deformation vibrations (in which the angles between bonds oscillate).

Naturally, every allowed vibrational state of a molecule is a linear combination of the states of a unique set of $3N - 6$ (or $3N - 5$) normal oscillators. The symmetry types of vibrations are fixed by the point symmetry group to which the molecule belongs.

One must bear in mind, however, that as the vibrational quantum number V grows, the concept of normal vibrations becomes less and less suitable for describing vibrations in a polyatomic molecule. Nevertheless, a vibrational state of a molecule is characterized by a set of vibrational quantum numbers V_i in all the normal vibrations or, as is sometimes said, in all the molecule's vibrational modes, whose number, as we know, is equal to the number of vibrational degrees of freedom of the molecule.

The modes of a molecule are usually denoted by ν_1, ν_2, ν_3, etc. up to ν_r

(where $r = 3N - 6$ or $r = 3N - 5$). The anharmonicity caused by the dissociation of the molecule leads to a violation of the selection rule $\Delta V = \pm 1$ and, hence, to the appearance of harmonics of the fundamental frequencies in the vibrational spectrum of the molecule. The respective states are denoted by $2\nu_1$, $3\nu_2$, $2\nu_3$, etc. The radiative transitions between the ground state and the second, third, etc. excited states of a mode occurs at double, triple, etc. frequency along with a corresponding change in the vibrational quantum number: $\Delta V_1 = 2$, $\Delta V_2 = 3$, $\Delta V_3 = 2$, etc. Besides, in polyatomic molecules anharmonicity may lead to the appearance in the vibrational spectrum of so-called combination frequencies, which appear when two or more vibrational quantum numbers change simultaneously, that is, involving transitions to so-called composite vibrations. For instance, the transition from the ground state to the composite vibration $\nu_1 + \nu_2$ corresponds to ΔV_1 and ΔV_2 being equal to unity simultaneously, and the transition from the ground state to the composite vibration $2\nu_2 - \nu_1$ to $\Delta V_1 = -1$ and $\Delta V_2 = \pm 2$. In other words, anharmonicity disrupts the independence of normal vibrations, coupling the corresponding near-harmonic oscillators. Since the strength of the molecule along different bonds is different, distinct modes have distinct dissociation energies and, hence, distinct intramode anharmonicity constants.

The possibility of vibrational transitions depends on the symmetry of the states. If in a certain vibration of a molecule the molecule's dipole moment does not vary, the corresponding transition is forbidden in the dipole approximation. An example is the symmetric stretching vibrations of molecules, vibrations that are active in the Raman scattering spectra and inactive in the IR absorption spectrum. Molecules of this kind include all homonuclear diatomic molecules of the "symmetric dumbbell" type, such as O_2, N_2, and H_2. Vibrations of this kind are also possible in polyatomic molecules.

The presence of several vibrational modes in polyatomic molecules and the difference in their frequencies and localization ensure the existence of different relaxation channels for the excitation energy. Relaxation proceeds at different rates in different channels, which makes it possible to facilitate the conditions for creating and maintaining population inversion by properly selecting the pressure and composition of the gas mixture. In addition, the fact that the vibrational spectrum is multimode enables separating excitation and oscillation channels, which also facilitates lasing.

An example of how the requirements on the active substance of high-efficiency high-power gas lasers are realized in molecular lasers is the carbon dioxide laser.

The carbon dioxide laser, that is, a laser whose lasing component of the active media is the carbon dioxide CO_2, occupies a special place in the diverse family of lasers. This unique type of laser is characterized primarily by a high power output combined with high efficiency. Exceptionally high outputs, several dozen kilowatts, have been generated in the cw mode, and the pulsed power output reaches several gigawatts, with the energy per pulse measured in kilojoules. The efficiency of the carbon dioxide laser (about 0.3) exceeds the efficiency of other types of lasers. The repetition frequency in the pulse-periodic mode can amount to several kilohertz. The wavelengths of the radiation emitted by the carbon dioxide laser fall into the 9 to 10-μm range (the medium IR range) and land in the spectral window of the Earth's atmosphere. For this reason the light of the carbon dioxide laser can be very useful for intense action on substances, say, for technological purposes. In addition, the resonance absorption frequencies of many molecules land in the wavelength range of carbon-dioxide laser radiation, which allows for intense resonance action of laser light on substances. Here one must keep in mind that both discrete and continuous wide-range tuning of the lasing frequency of the carbon dioxide laser are possible, which considerably broadens the possibility of the laser. The carbon dioxide laser can also operate in the single-frequency mode. All this explains the interest in this laser.

The CO_2 molecule has three normal vibrational modes: the symmetric stretching mode ν_1, the deformation mode ν_2, and the asymmetric stretching mode ν_3. The deformation mode is twofold degenerate. Accordingly, the population of the vibrational levels of the CO_2 molecule, including not only the normal modes ν_1, ν_2 and ν_3 but also their harmonics and the combination vibrations, determines the set of vibrational quantum numbers V_1, V_2, and V_3 describing the vibrational state of the molecule. Each vibrational level is designated by a triple of quantum numbers, V_1, V_2^l, and V_3, with the superscript l introduced because of the degeneracy of the deformation vibration ν_2. We are interested here in the lower vibrational levels of the ground electronic state, which are depicted in Figure 15.4 together with a schematic representation of the vibrations of the CO_2 molecule. The accidental coincidence of the levels ν_1 and $2\nu_2$ due to a Fermi resonance leads to a mixing of these levels, and in kinetic processes they often act as a single state. This has proved very important. The lower laser level 10^00 in the $00^01 \rightarrow 10^00$ lasing transition has a short relaxation time owing to the strong coupling with the deformation vibration 02^00 and, hence, the 01^10 vibration. Owing to the smallness of the energy gap, collisional energy exchange inside a single vibrational mode occurs very fast, practically in the course of a single gas-kinetic collision. Vibration 01^10, like any deformation vibration, has a large collisional-deactivation cross section.

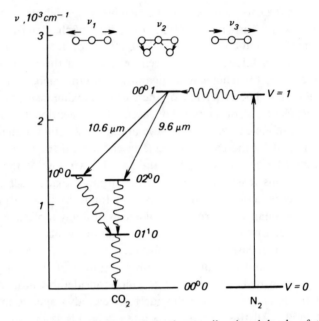

Figure 15.4. Diagram of the lower vibrational levels of the CO_2 and N_2 molecules in their ground electronic states. The slanted straight arrows correspond to the laser transitions, the straight arrow pointing up corresponds to nitrogen excitation, the horizontal wavy arrow corresponds to excitation-energy transfer from nitrogen to CO_2, and the other wavy arrows correspond to relaxation depopulation of the lower laser levels.

Simple geometric considerations indicate that the inelastic collision cross section for a stretching vibration localized along the molecule's axis is much smaller than that for a deformation vibration, which may be excited (deexcited) by a particle incident on the molecule at practically any angle to the molecule's axis.

If we also allow for the accidental coincidence of the energy of the first vibration of the N_2 molecule ($\Delta E = 18 \text{ cm}^{-1} \ll kT$) with the 00^01 level of the CO_2 molecule, we conclude that the CO_2 molecule fully meets the demands on an ideal gas-discharge laser discussed at the beginning of this lecture. Symmetric homonuclear nitrogen molecules have a large natural lifetime in a vibrationally excited state, are easily excited by electron impact, and readily give off excitation energy to CO_2.

In a glow discharge with a reduced electric field strength in the discharge plasma E/p = 5 V/cm · torr, an electron energy of 2-3 eV (resonant excitation of N_2 molecule in the 1-to-8 range of values of the vibrational quantum number), and an electron number density of $(0.5-5) \times 10^{10}$ cm^{-3}, from 40 to 80% of the nitrogen molecules are excited. The excitation cross section for nitrogen is 3×10^{-16} cm^2. The rate of collisional excitation-energy transfer from nitrogen to CO_2 is $(1-2) \times 10^4$ s^{-1} · torr^{-1}. This transfer proceeds effectively between the 00^0n harmonics of the CO_2 and N_2 molecules up to values V = 4-5 of the N_2 molecule. In this way the upper laser level of the carbon dioxide laser becomes populated.

As for the depopulation of the lower laser level, it has been found that the first excited level of the deformation mode ν_2, 01^10, effectively relaxes in collisions with He atoms. Helium depopulates the 01^10 level of CO_2 at a rate of 4×10^3 s^{-1} · torr^{-1} but has practically no effect on the 00^01 level of the ν_3 mode.

The pumping cycle of a carbon dioxide laser in steady-state conditions can, therefore, be represented in the following way. The electrons of the glow-discharge plasma excite nitrogen molecules whose excitation energy is then transferred to the asymmetric stretching vibration of CO_2 molecules, which has a greater lifetime and forms the upper laser level. The lower laser level is usually formed by the first excited level of the symmetric stretching vibration, which is strongly coupled via a Fermi resonance with the deformation vibration and, hence, rapidly relaxes with this vibration in collisions with helium. Obviously, the same relaxation channel is effective when the lower laser level is formed by the second excited level of the deformation mode. Thus, the carbon dioxide laser uses a mixture of carbon dioxide, nitrogen, and helium, with CO_2 ensuring lasing, N_2 the pumping of the upper level, and He the depopulation of the lower level.

Figure 15.5 shows the pumping cycle of a concrete carbon dioxide laser, with the diameter of the discharge tube 15 mm, the discharge current 40 mA, the reduced electric field strength 5 V/cm · torr, the gas temperature at the tube's axis 450 K, the total pressure of the gas mixture 15 torr, and the $CO_2 \div N_2 \div$ He ratio $1 \div 1 \div 8$. In conditions close to those shown in Figure 15.5 the weak-signal gain is 3-4 dB/m, the saturation flux 30-60 W/cm^2, and the power output per unit length at optimal coupling 50-100 W/m. For a laser 200-cm long (the common laboratory device), the optimal transmittance of the output mirror, T_{opt}, is 0.4.

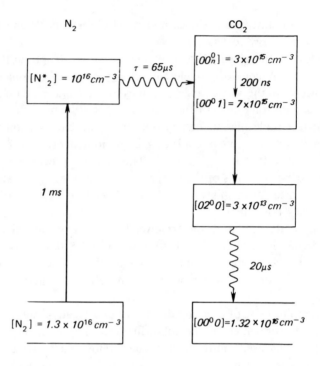

Figure 15.5. Pumping cycle of a low-pressure carbon dioxide laser with longitudinal discharge and slow longitudinal gas pumping.

Note that the depopulation of the lower laser level is not the only task that helium performs in the carbon dioxide laser. It also facilitates the emergence and maintenance of the glow discharge. It helium-rich gas mixtures the discharge develops in practically the same way as it does in pure helium, the classical object of plasma studies. Helium also has a high thermal conductivity, which is important in cooling CO_2 in the discharge area. Preventing CO_2 from overheating is necessary so as to avoid thermal population of the low-lying 01^10 level (see Figure 15.4). It must be added that a large amount of helium in the discharge area obstructs dissociation of CO_2 by electrons forming the discharge.

Medium-power carbon dioxide lasers (power outputs ranging from dozens to hundreds of watts) are designed in the form of fairly long tubes with a longitudinal discharge and longitudinal gas pumping. A typical design is shown in Figure 15.6. A longitudinal discharge is the simplest for realization: one needs only to introduce a large electrical resistance in series with the discharge so as to limit the discharge current and compensate for the effect of the falling segment of the volt-ampere characteristic. Longitudinal gas pumping serves to remove the products of dissociation of the gas mixture from the discharge region. The cooling of the active gas in such systems occurs thanks to diffusion to the discharge-tube wall, which is cooled from the outside. This means that the thermal conductivity of the wall's material is an essential factor. Hence, the use of tubes made of corundum (Al_2O_3) or beryllium (BeO) ceramics has proved expedient.

With both the discharge and the gas pumping longitudinal, the maximum power output per unit length (50-100 W/m) is independent of the diameter of the discharge tube. Indeed, when the self-excitation threshold of the laser is considerably exceeded, the power output is determined by the product of the pumping rate Λ and the volume $\mathscr{V}$ of the active medium:

$$P = \Lambda \mathscr{V} h\nu. \tag{15.20}$$

For a cylindrical geometry, $\mathscr{V} = \pi D^2 l/4$, where l is the discharge length and D the discharge tube diameter. The pumping rate $\Lambda = dN(00^01)/dt$ is deter-

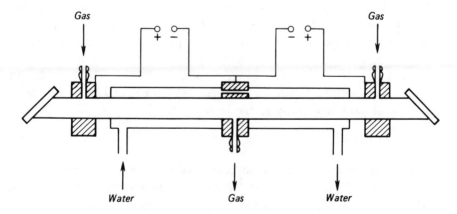

Figure 15.6. Schematic of a carbon dioxide laser with longitudinal discharge and longitudinal gas pumping. The coolant (water) passes through the cathode.

mined by the number density of molecules in the ground state, N_0, times the electron number density N_e times the cross section σ of excitation by electron impact times the average relative velocity u:

$$\Lambda = N_0 N_e \sigma u. \tag{15.21}$$

But the product $N_e \sigma u$ has the meaning of the discharge current density $\mathscr{J}$; hence, $\Lambda = N_0 \mathscr{J}$. The product $\mathscr{J} \pi D^2 / 4$ gives the value of the total discharge current $\mathscr{I}$, and we find that

$$P = \mathscr{I} N_0 l h \nu. \tag{15.22}$$

Since N_0 is proportional to the total pressure of the gas mixture, p, we obtain

$$P \propto \mathscr{I} p l. \tag{15.23}$$

As is known, the product of current by pressure is an important characteristic of plasma processes in long tubes. The conditions in steady-state glow-discharge plasma are determined by pD, and if pD is constant so are the plasma conditions (see Lecture 13). If for a certain tube diameter the optimal pressure is fixed, the optimum will be retained provided that pD remains fixed. Hence, $p = \text{const}/D$, and since $\mathscr{I} = \mathscr{J} \pi D^2 / 4$, we find that $\mathscr{I} p \propto \text{const} \times \mathscr{J} D$.

On the other hand, thermal conditions are extremely important for CO_2 lasers. Heat release per unit volume is proportional to the current density $\mathscr{J}$, while in the cylindrical geometry heat transfer from the central part of the discharge channel to the periphery is proportional to $1/D$. The requirement that thermal conditions remain constant and optimal leads to the requirement that $\mathscr{J} D$ be constant. Hence, the product $\mathscr{I} p$ is a constant, and we arrive at the following important conclusion: in optimal conditions the power output of a carbon dioxide laser with longitudinal gas pumping and longitudinal discharge is proportional solely to the laser length:

$$P \propto l. \tag{15.24}$$

In our discussion we ignored nonresonance losses by using the well-known formula $I = (\alpha/\beta) I_{sat}$ (see (6.44)), which limits the maximum power output of a laser. This is why, for one thing, the maximum power output achieved for the longitudinal configuration is just over 1 kW (the total discharge length is about 20 m).

Since the reason for pumping the gas mixture through the discharge tube is to replace the gas so as to remove the products of CO_2 dissociation, adding to the active media some admixture that facilitates the oxidation of CO in the glow-discharge plasma down to CO_2 makes it possible to create sealed-off carbon dioxide lasers. Water molecules at a concentration below 1% usually serve as such a regenerating admixture. Sealed-off lasers have proved very useful in laboratories, but their power outputs do not exceed several dozen watts.

Problems to Lecture 15

15.1. Estimate the gas temperature on the axis of the tube of a carbon dioxide laser (see Figures 15.4 and 15.5) at which the effect of thermal population of the 01^10 level (667.3 cm^{-1}) will begin to manifest itself.

15.2. Estimate the maximum power output of the carbon dioxide laser (see Figure 15.5) using formula (6.44).

15.3. Determine the efficiency of the carbon dioxide laser (see Figure 15.5).

15.4. Determine the Doppler amplification linewidth of the carbon dioxide laser.

15.5. Knowing the distance between the rotational components of the lasing line, $2B \approx 1 \text{ cm}^{-1}$, calculate the rotational quantum number of the most populated sublevel, 00^01, at 450 K.

LECTURE 16

The Carbon Dioxide Laser II

Spectral properties of carbon dioxide lasers. Rotational structure. The 00^01-10^00 and 00^01-02^00 bands. Rotational competition. Laser tuning. Continuous tuning. Pulse discharge. TEA carbon dioxide lasers. Self-maintained and semi-self-maintained discharges. Gasdynamic lasers

Let us now turn to the spectral properties of the radiation emitted by the carbon dioxide laser. At low pressures the lasing transition linewidth is determined by Doppler broadening and amounts to 50-60 MHz. This corresponds to a gas temperature of 400-450 K (see Lecture 2). Broadening caused by collisions when the pressure changes by 1 torr varies between 4.5 and 6 MHz/torr, depending on the composition of the gas mixture. Only at pressures exceeding 20-25 torr can the line be assumed homogeneously broadened. At pressures amounting to several torr, which is characteristic of carbon dioxide lasers with longitudinal gas pumping, the laser line remains so narrow that the laser automatically operates on a single longitudinal mode.

The rotational structure of vibrational levels considerably broadens the range of possible lasing frequencies in the event of vibrational population inversion. The P- and R-branch are observed in each vibrational band in lasing. The corresponding lasing lines are denoted $P(J)$ and $R(J)$, where J is the number of a rotational sublevel of the lower laser level. Transitions in the P-branch take place from a state with a smaller value of J to that with a larger value, and this is usually less populated in keeping with distribution (15.17). Hence, the gain achieved on transitions in the P-branch is somewhat higher than that on transitions in the R-branch. The distance between separate rotational lines is somewhat smaller than $2 \, \text{cm}^{-1}$. The rate at which equilibrium sets in in the system of rotational sublevels is high (of the order of $10^7 \, \text{s}^{-1} \cdot \text{torr}^{-1}$). For this reason in steady state conditions the lasing that emerges on the rotational line where the self-excitation conditions are best continues on the frequency of this line, and the lasing power is determined by the pumping into all the rotational sublevels of the asymmetric vibration 00^01. Intensive rotational relaxation supplies energy to the sublevel that is depopulated by radiation. The energy stored by all the sublevels is emitted by a single one. This constitutes what is known as the rotational competition effect, which in many respects is responsible for the high efficiency of the carbon dioxide laser.

In a nonselective cavity the rotational line emitted is the one whose initial sublevel in the vibrational state 00^01 is most populated. These are usually the $P(20)$ and $P(22)$ lines of the $00^01\text{-}10^00$ band with the frequencies 944.2 and 942.4 cm^{-1} (wavelengths 10.59 and 10.61 μm). In the R-branch of this band the lines $R(18)$ and $R(20)$ with the frequencies 974.6 and 975.9 cm^{-1} (wavelengths 10.26 and 10.25 μm) have the highest gain. If we turn to Figure 15.4, we see that lasing may occur not only on transitions of the $00^01\text{-}10^00$ band but also on transitions of the $00^01\text{-}02^00$ band (9.3 μm in the R-branch and 9.6 μm in the P-branch). However, transitions of the $00^01\text{-}10^00$ band have a higher gain, and since the two bands have a common upper level, lasing in a nonselective cavity usually occurs on one of the rotational lines belonging to the P-branch of this band (10.6 μm).

The rich rotational structure, the presence of two vibrational bands with population inversion, and the rotational competition make it possible, by using a tunable selective cavity, to obtain laser radiation with a high efficiency on practically every one of the lines belonging to the P- and R-branch of the $00^01\text{-}10^00$ and $00^01\text{-}02^00$ bands. The corresponding frequency ranges are shown in Figure 16.1. At low pressures only discrete tuning (from line to line) is possible, with an interval equal to the distance between the lines. This distance is different in different branches of different bands and varies from 0.8 to 1.5 cm^{-1} in R-branches and from 1.5 to 2.2 cm^{-1} in P-branches. At low pressures the width of the lines is much smaller than the line separation (we recall that 100 MHz is equivalent to 0.003 cm^{-1}). It is possible, however, to broaden the lines considerably by drastically increasing the pressure of the active gas mixture. Indeed, the rotational lines then merge into a continuous spectrum when their collision broadening becomes comparable to the line separation.

If we set the value of the spectral interval at 2 cm^{-1}, which is equivalent to 60 GHz, and the value of collision broadening at 6 MHz/torr, we find that the rotational lines overlap within a single branch of one vibrational band at a pressure of 10^4 torr $\approx$ 14 atm. At this pressure practically homogeneous overlapping of the lasing lines is achieved. Actually, if a selective cavity with a fairly high Q-factor is used, there is no need to require such complete overlapping and high-pressure carbon dioxide lasers guarantee the existence of fairly broad regions of continuous tuning of the lasing frequency in the P- and R-branch of the $00^01\text{-}10^00$ and $00^01\text{-}02^00$ bands at pressures of about 6-7 atm (see Figure 16.1).

Here the serious problem arises of maintaining a glow or like discharge at such high pressures. Longitudinal discharge in long tubes cannot be realized at these pressures. In a steady-state plasma the discharge conditions remain

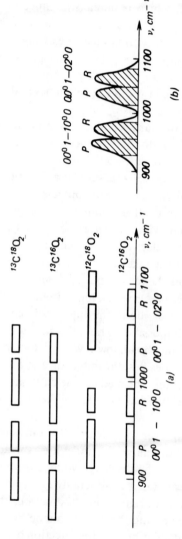

Figure 16.1. (a) Frequency ranges of discrete tuning of carbon dioxide lasers with different isotopic compositions of CO_2; (b) the regions of continuous tuning of the lasers (hatched regions).

unchanged if the product pD remains constant, a point made many times in these lectures. In gas mixtures common to carbon dioxide lasers the optimal conditions correspond to a pD value of approximately 25 torr · cm. At 10^4 torr we have a diameter of 25 μm, which is unrealistic. Since high pressures are promising not only because continuous tuning becomes possible but also because a higher specific power output becomes feasible owing to the increase in the number density of the emitting particles in the active media, high priority has been given to the problem of creating a homogeneous plasma in the gas mixtures of carbon dioxide lasers in large volumes and at high pressures.

Without going into the details of plasma physics, we note that what limits the possibility of obtaining a uniform discharge in gases at high pressures and in noncapillary geometries at static breakdown voltages and higher is the formation of a spark channel. The glow discharge present at low pressures transforms into an electric arc, which disrupts the lasing process. However, the time it takes an arc (or arcs) to form in a self-maintained discharge is finite. For this reason a homogeneous discharge can, at least theoretically, be obtained in any gas in the region between two electrodes if we ensure that the discharge time is small compared to the time it takes an arc to form. This leads us to the idea of a pulse discharge and, hence, to pulsed lasers.

In the traditional scheme of a gas laser with a long discharge tube, in which gas flow and electric discharge occur along the tube's axis coinciding with the optical axis of the cavity, it is impossible to raise the gas pressure considerably since this would bring up the breakdown voltage and require using megavolt sources of pulsed voltage. Besides, the inductance of a long discharge contour is high and the discharge cannot be made sufficiently short. Increasing the diameter of the gas-discharge tube has no effect either because this only facilitates constriction.

A solution was found by going over to systems with transverse discharge. The idea is quite simple. If the gas-discharge electrodes are "spread out" along the laser's optical axis (Figure 16.2), the discharge voltage is applied at right angles to the optical axis. This results in a considerably lower breakdown voltage compared to the longitudinal configuration. In addition, there appears the possibility of considerably increasing the active volume by extending the electrodes in the direction perpendicular both to the direction of the discharge-current flow and to the system's optical axis. The conditions of gas pumping are also less stringent in this direction, which facilitates the cooling of the gas in the case of pulsed operation with a high repetition rate (transverse gas pumping). In systems with transverse gas pumping and transverse discharge

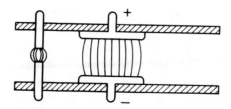

Figure 16.2. Schematic of a transverse discharge in which the spark discharge is pre-ionized with UV light.

the three basic directions that specify a laser's design, the direction of propagation of laser light (the z axis), the direction of flow of the discharge current exciting the active media (the y axis), and the direction in which the gas creating the active media is pumped (the x axis), are uncoupled, that is, made completely independent.

Let us now return to transverse excitation. For the transverse discharge to burn more or less homogeneously over the cross section and over the length of the laser chamber, fast power deposit is essential in the first place. Hence, great attention has been paid to various methods of organizing rapid discharges, to capacitance generators of pulse voltages based on Arkad'ev-Marx circuits, and to pulse transformers. It is also important to prepare the gas for fast power deposit. The best way is to pre-ionize the gas over the entire volume uniformly. Illumination of the active volume with UV light is widely used for this purpose, with the sources of UV pre-ionization chosen in the form of various types of spark gaps, gliding plasma discharges, etc. placed directly in the laser chamber and ignited with a small (of the order of several dozen nanoseconds) advance in relation to the discharge of the main power deposit. UV radiation in the 200-nm range ionizes the admixtures that are always present in the active gas mixture. In the presence of especially pure gases, controlled adding of easily ionized admixtures of the tripropylamine type to the mixture helps.

Thus, auxiliary discharges pre-ionize at least the region of the active volume near the main electrodes, which ensures the creation of a uniform discharge over the entire surface of an electrode. Pre-ionization reduces the time necessary for the electrons initiating the discharge to reach the discharge gap at the moment when the supply voltage of the main discharge is applied to the main discharge gap. This makes it possible to supply the gas with energy before arc formation. In turn, the rate of arc formation is affected by the finish of the electrodes and their configuration and geometry. In order to postpone arc formation as long as possible, the distance between the electrodes must be maintained constant with a high accuracy over the entire volume, and the electrode surfaces must be thoroughly polished and must have a configuration that ensures a constant electric field strength in the gap over the entire surface, say, the so-called Rogowski or Bruce profiles. The plasma of a gliding discharge, that is, a discharge that slides along the surface of a dielectric, may serve as the electrodes of the main discharge; the UV radiation emitted by the gliding discharge can be used simultaneously for pre-ionizing the active volume. In any case the inductance of the leads must be low and voltages must be switched on rapidly.

Lasers of this type (see Figure 16.2) operating at atmospheric pressure have become known as TEA (Transversely Excited Atmospheric) carbon dioxide lasers. Their moderate dimensions and simple design make TEA carbon dioxide lasers a useful laboratory device. The pulse energy amounts to 1-10 J and the pulse length 100-1000 ns depending on the composition of the gas mixture. The large amount of nitrogen not only increases the energy stored by the active medium but also prolongs the pumping of the emitting state 00^01 of the carbon dioxide molecules.

At atmospheric pressure the time it takes the energy to transfer from nitrogen to carbon dioxide is about 100 ns. Hence, the presence of nitrogen has no effect on the radiant energy over time intervals shorter than 100 ns. The energy stored in the asymmetric stretching vibration 00^0n of the carbon dioxide is emitted on just such a time scale. Gain switching, similar to Q-switching and discussed in Lecture 14, occurs during fast discharge. When there is little or no nitrogen or when time intervals are shorter than, or of the order of, 100 ns, the radiated pulse has the shape of a typical switch-on pulse. The presence of nitrogen, which acts as a fairly inertial energy reservoir, stretches out the radiated pulse, violates the monotonic nature of the pulse's tail, and leads to the appearance of a secondary peak (true, a more spread-out one). Figure 16.3 gives a typical shape of a pulse of radiation emitted by a TEA carbon dioxide laser. The overall length of this pulse increases to 1 μs, which leads to a corresponding increase in the radiant energy. Usually the first spike carries 1/3 to 1/2 of the total pulse energy.

The total energy emitted in a single pulse by a TEA carbon dioxide laser is easily estimated. Suppose that the self-excitation conditions are more than satisfied. Then the total pulse energy is determined by the energy of the asymmetric stretching vibration of carbon dioxide molecules and the vibrational energy of the N_2 molecules in the laser cavity. In the course of the full pulse

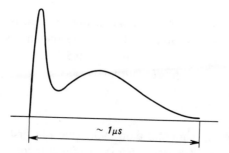

Figure 16.3. Typical shape of a pulse of radiation emitted by a TEA carbon dioxide laser.

length the nitrogen's vibrational energy is transferred to CO_2 and emitted. The sought energy, obviously, is

$$Q = \frac{1}{2} h\nu(n_1^* + n_2^*),$$ (16.1)

where n_1^* and n_2^* are the total numbers of CO_2 and nitrogen molecules, respectively. This formula can be given a more convenient form by introducing the coefficients $\varkappa_1$ and $\varkappa_2$ to characterize the degree of excitation, that is, the average number of vibrational quanta belonging to the ν_3 mode per CO_2 and N_2 molecule respectively, by going from particle numbers to partial pressures, and by recalling that for ideal gases $PV = NkT$:

$$Q = \frac{1}{2} \frac{\varkappa_1 + c_2\varkappa_2}{1 + c_2 + c_3} \frac{h\nu}{kT} VP,$$ (16.2)

where P is the total pressure of the gas mixture in the laser cell, V the active laser volume, and c_2 and c_3 the concentrations of nitrogen and helium, respectively, referred to the concentration of CO_2 in the mixture ($CO_2 \div N_2 \div He = 1 \div c_2 \div c_3$). The product of pressure and volume has, as is known, the dimensions of energy. If V is measured in liters, P in standard atmospheres, and Q in joules, for the reduced energy (the specific power output) we have the following simple relation:

$$q = \frac{Q}{VP} = 50 \frac{\varkappa_1 + c_2\varkappa_2}{1 + c_2 + c_3} \frac{h\nu}{kT} \quad J/(l \cdot atm).$$ (16.3)

As the temperature goes down, the specific power output increases. At room temperature $h\nu/kT \approx 4$. In a well-organized discharge the values of $\varkappa_1$ and $\varkappa_2$ are approximately equal and may vary from 0.4 to 0.5. Then for a $1 \div 2 \div 3$ mixture the value of the specific power output is

$$q = 40\text{-}50 \text{ J/l} \cdot \text{atm}, \tag{16.4}$$

which agrees well with the experimental data. Note that the first and second terms in the numerators of Eqs. (16.2) and (16.3) give the fractions of the energy carried in the first spike and in the tail of a pulse emitted by a TEA carbon dioxide laser (see Figure 16.3).

The above estimate is valid since at atmospheric pressure the rotational relaxation time is 0.1 ns, the time it takes equilibrium to set in in a single vibrational mode of CO_2 is also 0.1 ns, and the vibrational-translational relaxation of a CO_2 molecule proceeds through vibration 01^10 with a relaxation time of 500 ns.

Concluding this discussion of TEA carbon dioxide lasers, we note that because collision broadening is of the order of 3 GHz (or 0.1 cm^{-1}) the characteristic amplification linewidth is large for such lasers, which makes it possible, via mode-locking methods, to generate pulses that are about 1 ns long. We note also that these lasers have a tendency toward passive mode locking, because of which the emitted pulses are "dissected" in an irregular manner with a characteristic time scale of 1 ns. Figure 16.3 illustrates results obtained with a low temporal resolution.

The application of systems with double discharge in large-volume atmospheric-pressure carbon dioxide lasers yields good results thanks to preionization, primarily of the near-cathode region. This lowers the work function of the electrons at the cathode's surface and facilitates the formation of a homogeneous discharge. But initially there are only a few electrons (10^9 cm^{-3}) in the active volume, which means that the mechanism by which the main discharge develops remains avalanche-like and is inclined to form arcs. All this limits the deposit of energy into the active region. The fact that the discharge is self-sustained, which means that both the breakdown of the discharge gap (i.e. ionization of the gas in the gap) and the power input into this gap (i.e. a specific heating-up of the breakdown channel by an external electric field) are enforced by a single high-voltage source, obstructs the creation of sufficiently effective TEA carbon dioxide lasers with a large volume and does not allow raising the pressure significantly above the atmospheric.

Transition to the non-self-sustained discharge, where the ionization of the gas is done by an independent agent, say, a beam of fast electrons, and the deposit of energy into a pre-ionized medium is carried out in an optimal manner, ensures homogeneous excitation of the active gas mixture in the absence of breakdowns in the gas, that is, at $E/p < 10\,\text{kV/cm} \cdot \text{atm}$. As we know, $3\,\text{kV/cm} \cdot \text{atm}$ is the optimal value of this ratio from the viewpoint of excitation. Then for an electron number density of $10^{12}\text{-}10^{14}\,\text{cm}^{-3} \cdot \text{atm}^{-1}$ there is a fairly effective power deposit to the active gas and a high degree of excitation of nitrogen vibrations and the levels $00^0 n$ of CO_2 molecules. The external ionization source prepares the gas mixture for power input completely, linearizes the volt-ampere characteristic of the mixture, and makes it possible to excite the lasing volume in a controlled manner. Non-self-sustained discharges have been used to create high-pressure carbon dioxide lasers with continuous tuning of the laser frequency, and record-breaking radiant energies have been attained at atmospheric pressure.

Before we go on to nonelectric methods of creating population inversion in carbon dioxide lasers, let us discuss an extremely important point that requires attention in connection with the operation of carbon dioxide lasers. The point is the resistance to radiation of the optical materials used in lasers (windows, mirrors, lenses, prisms, etc.). Actually this matter is beyond the scope of this course of lectures; it is the object of study of the interaction of laser light with matter. But carbon dioxide lasers are sources of such strong laser light that the properties of materials, even taken as phenomenological constants, must be taken into account at the design stage of such lasers.

The best transparent materials for windows of the gas cells of carbon dioxide lasers are alkali-halide crystals of the NaCl or KCl type, wide-gap semiconductors of the ZnSe type, and thallium-based compound crystals (the KRS-5 and KRS-6 Soviet brands or the IRTRAN Western brand). Under pulses 0.1-to-1 μs long these crystals can withstand radiation loads (radiant energy densities) up to 10-20 J/cm^2. For a rough estimate we equate what the active material can produce (formula (16.2)) with what the window material can withstand and obtain the following simple condition:

$$pl = 2\text{-}4\,\text{m} \cdot \text{atm}, \tag{16.5}$$

which limits the length l of the laser's active medium and the pressure p in the medium. The optical resistance of the reflecting metallic mirrors and the laser's active medium proper is higher, with the result that the optical strength of transparent materials is the limiting factor.

Thermo-optical distortions pose a serious problem in the cw and pulse-periodic modes long before irreversible changes occur in the laser materials. On the whole we can assume that modern materials enable operating in the cw with intensities of the order of several kilowatts per square centimeter.

Now let us turn to a method of creating the active media of gas lasers that differs from those discussed in the previous lectures.

We start with gasdynamic lasers. In these the source of radiant energy is the thermal energy of a molecular gas heated in an equilibrium manner to a high temperature. In thermal equilibrium there can be no population inversion, no matter how high the temperature is. The thermal energy is equally distributed over all degrees of freedom of the molecules, including the vibrational degrees. But different vibrational modes of polyatomic molecules may relax at different rates. This means that if suddenly the conditions are changed, different modes may approach new equilibrium states at different times. Hence, when the thermodynamic parameters of the gas change suddenly as the gas goes over from one equilibrium state to another, there may be a period in which the nonequilibrium state may take the form of population inversion, provided that the higher energy level relaxes more slowly than the lower. This is just the situation with CO_2 molecules in relation to the 00^01 (the upper laser level) and 10^00-02^00-01^10 (effectively the lower laser level) vibrations. Hence, population inversion produced by the energy of the heated gas may form during vibration relaxation when the gas is quickly cooled.

The simplest method of cooling large masses of flowing gas quickly is gasdynamic cooling, that is, letting the compressed and heated gas supersonically escape to a region of space that for all practical purposes is a vacuum. Supersonic expansion must lower the temperature and pressure of the gas mixture in the course of a time interval that is short compared to the lifetime of the upper laser level and long compared to the lifetime of the lower laser level. This requires organizing the expansion of the gas through a supersonic nozzle with a small critical-section height (0.3-1.0 mm). Then the gas mixture changes its parameters over a path 1-2 cm long, along the flow, which at supersonic speed ($M = 4$) leads to a situation when the population of the upper level remains the same as that of the hot gas while the population of the lower level corresponds to the low temperature of the gas after expansion. An admixture of nitrogen plays an important role in realizing this mode since nitrogen has a long lifetime and pumps the 00^01 level of CO_2 molecules. Figure 16.4 illustrates what has been said.

Obviously, population inversion exists on a finite segment in a region downstream beyond the nozzle; it is here that we must place the mirrors. Note that the cavity of a gasdynamic carbon dioxide laser operates in peculiar con-

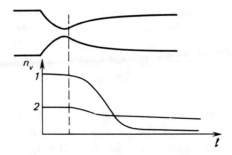

Figure 16.4. Diagram illustrating the idea of the gasdynamic laser. The upper part of the figure shows the profile of the nozzle and the lower the curves representing the dependence of the population number on distance (in arbitrary units): curve *1* represents the dependence for the lower level and curve *2* for the upper. The dashed line indicates the position of the critical section of the nozzle.

ditions: the active substance is introduced into the cavity at a high speed directed at right angles to the cavity's optical axis.

In a gasdynamic laser the thermal energy of the molecules of the gas is directly transformed into the energy of coherent electromagnetic radiation. This laser is a kind of heat engine, whose efficiency is determined by the temperatures of its working medium, the upper and lower temperatures. However, the upper temperature cannot be raised too high lest there be CO_2 dissociation and the lower temperature cannot be lowered too much lest there be CO_2 condensation. These considerations restrict the efficiency of gasdynamic lasers, which in reality amounts to several percent.

The main merit of gasdynamic lasers is their high output power in the cw mode, which reaches the megawatt range owing to the fact that a supersonic gas flow can carry through the laser cavity a large quantity of excited molecules per unit time. Naturally, such lasers are a complex engineering facility.

The gas that has passed through the cavity is usually ejected into the atmosphere. In the case of the ordinary laser mixture $CO_2 \div N_2 \div He$ and high output power this would lead to large irreplaceable losses of helium, which is inadmissible. Helium is, therefore, usually replaced by water vapor, which in moderate concentrations (1-2%) facilitates the collisional depopulation of the lower level 01^10. A higher concentration of water accelerates the relaxation of the 00^01 level of CO_2 molecules and thereby deactivates the main carrier of the excitation energy, molecular nitrogen, reducing in this way the efficiency of the laser.

Heating the gas with external sources of heat has proved inefficient. In large-scale devices the gas mixture of the required composition heated to 1300-1400 K is obtained by burning specially selected hydrocarbon fuels in an air atmosphere. This circumstance reemphasizes the fact that the thermal energy in gasdynamic lasers is transformed directly into the energy of coherent electromagnetic radiation.

The gasdynamic method of creating population inversion has been developed largely in connection with carbon dioxide lasers. Many variants have been suggested and developed of heating the gas and reducing the hazard of CO_2 dissociation at high temperatures by heating only the nitrogen and admixing CO_2 to vibrationally excited nitrogen after the nitrogen has passed the critical nozzle section and been "translationally" cooled.

Obviously, the gasdynamic method can be applied as well to molecules (or mixtures of these with other molecules) for which "freezing" of the high vibrational temperature of the higher vibrational level and rapid maxwellianization of the lower vibrational level in the process of rapid cooling of the gas is possible (see Figure 16.4).

Thus, in electric-discharge molecular lasers the relaxation channels are allocated to different vibrational modes, and the long-living mode is excited either directly by electron impact or indirectly by transfer of the excitation energy from the gas that carries the energy and is excited by electron impact. In gasdynamic lasers thermodynamically equilibrium heating is used for excitation, and this results in a thermodynamically nonequilibrium population distribution establishing itself in the course of supersonic outflow of the gas owing to the difference in relaxation rates in the separated relaxation channels mentioned above.

Another way of exciting the active gas, similar to the method just discussed, is to use exothermic chemical reactions, a fraction of whose energy may be released in the form of the vibrational energy of molecules. If this vibrational energy is then transferred, say, to CO_2 molecules, the resulting chemical carbon dioxide laser is very similar to a gasdynamic carbon dioxide laser with a high nitrogen content. At the same time the excitation energy of chemical origin can be directly used to create active laser media. For this reason chemical lasers form a separate class of laser systems and are of great interest in themselves. Yet it must not be forgotten that from a fairly general viewpoint the nature of gasdynamic and chemical lasers appears to be the same: in the active medium of these lasers the transition from one equilibrium state to another occurs in such a way that in the course of a finite time interval an essentially nonequilibrium state with population inversion is realized.

We note also that in chemical lasers the chemical energy is transformed directly into the energy of coherent electromagnetic radiation. What makes these lasers attractive is the large number of chemical reactions with colossal energy release.

Problems to Lecture 16

16.1. Can continuous tuning of the laser wavelength be achieved in a TEA carbon dioxide laser?

16.2. What type is the amplification line broadening in carbon dioxide lasers?

16.3. According to what scheme, the three-level or the four-level, do the different types of carbon dioxide lasers operate?

16.4. From the energy viewpoint why is the channel for the pumping of level 00^01 through level 00^02 attractive?

16.5. When must rotational competition be avoided?

16.6. How is rotational competition to be excluded?

LECTURE 17

Chemical Lasers

Exothermic reactions and vibrational energy. Vibrational-translational, vibrational-vibrational, and rotational-translational relaxations. Complete and partial inversion. Chemical pumping rate. Chain reactions. Initiation efficiency and chemical efficiency. Pulsed and cw chemical lasers. Laser parameters. The iodine photoionization laser

Thus, lasers whose population inversion in the active medium is created because of the nonequilibrium distribution among the chemical reaction products of the energy liberated in the course of the reaction right in the reaction volume are known as chemical lasers. In other words, the emission of light by a chemical laser is the direct result of a chemical reaction and not of some secondary effect of a chemical reaction, say, heating or an explosive-like increase in pressure, as in the case of gasdynamic lasers with an initial chemical energy input (steady-state or explosive combustion of fuel).

The emission of light by a chemical laser is the reverse of the photochemical effect, in which the chemical reaction emerges as the direct result of absorption of a photon by an atom or molecule and not of a secondary action of light, say, the heating of the reactants in the course of light absorption. Photochemical stimulation is carried out for inducing endothermic reactions. Obviously, only exothermic chemical reactions can be used for creating chemical lasers.

The interest in chemical lasers can be explained by the fact that in many exothermic chemical reactions a large amount of energy is liberated per unit reactant mass. There are the well-known examples of organic fuel and gun powder.

Analysis of reactions proceeding in the gaseous phase readily suggests that the vibrational degrees of freedom of molecules are good accumulators of the energy liberated in the chemical reaction. First, the chemical bonds of molecules are restructured in the course of molecular chemical reactions, that is, some bonds are broken, new ones are formed, some nuclei linked by the remaining bonds are replaced by other nuclei, and so on. The vibrational modes of molecules are localized, as we know, along bonds in the molecular structure. Hence, the energy is liberated in the course of an exothermic reaction through the chemical bond restructured in the reaction and is localized in the form of molecular vibrational energy. Second, vibrational (more precisely, vibrational-translational) relaxation proceeds rather slowly in the gaseous phase.

In view of what has been said, chemical lasers operate primarily on transitions between the vibrational levels of molecules. It is important to know, therefore, how chemical energy is distributed among the vibrational levels of molecules as a result of the reaction and in what reactions and which type of molecules an inverse distribution of the energy is possible.

The simplest example is the substitution reaction

$$A + BC \rightarrow AB(V) + C = AB^* + C. \tag{17.1}$$

A number of elementary reactions of the (17.1) type have been thoroughly studied. For instance, spectroscopic methods have revealed that for the reaction

$$F + H_2 \rightarrow HF^* + H \tag{17.2}$$

the ratio of the vibrational energy to the total energy liberated in the reaction, E_{vib}/E_{tot}, is 0.7. The relative populations on the levels $V = 1$, $V = 2$, and $V = 3$ are 0.31, 1.0, and 0.48, respectively. About the same situation occurs in the reactions $F + D_2$, $H + F_2$, $H + Cl_2$, and $H + Br_2$. As a rule population inversion emerges in all reactions that form hydrogen halides according to the (17.1) scheme as a result of exothermic substitution.

Once complete vibrational inversion has emerged in the restructuring of an intramolecular bond, that is, once

$$N_{V'}/g_{V'} > N_V/g_V, \quad V' > V, \tag{17.3}$$

it can exist only during a certain finite time interval. Relaxation processes drive a nonequilibrium system to an equilibrium that corresponds to a Boltzmann population distribution with a temperature corresponding to the energy liberated in the reaction. As we know, a molecule has vibrational, rotational, and translational degrees of freedom. The maxwellianization of the energy release of an exothermic reaction assumes that this energy is evenly distributed among all degrees of freedom of a molecule. In other words, in equilibrium all types of molecular motion are characterized by the same temperature T. The physical process responsible for this equilibrium is the intermolecular gas-kinetic collisions. We know that in collisions quasiresonant energy exchange is most effective. In vibrational-translational (V-T) relaxation, when

$$h\nu_{vib} > kT, \tag{17.4}$$

the quasiresonance conditions are strongly violated. Hence, V-T relaxation requires many hundreds and even thousands of gas-kinetic collisions. This explains why the V-T relaxation time is so great.

Along with this, exchange of vibrational energy quanta, $h\nu_{vib}$, is also possible between vibrational levels with different values of the vibrational quantum number V in the course of collisions. If the anharmonicity of a molecule is low, the gap energy in such a transfer (see Eq. (15.7)) is low, too,

$$2\chi(V + 1) < kT, \tag{17.5}$$

and vibrational-vibrational exchange (vibrational-vibrational, or V-V, relaxation) occurs practically at the rate of gas-kinetic collisions. In V-V relaxation the overall number of vibrational quanta is conserved, that is, the energy received by a system of molecules as a result (in our case) of a chemical reaction in the form of vibrational energy is conserved as vibrational energy. Only the distribution of this energy over the levels changes. If the anharmonicity of a molecule is low, the distribution obtained in the process of V-V relaxation is close to a Boltzmann distribution with an effective vibrational temperature $T_{vib} \neq T$. Obviously, in this case complete vibrational inversion (17.3) ceases to exist.

During V-V relaxation there also occurs rotational-translational, or R-T, relaxation, driving to equilibrium the rotational and translational degrees of freedom, which are characterized, as a result, by a unique temperature T. The fact that equilibrium sets in rapidly in the rotational-translational reservoir is due to the quasiresonant nature of energy exchange in the appropriate collisions owing to the smallness of the rotational energy:

$$BJ(J + 1) < kT. \tag{17.6}$$

The cross sections of V-V and R-T relaxations are close to the gas-kinetic value. Usually R-T relaxation proceeds somewhat faster than V-V relaxation. But V-T relaxation, as has been mentioned many times, proceeds much more slowly, with the result that $T_{vib} \neq T$ during a long time interval equal to many intervals (hundreds or thousands) between gas-kinetic collisions. If such reactions of the (17.1) type are chosen that their exothermic nature is realized primarily in vibrational energy release, then in the course of a fairly long time interval

$$T < T_{vib}. \tag{17.7}$$

In this situation what is known as partial inversion is possible. For a level

V', J' lying higher than another level V, J the condition for partial inversion is

$$N_{V',J'}/g_{V',J'} > N_{V,J}/g_{V,J}. \tag{17.8}$$

This condition is met only for some values of J and J', in contrast to the condition for complete inversion (17.3).

Let us consider this problem in greater detail. The Boltzmann distribution over vibrational-rotational levels at quasi-equilibrium vibrational and rotational temperatures, T_{vib} and T, assumes the form

$$\frac{N_{V,J}}{g_{V,J}} = \frac{1}{\Sigma_{\text{vib}}} \exp\left(-\frac{E_V}{kT_{\text{vib}}}\right) \frac{1}{\Sigma_{\text{rot}}} \exp\left(-\frac{BJ(J+1)}{kT}\right), \tag{17.9}$$

the energy $E_{V,J}$ of the rotational sublevel J of the vibrational level V being defined as

$$E_{V,J} = E_V + BJ(J+1) \tag{17.10}$$

(this is true for diatomic or linear molecules), and Σ_{vib} and Σ_{rot} stand for the vibrational and rotational partition functions, respectively. The condition of partial inversion is met if

$$\frac{T_{\text{vib}}}{T} > \frac{E_{V'} - E_V}{B} \frac{1}{J(J+1) - J'(J'+1)}, \quad J' < J. \tag{17.11}$$

The selection rules for vibrational-rotational transitions, $\Delta V = \pm 1$ and $\Delta J = 0, \pm 1$, make it possible to simplify (17.11) using the relations $E_{V'} - E_V = h\nu$ and $J = J' + 1$:

$$\frac{T_{\text{vib}}}{T} > \frac{h\nu}{2BJ}. \tag{17.12}$$

In other words, partial inversion is realized only in the P-branch (the value of J for the rotational sublevel of the upper laser level is by one unit smaller than that for the rotational sublevel of the lower laser level) and only for values of J of the lower level that are determined by the amount of stored vibrational energy, that is, the value of T_{vib}:

$$J > \frac{h\nu}{2B} \frac{T}{T_{\text{vib}}}. \tag{17.13}$$

Complete vibrational inversion, on the other hand, allows for gain in all three branches, the P-, Q-, and R-branch.

The possibility of partial inversion is an important feature of molecular systems. Population inversion may manifest itself in vibrational-rotational spectra not only because of differences in the populations of vibrational levels. What is important is the population distribution in a vibrational level over the rotational sublevels.

For transitions belonging to the P-branch that proceed from the $(J + 1)$st sublevel of the lower vibrational level to the Jth sublevel of the upper in an absorption act and from the $(J - 1)$st sublevel of the upper vibrational level to the Jth sublevel of the lower in an emission act a peculiar situation emerges. Let us imagine, for the sake of simplicity, a situation in which the population numbers of two vibrational levels V and $V + 1$ are equal. No complete inversion is present. The population distributions over the sublevels of these two vibrational levels are identical and Boltzmann in nature. Sublevel $J - 1$ is more heavily populated than sublevel J, which in turn is more heavily populated than sublevel $J + 1$, and so on. If the vibrational population numbers of levels V and $V + 1$ coincide, so do the population numbers of sublevels J_V and J_{V+1}. The same is true of sublevels $(J - 1)_V$ and $(J - 1)_{V+1}$, and so on. Hence, sublevel $(J - 1)_{V+1}$ is more heavily populated than sublevel J_V. The lowest sublevel of the upper level lies lowest on the scale of rotational energies because on its own vibrational level its rotational energy is lower. But it lies higher on the scale of vibrational energies, and, therefore, for a vibrational-rotational transition in the P-branch there is the possibility of population inversion (we have called this partial inversion). Usually the symbol J is used to denote a sublevel of the lowest vibrational level. When vibrational population inversion is absent from the $V + 1 \rightarrow V$ transitions, partial inversion can manifest itself only in $J - 1 \rightarrow J$ transitions, that is, in the P-branch for values of J specified by condition (17.13).

Thus, until V-V relaxation ceases, there is complete vibrational inversion caused by the manner of distribution of the energy of an elementary exothermic chemical act over the vibrational states of the molecules of the reaction products. Until V-T relaxation ceases, there is partial rotational inversion in the P-branch. These two types of inversion can easily be distinguished by the spectra of possible lasing frequencies.

When population inversion is achieved, the properties of chemical lasers coincide fully with those of all other molecular gas lasers, which, incidentally, is true for any other method of creating active media.

The most important problem in the case of chemical lasers is the choice of a proper chemical reaction that will produce population inversion.

Let us take the simple model depicted in Figure 17.1. Suppose that level 2 is populated in the process of the chemical reaction at a rate W. Then in the two-level approximation considered here the rate equations acquire the form

$$\frac{dn_2}{dt} = W - \left(\frac{1}{\tau_2} + \frac{1}{\tau_{21}}\right)n_2, \quad \frac{dn_1}{dt} = \frac{n_2}{\tau_{21}} - \frac{n_1}{\tau_1}, \tag{17.14}$$

where the meaning of the τ's is clear from Figure 17.1. The steady-state value of the population difference,

$$n_2 - n_1 = W\tau_2 \frac{\tau_{21} - \tau_1}{\tau_2 + \tau_{21}}, \tag{17.15}$$

reaches its maximum value

$$n_2 - n_1 = W\tau_{\text{eff}}, \tag{17.16}$$

where $1/\tau_{\text{eff}} = 1/\tau_2 + 1/\tau_{21}$, when $\tau_1 \ll \tau_{21}$. This maximum value must exceed a lasing threshold value n_{thr}. Hence, the chemical pumping rate must exceed a certain value determined by n_{thr} and the effective relaxation time:

$$W > n_{\text{thr}}/\tau_{\text{eff}}. \tag{17.17}$$

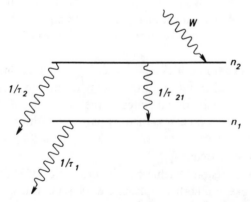

Figure 17.1. The kinetics of pumping a chemical laser.

Let us now assume that the chemical pumping rate is equal to the rate of the chemical reaction. Suppose that the reaction can be written as follows:

$$A + B \rightarrow AB^*. \qquad (17.18)$$

Then the rate at which the reaction product AB^* is produced is determined by the concentrations of the initial reactants, [A] and [B], and the so-called reaction rate constant K:

$$W = [A][B]K. \qquad (17.19)$$

The value of K obeys the Arrhenius equation

$$K = K_0 \exp(-E_a/kT), \qquad (17.20)$$

where E_a is the activation energy of the reaction, and the pre-exponential factor is a slowly varying function of T and always smaller than a value determined by the number of gas-kinetic collisions $\langle \sigma v \rangle$:

$$K_0 \leq \langle \sigma v \rangle. \qquad (17.21)$$

At 300 K and common values of the activation energy, $E_a \approx$ 10-20 kcal/mol, the reaction rate constant is extremely low, $K \approx 10^{-18}\text{-}10^{-26}$ cm^3/s, which cannot ensure that condition (17.17) is met. Chemical pumping requires reactions with an activation energy amounting to several kilocalories per mole (1 eV is equal to 23 kcal/mol). Only reactions involving free atoms or free radicals can have such low activation energies. However, the formation of free, chemically active atoms or free radicals requires a great deal of energy. The energy release of radical reactions at 300 K is of the same order of magnitude as the energy necessary for preparing the radicals.

Thus, when the issue is chemical pumping, there is the problem of lowering the expenditure of energy required to create radicals so that the initiation of chemical reactions of the (17.18) type is not energy-consuming. Then the power output in the form of radiation, which is approximately equal to the energy release (per unit time) of the chemical reaction, could considerably exceed the reaction initiation energy (per unit time). In the ideal limiting case we would expect a purely chemical laser effect, obtained without any expenditure of energy to reaction initiation.

The problem can be solved by using chain and self-sustaining chemical reactions. In these each chemically active center (radical or atom) is reproduced in the reaction. Multiple use of radicals transforms the chemical energy stored in the mixture into radiation. The fact that the same radicals are repeatedly engaged in chain chemical reactions may make up for the expenditure of energy required to create the small number of radicals needed for initiation of the chain process.

Indeed, to realize reactions of type (17.1) or (17.18), there must be atoms (radicals) A and/or B. In the simplest case we have diatomic molecules A_2 and B_2. Then the reaction product AB can usually be obtained in the reaction

$$A_2 + B_2 \rightarrow 2AB. \tag{17.22}$$

But this direct nonchain reaction has a high activation energy and proceeds at a rate too low to create population inversion above the threshold value.

Let us now employ an external initiating energy, say, subject the gas to an intense flash of fairly hard UV radiation. As a result a fraction of the molecules break up.

$$A_2 + h\nu_{UV} \rightarrow 2A, \tag{17.23}$$

and there may be the following fast chain reaction:

$$A + B_2 \rightarrow AB^* + B, \quad B + A_2 \rightarrow AB + A,$$
$$A + B_2 \rightarrow AB^* + B, \quad \ldots \; . \tag{17.24}$$

The second act in chain (17.24) restores the initial situation, and production of the sought reaction product AB^* goes on without the initiating UV radiation. But the real chain is not infinite, since the active centers finally disintegrate. A concept often used in analyzing chain reactions is the chemical chain length ν_{chem}, which is the ratio of the concentration of the molecular product AB to the concentration of the initial, or initiating, radicals A.

Actually, ν_{chem} is determined by the ratio of the rate of the slowest act in the chain to the active-center disintegration rate. Obviously, the longer the chain, the smaller the fraction of the first-initiation energy in the reaction energy balance and, hence, the higher the efficiency of a possible laser. But for a laser effect we need not simply a molecular product AB but excited molecules AB^*. The relaxation processes ever present in a reaction lead to the destruction of excitation, that is, the destruction of population inversion. Hence, the operation of a chemical laser can be characterized by introducing the concept of the laser chain length ν_{las}, which is defined as the ratio of the rate of production of excited molecules AB* to the inversion destruction rate. The reason why radicals disintegrate and population inversion relaxes is essentially the same, namely, gas-kinetic collisions. But relaxation proceeds faster, with the result that $\nu_{las} < \nu_{chem}$, and it is the laser chain length that determines the efficiency of the laser.

Suppose that the energy required to form an initial radical is E_d. Each radical has operated ν_{las} times. The energy expenditure of the initiation source (its efficiency is η_{ini}) per act of useful emission amounts to $E_d/\eta_{ini}\nu_{las}$. If the release in a single act of formation of an AB* molecule is ΔH and the fraction η_r of this energy (the reaction efficiency) is spent to excite the necessary vibration of molecule AB, the energy emitted in a single act is $\Delta H\eta_r$, and the initiation efficiency of a chemical laser, also often called the external efficiency, is

$$\eta_0 = \nu_{las}\eta_{ini}\eta_r\Delta H/E_d. \tag{17.25}$$

For large laser chain lengths this efficiency may exceed, and even considerably, 100%. This is the main merit of chemical lasers.

In addition to initiation efficiency, another quantity of interest is the chemical efficiency of the laser, η_{chem}, which by definition is the ratio of the laser radiation energy Q_{las} to the overall reserve of chemical energy in the system, Q_{chem}:

$$\eta_{chem} \equiv Q_{las}/Q_{chem}. \tag{17.26}$$

The quantity η_{chem} is important primarily because chemical energy is by no means gratuitous; it is contained in the initial reagents, whose manufacture may require excessive amounts of energy.

The above discussion suggests that

$$Q_{las} = \nu_{las}\eta_r\Delta H/[A]_0, \tag{17.27}$$

where $[A]_0$ is the initial number of radicals at the initiation stage.

The amount of chemical energy stored in the system is determined by the amount of energy released in the reaction of formation of a single AB* molecule and by the number of such molecules that can be formed from the given composition of the reactants of the initial mixture. The latter is equal to the number of molecules of the initial mixture that enter into the molecular product at a stoichiometrically lower concentration. This number, denoted by the symbol [N], is equal to the concentration $[A_2]$ of molecules A_2 in the case of a reaction of the (17.22) type or, more precisely, of reactions (17.23) and

(17.24). Let us consider the simplest case. Then

$$Q_{chem} = \Delta H[A_2],$$ (17.28)

$$\eta_{chem} = \nu_{las}\eta_r[A]_0/[A_2].$$ (17.29)

Chemical efficiency, just as initiation efficiency, is the greater the longer the laser chain and the higher the reaction efficiency η_r. At the same time, the initiation level, that is, the fractional number of active centers, $[A]_0/[A_2]$, created by the initiation source also determines η_{chem}. Unfortunately, further comparison of η_{chem} and η_{las} proves difficult because the laser chain length ν_{las} depends in a complicated manner on the kinetics of the chain reaction and on the initial concentration of radicals $[A]_0$. Often the laser chain length ν_{las} decreases as the initiation level increases, and η_{chem} and Q_{las} may increase, too, while η_0 drops. In real conditions a compromise is sought.

The hydrogen halide lasers mentioned at the beginning of this lecture operate in the medium IR range. The following wavelengths are characteristic of these lasers: 2.7 μm for the HF laser, 3.7 μm for the HCl, 4.2 μm for the HBr, and 4.3 μm for the DF. Here one must bear in mind that the complex structure of the vibrational-rotational transitions in these molecules makes it possible, under heavy vibrational excitation, to lase on many lines in a broad range in the vicinity of the above-mentioned wavelengths. All these wavelengths are interesting in themselves, but what is striking is the magnitude of the radiation quantum emitted by the DF laser, which coincides with the energy of the 00^01 vibration of the CO_2 molecule. Hence, chemically excited DF can be used successfully to create chemical carbon dioxide lasers by employing the effect of resonant excitation energy transfer.

Both pulsed chemical lasers and cw chemical lasers have been well developed.

In pulsed lasers pulsed initiation switches on the chain reaction in a mutually stable mixture of gases. The result is the emission of radiation. But the reaction must be fast, since in a slow quasistationary reaction relaxation processes lead to the destruction of inversion by equalizing all degrees of freedom. Here there is an analogy with the gasdynamic laser, where the cooling of the gas must also be rapid.

Various processes of photolysis and radiolysis are used for pulsed initiation, primarily UV dissociation and dissociation by electron beam. Spark discharges, exploding wires, gliding discharges, and quartz lamps are used as sources of UV radiation. The characteristic length of pulses of radiation emitted by these sources is about 1 μs. Electron beams can ensure shorter pulses (1-100 ns), but their overall effectiveness is lower.

To obtain population inversion it is better to use the simplest molecules (preferably, reaction products), namely, diatomic heteronuclear molecules of the AB type. Otherwise, V-V relaxation becomes extremely complex and in the initial act of the exothermic reaction chemical energy will be redistributed over the degrees of freedom of the molecule.

The best example is the HF laser. Lasing is observed primarily in the vibrational band of the $V = 2 \rightarrow V = 1$ transition in the 2.7-μm range. To produce first atomic fluorine and then hydrogen fluoride, mixtures are used of fluorine-carrier molecules of the NF_3, UF_6, SF_6, or WF_6 type or fluorinated hydrocarbons with hydrogen or hydrogen-containing molecules. For instance, in the mixture $H_2 \div F_2 \div O_2 \div SF_6$ with a ratio of $10 \div 36 \div 14 \div 10$ and an overall pressure of 1 atm and under initiation by electron beam, an energy release of 100 J/l in a pulse 30-ns long was produced. The initiation efficiency was found to reach 800%. Here it is worth noting an important feature of chemical lasers related to their chemical origin. A mixture of absolutely pure hydrogen H_2 and fluorine F_2 is unstable, so that its spontaneous explosion is a rule rather than an exception. Adding oxygen and SF_6 stabilizes the mixture. Selection of the initial gas mixture, therefore, constitutes a serious problem in chemical lasers from the standpoint not only of relaxation kinetics but also of the kinetics of chain reactions of the explosive type.

Considerable interest, especially for autonomous applications, has been aroused by the DF-CO_2 laser, whose operation is based on the energy transfer scheme, already mentioned, $DF^* + CO_2 \rightarrow DF + CO_2^*$. In pulses several microseconds long an energy release of up to 150 J/l has been obtained at $\lambda = 10.6\,\mu$m.

Continuous-wave operation of chemical lasers is possible by mixing gases, for instance, in molecular streams. Since mutually unstable reactants can be mixed in a stream and the reaction products can then be removed very rapidly, at a rate close to sonic velocity, a purely chemical laser (i.e. without initiation) becomes feasible. Here the most interesting laser is the DF-CO_2, in which the excited molecules DF^* emerge as a result of a chain reaction ignited when stable radicals NO are mixed with molecular fluorine, which produces atomic fluorine, and this serves as the active center of the chain reaction:

$$NO + F_2 \rightarrow NOF + F, \tag{17.30}$$

$$F + D_2 \rightarrow DF^* + D, \quad D + F_2 \rightarrow DF^* + F, \quad \dots \ . \tag{17.31}$$

Downstream, CO_2 is mixed to DF^*. The carbon dioxide is excited according to the well-known scheme

$$DF^* + CO_2 \rightarrow CO_2^* + DF. \tag{17.32}$$

Implementing the gasdynamic system that realizes this scheme poses certain difficulties because of the risk of explosion in the mixtures used and the high toxicity of the reaction products. Figure 17.2 shows a schematic of a gas-flow reaction vessel with a laser cavity attached to the vessel at the outlet. Essentially this scheme is close to the one used in a gasdynamic laser, and the designer encounters the same difficulties of constructing a laser cavity that must receive a powerful transverse influx of the active substance. The reduced power output of such a laser may reach several hundred watts recalculated for a flow rate of 1 g/s of the active substance.

Strictly speaking, lasers can be considered chemical if the active medium is excited by chemical means and the initiation efficiency (external efficiency) exceeds 100%.

An example of the opposite situation is the iodine photodissociation laser.

It is well-known from photochemistry that in some cases atoms enter into addition reactions more readily when they are electronically excited. Hence, dissociation may lead to the emergence of electronically excited atoms. This

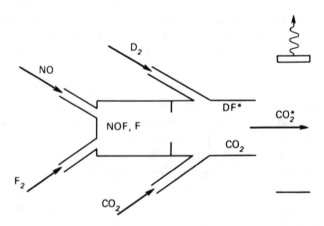

Figure 17.2. The schematic of a gas-flow reaction vessel of a chemical DF-CO_2 laser. The wavy arrow indicates the direction in which the radiation exits from the laser through a partially transparent mirror.

means that at least a fraction of the binding energy of the atoms in a molecule may be spent on exciting a fragment of the molecule. If the dissociation fragment is an atom, only its electron energy levels may be excited. But if the initial molecule is stable, the amount of energy that must be spent to break a bond is greater than the fraction of energy released in the form of electron energy of the atoms as a result of various types of redistribution.

In view of relaxation processes, dissociation must, obviously, be carried out rapidly. Any further use of the excited dissociation products for their chain breeding is impermissible, since this would lead to the destruction of excitation. For this reason direct pulsed dissociation is most suitable. This suggests that initiation efficiency in this case is lower than unity. Photodissociation lasers differ in this respect from the chemical lasers to which this lecture is mainly devoted. One must bear in mind, however, that within the framework of the more general definition given at the beginning of the lecture, photodissociation lasers are chemical lasers.

The best-known photodissociation laser is the iodine laser emitting radiation at $\lambda = 1.315\ \mu$m. The excited iodine atoms in this laser are produced by pulsed photolysis of CF_3I molecules. The reaction proceeds according to the scheme

$$CF_3I + h\nu \rightarrow I^*(^2P_{1/2}) + CF_3. \tag{17.33}$$

The excited iodine state $^2P_{1/2}$ is depopulated by the radiative $^2P_{1/2} \rightarrow {}^2P_{3/2}$ transitions within one and the same electronic configuration of iodine, $5p^5$. The radiation emitted by the nonmonochromatic source is absorbed in the broad molecular absorption band, while the excited atoms emit radiation in a narrow line characteristic of atomic spectra.

Since large volumes of the CF_3I gas can be subjected to photodissociation, iodine lasers of this type appear very promising for obtaining high levels of pulsed energy.

Pulsed quartz lamps, exploding wires, and open discharges can serve as sources of UV dissociation radiation.

In addition to CF_3I molecules, C_3F_7I and CH_3I molecules have found application as sources of excited iodine atoms.

Problems to Lecture 17

17.1. Make a qualitative estimate of the emission spectrum of a pulsed chemical laser. Can the spectrum change during the pulse?

17.2. According to what scheme do chemical lasers operate: the two-, three-, or four-level? Consider two cases, complete vibrational inversion and partial vibrational inversion.

17.3. How are the lower laser levels depopulated in chemical lasers? Consider the two cases of the previous problem.

17.4. Can a continuous-wave chemical laser be created without pumping the active medium through the laser?

17.5. Can a sealed-off chemical laser be designed?

LECTURE 18

The Carbon Monoxide Laser. Gas Lasers Operating on Electron Transitions in Molecules

The plateau in the vibrational population curve. Partial inversion. Distinctive features of the lasing spectrum of carbon monoxide lasers. Electron transitions in molecules. The Franck-Condon principle. The nitrogen laser. The hydrogen laser. Excimer lasers

Now to return to molecular lasers.

In the medium part of the IR range it is worth calling attention to the carbon monoxide laser, which operates in the 5-6.5 μm range. In many respects this laser is similar to the carbon dioxide laser. Bringing the two closer together are such features as high efficiency, high power output, the capability of operating in the continuous-wave and pulsed modes, a broad range of methods for creating population inversion (gas-discharge, gasdynamic, chemical, and electron-beam excitation), the relative closeness of the frequency ranges, and the possibility of selecting the lasing wavelength in a fairly wide range.

Like the carbon dioxide laser, the carbon monoxide operates on vibrational-rotational transitions in the ground electronic state. The mechanisms of vibrational excitation of carbon monoxide molecules are similar to those of carbon dioxide, namely, either direct excitation of the upper vibrational states of CO by electron impact, in a chemical reaction, or by other means, or excitation energy transfer from N_2 molecules excited by a well-known method. But vibrational excitation and population inversion are quite different matters. The mechanism of creating population inversion in the carbon monoxide laser differs considerably from the mechanism of creating population inversion in the carbon dioxide laser.

The behavior of polyatomic CO_2 molecule characterized by three vibrational modes and low anharmonicity, obeys the theory of relaxation of harmonic oscillators. The rapid V-V exchange proceeding along the modes establishes in each a specific vibrational temperature, whose value depends on the amount of energy received by the molecule and on the V-T relaxation, which proceeds at different rates in different modes (see Lecture 15).

This approach is invalid for the CO molecule, which is a diatomic molecule, that is, possesses only one vibrational mode. In the case of carbon monoxide the relaxation is of essentially anharmonic oscillators. This process is described by a system of rate equations that incorporate the populations of all the vibrational levels, the rate constants for all partial V-V and V-T relaxations, the excitation cross sections, donor concentrations, excitation energies, the concentrations of impurities, which open new relaxation channels, and the respective relaxation rate constants. Even if we are interested only in the steady-state case and assume that $dN_V/dt = 0$, there remains a system of approximately 60 nonlinear algebraic equations that cannot be reduced, in contrast to CO_2, to a two- or three-level system.

The results of computer calculations are as follows.

The system of vibrational levels of the CO molecule (Figure 18.1)

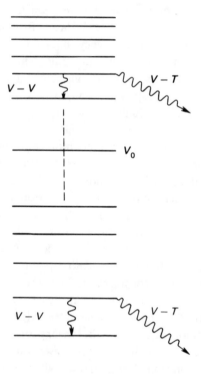

Figure 18.1. Vibrational levels of the carbon monoxide molecule. Level V_0 is indicated. Above it translational relaxation prevails and below it vibrational relaxation.

contains a level V_0 such that for levels $V > V_0$ the V-T relaxation proceeds faster than the V-V relaxation, while for levels $V < V_0$ the V-V relaxation proceeds faster than the V-T relaxation. The value of V_0 is temperature-dependent, decreasing as the gas temperature is raised. Hence, at low temperatures, when V_0 is high, V-V relaxation prevails. As a result, in vibrational excitation the distribution of particles over V differs from the Boltzmann distribution drastically.

For gas temperatures in the 150-300 K range the vibrational population curve exhibits a plateau. This extends from $V = 5$ to $V = 10$ at 200 K for gas discharge excitation in a mixture with nitrogen, but may reach $V = 30\text{-}35$ at lower temperatures and in carefully selected mixtures. The plateau disappears in the 150-500 K range, since V_0 drops, that is, the role of excitation of the upper vibrational levels diminishes in the process of V-V exchange and that of V-T relaxation grows while its rate increases. At high temperatures and in the presence of excitation the vibrational distribution tends rapidly to the Boltzmann equilibrium (Figure 18.2).

What has been said indicated the necessity of cooling the gaseous mix-

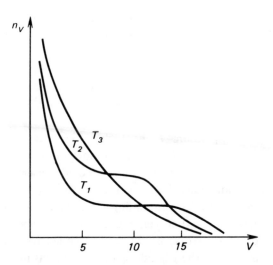

Figure 18.2. Time-independent population distribution over the vibrational levels of the ground electronic state of the carbon monoxide molecule in the presence of vibrational excitation ($T_1 < T_2 < T_3$).

ture in the carbon monoxide laser and the advisability of introducing helium into the mixture (helium is a gas with a high thermal conductivity).

The plateau on the vibrational population curve indicates the absence of complete vibrational inversion. However, the discussion of chemical lasers in Lecture 17 showed that in the absence of vibrational inversion there may be partial rotational inversion, in which case none but lines of the P-branch should appear in the lasing, a fact observed in experiments.

The exact formula for the gain at the center of the line corresponding to the Doppler broadened vibrational-rotational transition $V \to V - 1$, $J - 1 \to J$ yields

$$\alpha_{V, V-1}(J) = \left(\frac{2c}{\pi kT} \right)^{3/2} (2J - 1) A_{V, V-1} h\lambda^3 M^{1/2}$$

$$\times \left[N_V B_V \exp \left(- \frac{B_V J(J-1)hc}{kT} \right) \right.$$

$$\left. - N_{V-1} B_{V-1} \exp \left(- \frac{B_{V-1} J(J+1)hc}{kT} \right) \right]. \quad (18.1)$$

When vibrational population numbers are equal, $N_V = N_{V-1}$, and the rotational constants are approximately equal, $B_V \approx B_{V-1}$, the gain is positive since

$$\exp \left(- \frac{a(J-1)}{kT} \right) - \exp \left(- \frac{a(J+1)}{kT} \right) > 0, \quad (18.2)$$

and the lower the gas temperature T the higher the gain.

Formula (18.1) and Figure 18.2 lead to an important conclusion on the cascade nature of the lasing of the carbon monoxide laser. Figure 18.3 shows that successive transitions lead to successive population of the vibrational states. Lasing on the corresponding vibrational-rotational transitions occurs successively with a time lag. Figure 18.3 also shows how and why the carbon monoxide laser operates at many frequencies. At the same time it is clear that in view of the cascade nature of the lasing (in contrast to the lasing of the carbon dioxide laser, where there is a common upper level and, hence, competition of lasing lines is possible) it is impossible with the carbon monoxide laser at a single lasing frequency to transform into radiation all the energy stored in the nonequilibrium distribution of

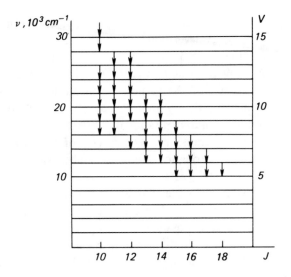

Figure 18.3. The cascade nature of lasing of the carbon monoxide laser.

particles among the vibrational levels. In the continuous-wave mode the carbon monoxide laser operates on many frequencies simultaneously, while in the pulsed mode successive switching of frequencies usually occurs. At high pressures and intense pumping (non-self-sustained discharge driven by an electron beam) lasing occurs on many frequencies simultaneously in the pulsed mode, too. Selective cavities, as well as external dispersing elements, make it possible to select any of the possible lasing frequencies.

Simple systems with transverse discharge, similar in design to carbon dioxide lasers, have gained the widest acceptance. An overall lasing power of several dozen watts has been reached in all lasing lines in the continuous-wave mode with systems that use a mixture of carbon monoxide with helium, nitrogen, and a small amount of oxygen, cooled to 150-200 K. Complex chemical reactions, which alter the composition of the mixture and the set of relaxation partners, take place in the discharge. The transition to room temperature and sealed-off systems became possible when mixtures containing xenon were employed. Mixtures of the $CO \div Xe \div He = 1 \div 1 \div 8$ type are used in TEA carbon monoxide lasers and carbon monoxide lasers with semi-self-maintained discharge.

Interest in the carbon monoxide laser arises not only from the importance of the wavelength range which the laser spans but also from the high energy parameters. The energy level diagram of the active medium of the carbon monoxide laser (Figure 18.3) is such that the total power-input efficiency over all the lasing lines from $V \geq 25$ to $V = 5$ may exceed 50-75%.

Concluding our discussion of the carbon monoxide laser, we note that its lasing lines are denoted by the symbol $P_{V, V-1}(J)$, with P standing for the P-branch, J the number of the rotational sublevel of the lower vibrational level in the $V \rightarrow V - 1$ transition. The spectra of the carbon monoxide molecule have been thoroughly studied; hence, such notation is sufficient.

Up to this point in studying molecular gas lasers we have had in mind the vibrational spectra of molecules. However, lasing that involves electron transitions in molecules is also possible.

The electron energy of molecules depends in a complex way on molecular structure. The system of levels is rich and complicated. The energy is distributed between many degrees of freedom. The axial symmetry of diatomic and linear polyatomic molecules makes it possible to classify the electronic states of a molecule according to the values of the quantum number Λ, which defines the absolute value of the projection of the total orbital angular momentum L on the molecule's axis. By analogy with the atomic states S, P, D, F, G, and so on, the molecular states with $\Lambda = 0$, 1, 2, 3, 4, etc. are denoted, respectively, by Σ, Π, Δ, Φ, Γ, etc. The multiplicity of the level, $\mu = 2s + 1$, is indicated by a superscript to the left of the symbol: the state with $\Lambda = 1$ and $s = 1/2$ by $^2\Pi$, that with $\Lambda = 0$ and $s = 1$ by $^3\Sigma$, etc. As a rule, the ground state of chemically stable molecules is the $^1\Sigma$ state.

For dipole transitions the following selection rule holds true: $\Delta\Lambda = 0$, ±1; this means that in the dipole approximation the allowed transitions are $\Sigma \rightarrow \Sigma$, $\Pi \rightarrow \Sigma$, etc.

The intensity distribution in the system of bands of electron transitions of a molecule is determined by the properties of the potential-energy curves in the ground and excited electronic states, $E'(r)$ and $E''(r)$, in accordance with the Franck-Condon principle: in any molecular system the transition from one electronic state to another is so rapid that the nuclei of the atoms involved can be considered stationary during the transition. In other words, the Franck-Condon principle allows for transitions from one electronic state to another only along the vertical line on the potential-energy diagram that connects the so-called turning points in the vibrational motion of a molecule between the walls of a potential well (Figure 18.4).

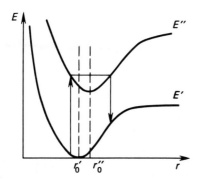

Figure 18.4. Illustration of the Franck-Condon principle.

As a rule, an excited electronic state corresponds to a looser molecule. In the case of a diatomic molecule the horizontal coordinate of the minimum of an electronic state (term), obviously, corresponds to the equilibrium distance between the nuclei, r_0. What has been said means that usually $r_0'' > r_0'$ (Figure 18.4). This has proved decisive for operation of the nitrogen UV laser ($\lambda = 337.1$ nm).

The energy level diagram for the nitrogen laser is shown in Figure 18.5. The potential energy curves of the upper state ($C^3\Pi_u$), the lower state ($B^3\Pi_g$), and the ground state ($X^1\Sigma_g^+$) are positioned in such a way that under electronic excitation the levels $V = 0$ and $V = 1$ of state C are effectively populated in accordance with the Franck-Condon principle. State B is not populated in the process. Generally speaking, electrons collisionally "shift" N_2 molecules vertically upward on the potential-energy vs. internuclear-distance diagram without touching the states shifted to the right and, hence, not populating the lower laser levels. At the same time the Franck-Condon principle allows radiative transitions from state C to state B, but only vertically downward and from the right turning points of the $C^3\Pi_u$ state.

In this way in the case of the nitrogen laser the Franck-Condon principle makes it possible to separate the excitation and lasing channels. This separation manifests itself even more strongly in the case of the hydrogen laser (the 116-126 nm wavelength range, vacuum ultraviolet), whose level diagram is shown in Figure 18.6.

Since the upper states have a short lifetime, pulsed gas-discharge systems with low inductance are used to excite hydrogen and nitrogen lasers. These

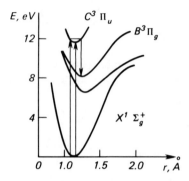

Figure 18.5. Energy level diagram for molecular nitrogen.

two types of lasers, it must be emphasized, are self-terminating because their lower levels have a longer lifetime than the upper.

The radiation lifetime of the upper active state of the nitrogen laser (38 ns) is almost 50 times longer than that of the hydrogen laser (0.8 ns), which means that the design of the excitation system of the nitrogen laser is simpler. However, this case also requires a transverse discharge. More than that, because of the short lifetime of the upper state, the sudden way in which population inversion is switched on, and the self-terminating nature of lasing, the time of nonzero of gain is very short (see Lecture 14). For

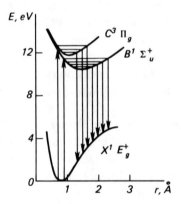

Figure 18.6. Energy level diagram for molecular hydrogen.

the nitrogen laser this time interval is approximately 3-10 ns, and the pumping pulse length must be no longer. But in 3 ns light travels about 1 m. Hence, when a pulse is generated at one end of a long laser and travels the full length of the laser, gain vanishes if a transverse pumping discharge is ignited simultaneously along the entire length of the laser. Therefore, we need a traveling excitation wave that propagates along the axis of the laser tube and is synchronized with a pulse of laser light. In this case high unidirectional gain is achieved, and the pulse of spontaneous radiation emitted in the process becomes stronger as it propagates and acquires directivity and monochromaticity.

The nitrogen, hydrogen, and similar short-wave (ultraviolet, vacuum ultraviolet) lasers are known as superluminescence lasers because population inversion lasts a very short time (less than 1 ns). This means that negative feedback has no time to develop while population inversion lasts. Hence, such lasers are not self-oscillating systems with positive feedback, they emit amplified noise, and owing to the properties of coherent amplification of a traveling wave in the acts of stimulated emission the radiation emitted is spectrally and spatially "purified" (see Lecture 5), that is, practically monochromatic and highly directional.

The gas pressure in nitrogen and hydrogen lasers usually amounts to 10-100 torr. At higher pressures collisional quenching of excited molecular states begins to manifest itself.

At pumping-power densities of up to 5 kW/cm^3 the characteristic peak lasing power in a pulse 10-ns long is about 1 MW for a nitrogen laser and 10 kW in a pulse 1-ns long for a hydrogen laser. The efficiency of these lasers is not very high and, depending on the excitation conditions, lies in the 0.01-to-1% interval. Pulsed power sources for these lasers, characterized by multi-kiloampere currents at multi-kilovolt voltages in pulses of several nanoseconds, are designed on the basis of Marx and Blumlein circuits and similar technical devices of high-current electronics and require thorough tuning. It must be emphasized that mastering the ultraviolet range by quantum electronics is extremely important for application, primarily in photochemistry and photobiology.

The nitrogen laser operates at a repetition frequency of up to 100 Hz. Transverse gas pumping, which effectively cools the laser's active medium, can create the conditions for raising the repetition frequency to 10^5 Hz, which is the inverse lifetime of the lower laser level $B^3\Pi_g$. From the user's standpoint a pulsed laser becomes equivalent to a continuous-wave laser at a repetition frequency of 1 kHz. Pulsed-periodic nitrogen lasers have found application in pumping dye lasers. The design of pulsed-periodic hydrogen lasers has been hindered by the much more complicated power source required.

Let us return to the question of mastering the ultraviolet range. The difficulties are obvious. They are fundamental as well as technical. The crux of the matter lies in shortening the wavelength. As we already know (see Eqs. (3.13) and (4.25)), the gain is proportional to inversion Δn, the square of the matrix element $\langle \mu \rangle$, and the ratio of the transition frequency ν to the transition linewidth $\Delta \nu$:

$$\alpha \propto \langle \mu \rangle^2 \Delta n \nu / \Delta \nu. \tag{18.3}$$

In turn, population inversion is determined by the pumping rate Λ, the total number N of particles, and the lifetime τ in the following manner:

$$\Delta n = N \Lambda \tau. \tag{18.4}$$

Hence,

$$\alpha \propto \Lambda \langle \mu \rangle^2 N \tau \nu / \Delta \nu. \tag{18.5}$$

The frequency dependence of $\tau \nu / \Delta \nu$ has a strong effect on the α vs. ν dependence.

Here are some particular cases.

Suppose that collision broadening determines the lifetime and the linewidth. Then $N \propto p$, $\Delta \nu \propto p$, $\tau \propto 1/p$, and $\alpha \propto \nu/p$, where p is the pressure of the active gas. But if the collision broadening determines the lifetime and the linewidth is Doppler, then $N \propto p$, $\Delta \nu \propto \nu$, and $\tau \propto 1/p$ and the gain is frequency-independent. However, as the frequency grows, the role of spontaneous emission becomes ever greater owing to the cubic dependence of its probability on frequency, and sooner or later, depending on the values of the transition matrix element, pressure, and temperature of the gas (as well as other factors), spontaneous emission becomes predominant.

These particular cases belong to the IR range and, as a rule, to the long-wave part of the visual range. In the shorter-wave range the lifetime is determined by spontaneous decay. If the linewidth is Doppler and the lifetime natural, then $N \propto p$, $\Delta \nu \propto \nu$, $\tau \propto 1/\nu^3$, and $\alpha \propto p/\nu^3$. If both the linewidth and the lifetime are determined by spontaneous emission, $N \propto p$,

$\Delta\nu \propto \nu^3$, $\tau \propto 1/\nu^3$, and $\alpha \propto p/\nu^5$. Summing up the discussion in the form of the table

$$\alpha \propto \Lambda\langle\mu\rangle^2 \times \begin{cases} \nu/p, & \text{collision broadening,} \\ p/\nu^3, & \text{Doppler broadening and natural lifetime,} \\ p/\nu^5, & \text{natural broadening,} \end{cases} \qquad (18.6)$$

we see that a sharp drop in gain as frequency grows can be compensated for, at least to some extent, by increasing the pumping rate Λ and the gas pressure p and by the proper selection of transition with a large value of $\langle\mu\rangle^2$. This last fact almost entirely excludes atomic gases from the present discussion, since for atoms ulraviolet transitions are usually forbidden. Molecules remain.

The examples just discussed of nitrogen and hydrogen molecular lasers showed the need for intense pumping, which grows from the nitrogen laser (337.1 nm) to the hydrogen laser (116 nm). Unfortunately, lasing transitions between electronic states of stable molecules are characterized by the fact that the lifetime of the upper laser level is shorter than that of the lower. This considerably restricts the potentials of the respective lasers.

The solution has been found in replacing stable molecules with molecules that do not exist in the ground state. One example is noble gases, which do not form dimer molecules of the A_2 type in the ground state because the electron orbits of each atom A are filled in a symmetric manner. The interaction of atoms A with each other during gas-kinetic collisions is repulsive. However, when in collision two atoms are close to each other and form a dimer A_2 during the time of transit. As pressure increases, the fraction of atoms in dimer states, which exist only temporarily and do not have constant parameters, increases.

As is known, the time between gas-kinetic collisions is

$$\tau = 1/n\sigma u, \qquad (18.7)$$

where n is the number of gas particles per unit volume, σ the gas-kinetic cross section, and u the average thermal velocity. The effective time of transit, $\Delta\tau$, can be estimated by using the following formula:

$$\Delta\tau = \sqrt{\sigma}/u. \qquad (18.8)$$

At $\sigma = 10^{-16}$ cm^2 and $u = 10^5$ cm/s this time interval is 10^{-13} s, which amounts to several periods of IR vibrations. The fraction of atoms in dimer state can be estimated by the ratio

$$\Delta\tau/\tau = n\sigma\sqrt{\sigma}. \tag{18.9}$$

This number can be seen to grow with pressure and is about 10^{-5}-10^{-4} at atmospheric pressure. But if at least one of the colliding atoms is in an excited electronic state, the symmetry of the electronic shell is violated and the shell becomes loose. Then there is the possibility that an electron may "jump" in the collision process to an orbit surrounding both atoms, which results in the formation of a chemical bond and stabilization of the dimer. Since the lifetime of the electronic excitation noticeably exceeds the time of transit $\Delta\tau$, the fraction of excited dimers greatly exceeds the number given by (18.9). After electronic excitation has ceased, the atoms return to the ground state and the dimer disintegrates.

Lasing that employs transitions from a stable upper state to a repulsive molecular lower state has been achieved using dimers and halogenides of noble gases (Xe_2, Kr_2, Ar_2, $XeCl$, XeF, $KrCl$, KrF, $ArCl$, and ArF) at high gas pressures combined with excitation of the active medium by a beam of high-energy electrons or a powerful gas discharge. Lasers of this type have became known as excimer lasers since the base of their active medium is formed by dimer molecules, which exist stably only in an excited electronic state and for this reason are called excimers (from *exci*ted di*mers*).

Excimer lasers, a new class of laser systems, open up the UV range to quantum electronics. Their discovery was as important to quantum electronics as the discovery of carbon dioxide lasers or dye lasers. They will now be discussed in greater detail.

The principle underlying the operation of excimer lasers can be conveniently explained using the example of the xenon laser ($\lambda = 172.5$ nm). The ground state of the Xe_2 molecule is unstable (Figure 18.7). The unexcited gas consists primarily of atoms. The upper laser state becomes populated, that is, excited stable Xe_2^* molecules are formed, owing to the action of the high-energy electron beam in the course of a complicated sequence of collision processes. An essential role in this process is played by the ionization and excitation of xenon atoms by electrons. Excited molecules are formed in triple collisions of excited and unexcited xenon atoms:

$$Xe^* + 2Xe \rightarrow Xe_2^* + Xe. \tag{18.10}$$

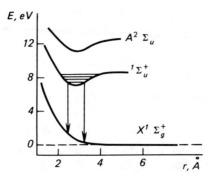

Figure 18.7. Potential energy curves of the excimer Xe*. The $^3\Sigma_u^+$ state, adjoining $^1\Sigma_u^+$, is not shown.

In the overall balance an important role is played by the conversion of atomic xenon ions into a molecular ion,

$$Xe^+ + 2Xe \rightarrow Xe_2^+ + Xe, \tag{18.11}$$

followed by dissociative recombination,

$$Xe_2^+ + e \rightarrow 2Xe^* + e, \tag{18.12}$$

which supplies excited atoms for further combination into excited molecules.

What is important here is the three-particle nature of collisions leading to the formation of an excimer molecule, a process that requires high pressures. The xenon laser operates at pressures exceeding 10 atm.

An excited molecule (see Figure 18.7) loses its excitation energy in the radiative processes

$$Xe_2^*(^1\Sigma_u^+) \rightarrow 2Xe + \hbar\omega \tag{18.13}$$

and

$$Xe_2^*(^3\Sigma_u^+) \rightarrow 2Xe + \hbar\omega \tag{18.14}$$

with decay times of 5 and 40 ns, respectively. As soon as a molecule proves to be in the ground state as a result of lasing transitions, it disintegrates. This leads to automatic depopulation of the lower laser level. A characteristic feature of excimer lasers is that the lower laser level is not populated.

Because of its repulsive nature the lower state contains no pronounced vibrational-rotational states and a lasing transition is broad-band, which makes tuning possible. For the xenon laser the amplification linewidth is 5 nm at a central wavelength of 172.5 nm.

The high pressure, large amplification linewidth, and short lifetime of the upper states require intense pumping with an energy input of 0.2 J/cm^3 in the course of 0.1-1 ns. The xenon laser has a high efficiency (up to 20%) of transformation of the electron-beam energy into laser light. The attained lasing power amounts to several hundred megawatts.

Thus, what makes it possible to design excimer lasers is the special shape of the potential-energy curves of the ground and excited states of the quasimolecule that forms when atoms are brought close together. For a dimer, formed by two similar atoms of noble gases, the depth of the potential well of the ground state is much less than kT at room temperature; hence, such dimers are not formed in ordinary conditions.

Noble gas halogenides excimers (monohalides of noble gases) have attracted much attention primarily because unlike noble gas dimers the corresponding lasers operate by gas-discharge excitation as well as electron-beam excitation.

There is no reliable data on the structure of the repulsive state of halides of noble gases. Since at room temperature no stable dimers of this type have been observed, the corresponding potential wells are very shallow, if they exist at all. In many respects the mechanism of formation of the upper states of the laser transitions in these excimers remains unclear. Quantitative reasoning suggests that these dimers are much more easily formed than noble gas dimers.

The fact is that there is profound analogy between excited molecules comprised of atoms of a noble gas and a halogen, on the one hand, and molecules comprised of atoms of an alkali metal and a halogen, on the other. The atom of a noble gas in an excited electronic state resembles an atom of the alkali metal that follows it in the Periodic Table. This atom is easily ionized because the binding energy of the excited electron is low. Owing to its high affinity of the halogen to the electron, this etcited "noble" electron easily tears off and in the collisions of the respective atoms readily jumps to a new orbit that unites the atoms, realizing in this way the so-called harpooning reaction.

The harpooning process proves so effective that it is realized not in triple collisions, as is the case with reaction (18.10), but in binary collisions, and not even with the atom of the halogen but with the molecule:

$$R^* + X_2 \rightarrow RX^* + X, \tag{18.15}$$

where R is the atom of the noble gas, and X the halogen. The reaction rate constant R reaches 10^{-11}-10^{-10} cm^3/s, which is fairly high. An important role in the overall balance of processes that lead to the formation of RX* excimer molecules is also played by the reaction

$$R^+ + X_2 + e \rightarrow RX^* + X. \tag{18.16}$$

Since halogens, primarily fluorine, are extremely active chemically, the question of the halogen-carrier has proved important. Fortunately, reactions of the (18.15) type also proceed at a high rate when for the halogen containing molecule we take not only the halogen molecule X_2 but also any other molecule, say, NF_3, SF_6, BF_3, and fluorinated hydrocarbons.

The binary nature of processes (18.15) and (18.16) leads to a situation in which excimer lasers using RX* molecules operate at noticeably lower pressures than lasers using R_2^* molecules. However, they require a buffer gas in the active mixture, usually argon or helium at a high partial pressure. In electron-beam excitation the buffer gas serves to multiply the electrons of the beam and transform their high energy into the energy corresponding to effective excitation of the molecules, while in gas-discharge excitation it serves to ensure rapid development of volume ionization and attainment of the required electron number density $n_e = 10^{14}$-10^{15} cm^{-3}. The difference in pressures and mixtures between lasers of the R_2^* and RX* types is considerable. Typical pressure values for Xe$_2$-, Kr$_2$-, and Ar$_2$-lasers are about 10^4 torr, and the active media contain no impurities. Laser mixtures for RX*-lasers contain several hundred torr, or 1.5 atm, of argon or helium, several dozen torr of the noble gas R, and several torr of a halogen or a halogen-carrier.

For the sake of reference here are the wavelengths of the radiation emitted by the most widespread lasers:

126.5 nm, Ar$_2$-laser; 145.4 nm, Kr$_2$-laser; 172.5 nm, Xe$_2$-laser; 192 nm, ArF-laser; 222.0 nm, KrCl-laser; 249.0 nm, KrF-laser; 308.0 nm, XeCl-laser; 352.0 nm, XeF-laser.

We must note the importance that KrF-, XeCl-, and XeF-lasers have gained.

We note also that the formation of a homogeneous nanosecond discharge in a gas at atmospheric pressure with an electron number density of 10^{14}-10^{15} cm^{-3} constitutes a complicated engineering problem. In developing excimer lasers based on monohalides of noble gases important use has been made of the experience gained in designing high-pressure pulsed lasers based on CO_2, CO, N_2, and H_2. It has proved possible in many cases to use already tested designs and simply replace the gas.

Problems to Lecture 18

18.1. How can the energy level diagram help to explain the high efficiency of carbon monoxide lasers?

18.2. In what respect do the mechanisms of achieving high quantum efficiencies in the carbon monoxide and carbon dioxide laser differ?

18.3. Why does a "vertical" transition of a particle appear primarily at turning points?

18.4. Use Figure 18.5 to estimate the quantum yield of the pumping process in the nitrogen laser.

18.5. Why has there been no discussion in this lecture of the depopulation of lower laser levels in the nitrogen and hydrogen lasers?

18.6. What is the width of the Lamb dip in the excimer, nitrogen, and hydrogen lasers?

Auxiliary Pumping Radiation in Systems with Many Energy Levels

Lasers using condensed media with many energy levels. The auxiliary pumping radiation method. Three- and four-level schemes. Nonradiative relaxation in a solid body. The host matrix of a solid-state laser. Ruby. Electronic configurations of atoms and ions of the transition groups. The ground states of the trivalent ions of chromium and neodymium

A feature that manifests itself as we go from gases to condensed media in general and to the solid state in particular is the sharp increase in population inversion density (up to a thousand times). Hence, if we create active media in the condensed phase, a considerably higher population inversion density is possible and, therefore, a considerably higher output. But for the solid state it does not seem possible to realize all the methods of creating active media characteristic of gases (see Lecture 13) and based in effect on the transfer of charges and particles or on organizing nonequilibrium flow in the medium.

In the case of solid dielectric media, practically the only way to act on a solid without destroying it but perhaps altering the thermodynamic equilibrium in it in a large volume is to irradiate it with light. If the solid has conductivity, electric current can be sent through it, but this will be discussed in Lectures 24 and 25, when we consider semiconductor lasers. The presence of broad absorption bands in the spectrum of a solid makes it possible to absorb large amounts of energy by irradiation, employing powerful sources of nonmonochromatic light. Hence, if generally speaking, a transparent insulator contains impurity centers that create an appropriate system of energy levels, population inversion in this system can be created only by absorption of the energy of the light incident on it. What is important in this process is the presence of at least three energy levels.

The problem comes down to the following. As we know, fairly strong resonance radiation can considerably change, up to saturation, the population distribution among the levels "linked" by the radiation. The population of the upper level increases while that of the lower decreases. If there is a third, nonresonance, level between these two levels on the energy scale, population inversion may emerge on a transition for which this third level is either the upper or the lower. Figure 19.1 illustrates this and simplifies the idea of the method of auxiliary pumping radiation used to create active media in multilevel systems. The method was suggested by Basov and Prokhorov in 1955.

Let us consider more thoroughly the level diagram depicted in Figure 19.2. We assume that the degeneracy multiplicities of levels 1, 2, and 3 are equal: $g_1 = g_2 = g_3$. In addition, we suppose that all the $h\nu_{ik}$ are much higher than kT (the optical range). Levels 1 and 2 are coupled by the radiation stimulating transitions with a probability $W_{12} = W_{21} = W$. The decay of level 2 proceeds through two channels ($2 \rightarrow 1$ and $2 \rightarrow 3$) with probabilities w_{21} and w_{23}, respectively. Level 3 decays with a probability w_{31}. Since $h\nu_{ik} \gg kT$, the thermal population of levels 2 and 3 can be ignored by setting relaxation probabilities w_{12}, w_{13}, and w_{32} equal to zero. Then the rate equations for the populations of levels 1, 2, and 3 can be written, respectively, as follows:

$$\frac{dn_1}{dt} = (n_2 - n_1)W + n_2 w_{21} + n_3 w_{31}, \qquad \frac{dn_3}{dt} = n_2 w_{23} - n_3 w_{31},$$

$$n_1 + n_2 + n_3 = N, \qquad\qquad (19.1)$$

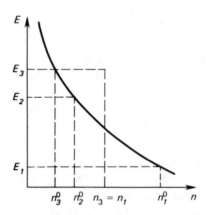

Figure 19.1. Illustration of the idea of the method of auxiliary pumping radiation in multilevel systems. The populations of levels E_3 and E_1 become equal when the $1 \rightarrow 3$ transition is saturated with radiation: population inversion has been created on the $3 \rightarrow 2$ transition.

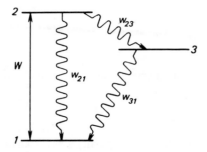

Figure 19.2. Determining the conditions for population inversion in a three-level system in the optical range. The numbering of the energy levels is the generally accepted.

where N is the number density of particles possessing the energy levels shown in Figure 19.2. In the time-independent case, where all the dn_i/dt vanish, we can easily find the following formula for the population difference:

$$n_3 - n_1 = N\frac{W(w_{23} - w_{31}) - w_{31}(w_{23} + w_{21})}{W(w_{23} + 2w_{31}) + w_{31}(w_{23} + w_{21})}. \qquad (19.2)$$

The condition for population inversion to set in on the $3 \to 1$ transition is the requirement that the numerator in (19.2) be positive, which is achieved if

$$W > \frac{w_{31}(w_{23} + w_{21})}{w_{23} - w_{31}}. \qquad (19.3)$$

This is possible only if

$$w_{23} > w_{31}. \qquad (19.4)$$

The meaning of the last condition is obvious: in the process of nonradiative transitions the upper level must get populated faster than it is depopulated. The level diagram in Figure 19.2 and the rate equations (19.1) correspond to the ruby laser. For ruby, level 3 is metastable, and the nonradiative transition probabilities w_{23}, w_{31}, and w_{21} obey the following inequalities:

$$w_{23} \gg w_{31}, \; w_{21}. \tag{19.5}$$

Then the condition for population inversion assumes the form

$$W > w_{31}. \tag{19.6}$$

The meaning of this condition is obvious: if particles are transferred from the upper, directly pumped, level to the metastable level much faster than they are removed from the latter ($w_{23} > w_{31}$), population inversion is created and maintained if the upper level is pumped faster than the metastable level is depopulated. In these conditions the population difference is given by the formula

$$n_3 - n_1 = N\frac{W - w_{31}}{W + w_{31}} \tag{19.7}$$

and tends to N as $W \to \infty$, which, of course, is not usually realized.

Since in the three-level scheme considered here population inversion is created with respect to the ground state, it takes some time for it to set in after pumping is switched on. Particles must gather on the metastable level as a result of the pumping over a finite time interval τ. To determine when the difference $n_3 - n_1$ will become positive, we must consider transient processes, that is, solve the system of differential equations (19.1). Without considering the general solution, let us turn to the ruby laser. In this case the nonradiative-transition probability w_{23} is the greater of all the rates of processes determining the distribution of particles among levels *1*, *2*, and *3*. Not only is condition (19.5) met for the ruby laser; for all reasonable pumping intensities we have

$$w_{23} > W. \tag{19.8}$$

This means that particles do not gather on the resonance level 2. There are no particles on this level since they are immediately transferred to the metastable level 3. In the laser pumping cycle shown in Figure 19.2 level 2 acts as the mediator that transfers excitation energy to the upper laser level 3, as nitrogen does in the carbon dioxide laser. Then, setting $n_2 = 0$ and, hence, assuming that $n_3 = N - n_1$, we reduce system (19.1) to the single first-order equation

$$\frac{dn_1}{dt} = -(W + w_{31})n_1 + Nw_{31}.$$ (19.9)

From this it is easy to find the equation for inversion $x(t) = n_3(t) - n_1(t) = N - 2n_1(t)$:

$$\frac{dx}{dt} = -(W + w_{31})x + (W - w_{31})N.$$ (19.10)

With the initial condition $x(0) = -N$ the solution to this equation is

$$x(t) = N\left\{\frac{W - w_{31}}{W + w_{31}} - \frac{2W\exp[-(W + w_{31})t]}{W + w_{31}}\right\},$$ (19.11)

which transforms into (19.7) as $t \to \infty$, as expected. Here and in the subsequent analysis of (19.11) we assume that pumping was switched on at time $t = 0$ and remains constant after that. The time dependence of the fractional inversion x/N is shown is Figure 19.3.

The bleaching time τ of the active transition is easily determined by setting $x(t)$ equal to zero in (19.11), which yields

$$\tau = \frac{1}{W + w_{31}}\ln\frac{2W}{W - w_{31}}.$$ (19.12)

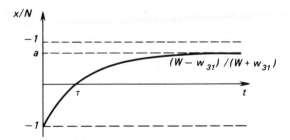

Figure 19.3. The time dependence of the fractional inversion in a three-level scheme with a pumping rate W, the pumping being switched on at $t = 0$.

In the practically unattainable case of maximum inversion ($W \gg w_{31}$, $n_3 - n_1 = N$), the product $W\tau$ is constant:

$$W\tau = \ln 2. \tag{19.13}$$

Since the probability of the pumping stimulated transitions is proportional to the pumping intensity, the above product determines the minimum energy necessary for bringing the system to a state with negative absorption. In the general case,

$$W\tau = \frac{W}{W + w_{31}} \ln \frac{2W}{W - w_{31}}. \tag{19.14}$$

The fact that $W\tau$ is constant for high pumping intensities, when all relaxation processes can be ignored, has a simple physical meaning: this product is proportional to the amount of energy necessary for transferring all the particles from the ground state to the upper laser level (see Eq. (19.7) and Figure 19.3).

Once more we note that the need to spend energy to create population inversion on the metastable level with respect to the ground state in the three-level optical system follows from the fact that at least half of all the particles must be transferred from level *1* to level *3* through level *2*. Population inversion in a three-level scheme is created with respect to the well-populated ground state. Hence the great interest in schemes in which optical pumping creates population inversion with respect to the thermally nonpopulated level, as is the case, for example, in gas lasers when excited by electron impact. Such schemes can be realized in four-level systems.

Let us write the rate equations for the population numbers in a system whose level diagram is depicted in Figure 19.4. We will take into account only downward nonradiative transitions. Then

$$\frac{dn_2}{dt} = (n_1 - n_2)W - n_2(w_{21} + w_{23} + w_{24}),$$

$$\frac{dn_3}{dt} = n_2 w_{23} - n_3(w_{31} + w_{34}), \quad \frac{dn_4}{dt} = n_2 w_{24} + n_3 w_{34} - n_4 w_{41},$$

$$n_1 + n_2 + n_3 + n_4 = N. \tag{19.15}$$

Now we consider the conditions for time-independent population inversion setting in on the $3 \rightarrow 4$ transition. Equations (19.15) have been written on

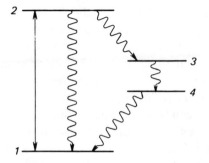

Figure 19.4. The four-level scheme.

the assumption that level *4* is not thermally populated (the same is true, therefore, for all levels lying higher). By introducing the notation $x = n_3 - n_4$ and carrying out simple algebraic transformations we find from (19.15) that x is positive if

$$w_{41} > w_{34} + \frac{w_{24}}{w_{23}} (w_{34} + w_{31}). \qquad (19.16)$$

This condition for population inversion differs greatly from that in the three-level case (see (19.3)) in that it is independent of W. But if inequality (19.16) holds true, then $n_3 - n_4$ is always positive. Usually (cf. (19.5))

$$w_{23} \gg w_{24}, \; w_{34}, \; w_{31}. \qquad (19.17)$$

Then the condition for inversion assumes the simple form

$$w_{41} > w_{34}, \qquad (19.18)$$

whose meaning is simple: the lower laser level must be depopulated through nonradiative transitions to the ground state faster than it is populated through transitions from the upper laser level. The inversion-favorable condition (19.17) means that the channel of nonradiative population of the upper laser level through $2 \rightarrow 3$ transitions is the most effective relaxation process in this level diagram.

Next we consider why there is no threshold in pumping when population inversion is created in the four-level scheme, in contrast to the three-level case. The fact is that in the transport equations (19.15) we ignored w_{14}, the probability of nonradiative population of the lower level. But this quantity is not exactly zero. If the energy of level *4*, $E_4 = h\nu_{41}$, is considerably higher than kT, the value of w_{14} is exponentially small in comparison to w_{41} (see (3.25)) but finite. In the absence of pumping there are still some particles on level *4*. These explain the inversion threshold in pumping, and the stronger the inequality $E_4 \gg kT$ the lower the threshold. Actually, this inequality is the condition of applicability of the ideas just discussed and the criterion that the system considered is a four-level. Heating may reduce a four-level system to a three-level with an accompanying sharp rise in the threshold. So, the inversion threshold in pumping is low when the lower level in a laser transition is located above the ground state by a distance of

$$\Delta E \gg kT. \tag{19.19}$$

In accordance with the above discussion, time-independent population inversion in a four-level system depends on the pumping intensity in a way greatly different from that in the three-level scheme. If we assume that condition (19.17) is met, then Eqs. (19.15) yield

$$n_3 - n_4 = N \frac{W(w_{41} - w_{34})}{W(w_{41} + 2w_{34}) + w_{23} w_{34}}. \tag{19.20}$$

We see that the extent of population inversion is proportional to the pumping intensity for $W \ll w_{23}$, which is characteristic of the real situation, and the condition for inversion (19.18), which does not depend on the pumping intensity, is obvious. All this sets the four-level scheme apart from the three-level (see (19.7)). The measure of population inversion, $n_3 - n_4$, tends to N as $W \to \infty$ and $w_{41} \gg w_{34}$.

The above examples show how the auxiliary pumping radiation, by greatly changing the population distribution among the levels involved in the auxiliary transition, creates population inversion between levels linked by nonradiative transitions to auxiliary levels. The radiation pumping method applied to multilevel systems has proved a very powerful and fairly general method for creating active media. We will refer to it repeatedly in these lectures. Here we only note that in quantum electronics of the radio-frequency range this method is successfully used to create amplifying devices with exceptionally high sensitivity (paramagnetic amplifiers, or masers).

Another important fact is worth noting. Throughout our discussion of the auxiliary pumping radiation method the nonradiative transition probabilities (rates) w_{ik} have played an important role. In Lecture 2 we introduced phenomenologically the idea of nonradiative relaxation, which is caused by the interaction of an active center with the surrounding medium and inevitably leads the nonequilibrium-excited center to equilibrium with the ambient. We noted that the concrete mechanism of such interaction depends strongly on the type of system considered. In the case of gas lasers the dominating and practically only mechanism of nonradiative excitation-energy transfer is gas-kinetic collisions. The role of collisional energy transfer was thoroughly discussed in previous lectures. In a solid, particles collisions are excluded, but there remains the interaction with phonons, or lattice vibrations. The active centers, usually ions, dissipate their excess energy in the vibrational-energy reservoir of the solid's crystal lattice, a reservoir that acts as a thermostat (thermal bath) in which the active centers are immersed. The micromechanisms of the interaction of active centers with phonons are different for different states of different centers in different crystals. But roughly the essence of the matter is the same: energy exchange takes place in the dipole-dipole interaction of some order between a given electronic or electron-vibrational state of an active center and the electronic dipoles created by the vibrations of the lattice atoms in the solid.

Usually phonon energies do not exceed 250-500 cm^{-1}. The energies of electronic states are several dozen times higher. Hence, relaxation of the energy of the active centers into lattice vibrations is a multiphonon process, that is, many phonons are created simultaneously in a single relaxation act. The probability of this process falls as the ratio of the energy of the relaxing state to the energy of the phonon with the highest possible frequency in the solid increases. (As is known, the frequency of such a phonon and the Debye temperature θ_D are connected by the simple formula $h\nu_D = k\theta_D$.) The result is that, as a rule, higher energy levels of an active center relax directly to the lattice more slowly than lower levels. As we know, this facilitates the creation of population inversion.

Besides nonradiative relaxation of the energy of an impurity center to lattice vibrations, in the discussed three- and four-level schemes described by probabilities w_{21}, w_{31}, and w_{41} an important part is played, as we have seen, by the nonradiative energy transfer from an excited level of one center to an excited level of another center (probabilities w_{23}, w_{24}, and w_{34} in the same schemes). If the gap between the levels is small compared to kT, energy transfer takes place effectively in the dipole-dipole interaction. If the gap is great, as it is the case of interest to us, binary intercenter dipole-dipole interaction alone cannot ensure effective energy transfer. Phonons, acting as a third "body", become involved in the transfer process. The interaction becomes ternary, and in the process of multiphonon relaxation the excess energy goes to lattice vibrations, heating in this way the solid as a whole.

The mentioned ratio of the energy of the relaxing state to the energy of the phonon with the highest possible frequency in the solid is determined by the ratio of the gap energy to the energy of the phonons participating in this process. We are interested here in the transfer of energy from the directly excited level to the upper laser level. The closer the latter is to the former, the smaller the ratio and the more effective the energy tansfer. It is also obvious that as the ratio decreases, a smaller share of excitation energy is spent on parasitic heating of the medium. Besides, a smaller ratio explains why in the cases discussed the probability of nonradiative energy transfer may be the highest among those in Eqs. (19.1) and (19.15), w_{ik}.

In conclusion we note that this necessarily qualitative description shows the great resemblance between the processes of nonradiative relaxation of active centers and transfer of excitation energy between the centers in solids and in gas-kinetic collisions in gases and plasmas.

Now let us turn directly to solid-state lasers. Traditionally, these are lasers whose active media is a solid insulator, a crystal or glass, into which active centers have been introduced as isomorphous impurities. But any tradition is inconsistent. A semiconducting crystal is a solid to a far greater degree than a glass, an amorphous body of the supercooled-liquid type. Nevertheless, semiconductor lasers constitute a separate class of laser systems primarily because of the special way in which their active media are pumped, while solid-state lasers are those that use dielectric crystals and glasses for their active media.

The active media of a solid-state laser is a sort of matrix containing active centers or a collection of active centers of different types as an activator impurity. There exist a number of obvious requirements on matrices of laser active elements. First, the matrix must be capable of being easily activated, that is, the active impurity must enter into the matrix easily and uniformly in controllable amounts, without destroying the optical and mechanical properties of the matrix. In addition, the values of the probabilities of the nonradiative relaxation transitions for the impurity centers introduced into the matrix must be favorable for population inversion.

Second, the matrix must be optically homogeneous and transparent for the laser (amplified) radiation and the pumping radiation. In high-power lasers the active medium is subjected to intensive laser light. A noticeable fraction of the energy of this light transforms into heat. Hence, the material of the matrix must have a high thermal conductivity, and the matrix must be heat resistant and thermo-optically stable, that is, its optical parameters must change as little as possible under heating. The requirement that the matrix be highly resistant to mechanical and chemical stresses is obvious. Also, the matrix must be optically and photochemically strong, that is, must resist the impact of the most intensive light fluxes of the continuous-wave and pulsed laser modes in the spectral ranges of pumping and lasing (amplification) radiation.

Finally, the matrix of the active element of a solid-state laser must be technologically suitable for manufacture and optical treatment.

Lasers using more than 250 types of crystals and many dozens of types of glass have been built. But a remarkable example is the ruby laser, the first laser constructed in 1960 by Theodore H. Maiman of Hughes Research Laboratories (USA). Its active substance was a solid solution α-$Al_2O_3 : Cr_2O_3$, or ruby. Pure, that is, impurity-free, crystals of the α modification of corundum Al_2O_3 are known as leucosapphire. The crystals are transparent in a broad wavelength range extending from the vacuum ultraviolet to 5-6 μm. In common ruby the concentration of chromium ions reaches several percent, which gives ruby crystals their beautiful dark red color. Laser crystals contain about 0.05% of Cr^{3+} ions (what is known as a pink ruby). The absolute concentration of chromium ions amounts to 1.6×10^{19} cm^{-3} in such crystals.

The synthesis and growth of ruby crystals is usually realized by the Verneuil method of melting the powdered mixture of Al_2O_3 and Cr_2O_3 in a high-temperature flame and then letting it crystallize on a rotating seed. A technology has been developed for growing large specimens (20-25 mm in diameter and 250-300 mm in length), and it is now possible to manufacture large specimens that are flat or of complex configuration.

Corundum has excellent mechanical, thermal, dielectric, and optical properties. It is characterized by a high thermal conductivity: at 300-400 K its thermal conductivity is only ten times lower than that of metals, and at liquid helium temperatures close to the metal value. The crystal possesses rhombohedral symmetry, and the threefold axis coincides with the optical axis of the crystal (the c axis). For the ordinary wave the refractive index of the crystal is 1.769 and for the extraordinary 1.760 (the D-line of sodium).

The first to suggest the use of ruby crystals in quantum electronics was Prokhorov (1956) in connection with masers.

In ruby crystals, that is, chrome corundum, the chromium ions are the active impurity centers. In the Al_2O_3 lattice, the Cr^{3+} ions isomorphically replace Al^{3+} ions, so that each is surrounded by six O^{2-} ions, which form an octahedron. This octahedral environment generates a strong electric field that greatly influences the energy of the Cr^{3+} ion. The more distant Al^{3+} ions have a weaker effect on this ion. We have, therefore, arrived at the important question of the spectrum of impurity ions in crystals.

The active impurities in crystals used in quantum electronics are ions of elements that belong to the transition groups. A specific feature of atoms of these groups is the presence of partially filled inner electronic shells. The Periodic Table contains five transition groups: the iron, palladium, rare-earth, platinum, and actinide.

In the main-group elements the filling of the electronic shells proceeds in strict order: for each principal quantum number the s-shell is filled first, then the p-shell. This strict order breaks down on the d-shell. Filling of the d-shell proceeds in competition with filling of the s-shell of the next principal-quantum-number value, that is, the $3d$-subshell competes with the $4s$-subshell, the $4d$ with the $5s$, the $5d$ with the $6s$. In the iron, palladium, and platinum groups the subshells being filled are the $3d$, $4d$, and $5d$, respectively. Even more complicated is the filling of the $4f$-subshell of rare-earth atoms, which competes with $5d$- and $6s$-subshell.

By way of illustration we write the electronic configurations of several elements of the iron group and rare-earth group:

$$(Ar)3d^3 4s^2 \text{ for } {}^{23}V \qquad (Xe)4f^3 6s^2 \text{ for } {}^{59}Pr,$$
$$(Ar)3d^5 4s^2 \text{ for } {}^{24}Cr, \qquad (Xe)4f^4 6s^2 \text{ for } {}^{60}Nd,$$
$$(Ar)3d^5 4s^2 \text{ for } {}^{25}Mn, \qquad (Xe)4f^5 6s^2 \text{ for } {}^{61}Pm, \qquad (19.21)$$
$$(Ar)3d^6 4s^2 \text{ for } {}^{26}Fe, \qquad (Xe)4f^6 6s^2 \text{ for } {}^{62}Sm.$$

Impurity crystals and glasses used in quantum electronics incorporate additions of elements not in the form of neutral atoms but in the form of ions. In ion formation the electronic configurations and states of ions of the transition-group elements are not constructed as simply as in series of main-group elements. Since for transition elements the strict order in filling the shells is disrupted, the ion of the subsequent element loses all resemblance, generally speaking, to the atom of the preceding. A ruby crystal, as has been said, is the solid solution $Al_2O_3 : Cr_2O_3$, that is, trivalent

Cr^{3+} ions are contained in the crystal. The configuration of the neutral atom, $(Ar)3d^54s$ provides the configuration $(Ar)3d^3$ for the trivalent chromium ion. Thus, the spectrum of Cr^{3+} is determined by three $3d$-electrons. Here it is well to consider an important circumstance.

In the transition elements of the iron group the inner $3d$-subshell become filled and is screened by the outer $4s$-subshell. The optical and chemical properties of these elements are largely determined by the filling of the $3d$-subshell and, because this is screened, are to some extent the same in situations where a neutral atom is interacting with the ambient. Trivalent ions of these atoms lose the screening subshell. As a result the spectra of these ions isomorphically introduced into the crystalline matrices are far from being similar, differ from the spectra of free ions, and differ for the same ions introduced into different matrices.

The situation is quite different for the transition elements of the rare-earth group. The optical and chemical properties of rare-earth elements are determined by the far more deeply screened $4f$-subshell. The argon-atom configuration, which is designated in (19.21) by (Ar) in the configurations of the transition elements of the iron group, is fairly simple. It is written as $1s^22s^22p^6$ and contains no electrons with principal-quantum-number values higher than those of the electrons of the $3d$-subshell being filled and of the screening $4s$-subshell. This is natural since the iron group is the first transition group (in order of increasing atomic number) in the Periodic Table. The rare-earth elements are preceded by xenon, a heavy atom with a far more complicated electronic configuration. In addition to the configurations of all the preceding noble-gas atoms (helium, neon, argon, and krypton), this configuration contains $5s$- and $5p$-electrons and can be written as $(Kr)5s^25p^6$. Hence, the subshell of the rare-earth group of elements being filled is additionally screened by two $5s$- and six $5p$-electrons. The deep screening of the $4f$-subshell explains the similarity in chemical properties of rare-earth elements and in their spectra. In the trivalent ions of rare-earth elements the $4f$-subshell remains screened by the same two $5s$- and six $5p$-electrons. As a result any external perturbation has little effect on the spectra not only of neutral atoms but also of ions of elements of this group. When trivalent rare-earth ions are introduced into different crystalline matrices, their spectra are practically the same. This is true of the highly interesting Nd^{3+} ion, whose spectrum is determined by three $4f$-electrons $((Xe)4f^3)$.

The interaction of the $3d$-electrons of the Cr^{3+} ion and the $4f$-electrons of the Nd^{3+} ion with the electric fields of their ambient in the matrices of active laser materials determines the level diagrams of the respective lasers.

The ground states of free ions are determined by the well-known Hund rule: in a given electronic configuration the state with the lowest energy is that in which the total spin angular momentum S is the greatest possible and the total orbital angular momentum $L = k(2l - k + 1)/2$, with k the number of electrons in the shell that is less than half filled, is the greatest possible for the given value of S. In the case of Cr^{3+}, that is, three $3d$-electrons, the maximum values are $S = 3/2$, $k = 3$, $l = 2$, and $L = 3(4 - 3 + 1)/2 = 3$. Since $L = 3$, the state is designated by the letter F. Its multiplicity $2S + 1$ is equal to four. This means that the quantum number J can assume four values ranging from $L + S$ to $L - S$: 9/2, 7/2, 5/2, and 3/2. If the shell is less than half filled, the ground state has $J = L - S$; otherwise, it has $J = L + S$. The d-shell may contain 10 electrons; hence, in the Cr^{3+} ion the subshell being filled is less than half filled, which means that in the ground state this ion has $J = 3/2$. Thus, the ground state of the Cr^{3+} ion can be written as

$$^4F_{3/2}. \tag{19.22}$$

Similarly, the Nd^{3+} ion, which has three of the fourteen electrons allowed in the $4f$-subshell that is being filled in the rare-earth group, has in the ground state according to Hund's rule $S = 1/2$, $L = 3(6 - 3 + 1)/2 = 6$, and $J = L - S = 9/2$, which means that its ground state is written as

$$^4I_{9/2}. \tag{19.23}$$

However, for solid-state lasers the ground states of free ions are interesting only to the extent to which they determine the nature and strength of the interaction between the ion and the crystalline fields generated by the matrix containing the ion.

Problems to Lecture 19

19.1. Transform the rate equations (19.15) to describe the case where the upper laser level becomes populated almost instantly,

$w_{41} \gg w_{34}$, and the equilibrium populations of all the levels except the lowest are negligible.

19.2. Let us consider three- and four-level systems with equal concentrations of active particles and the same population inversion number (equal to $N/2$). We also assume that $w_{31}^{(3)} = w_{34}^{(4)}$ and $w_{23}^{(3)} = w_{41}^{(4)} = w_{23}^{(4)}$ (the superscript indicates the type of system). In the three-level system $w_{23} \gg w_{21}$ and in the four-level $w_{23} \gg w_{24}$, w_{34}, w_{31}. Show that the required values of W in both systems are the same. When will $W \to \infty$?

19.3. Using the solution of the previous problem, show that, all other things being equal, the energy necessary for creating population inversion in the three-level system is $\tau^{(3)}/\tau^{(4)}$ times greater than the energy necessary for creating population inversion in the four-level system, with $\tau^{(3)}$ and $\tau^{(4)}$ the bleaching times of the lasing transition in the two systems, respectively.

19.4. Suppose that a crystal contains two types of impurity ions, and the excited energy level E_n of one type is distant from the energy level E_m of the other type by $\Delta E = E_n - E_m = 330$ cm^{-1}. Will there be an effective energy transfer with dipole-dipole interaction between the ions if the crystal is cooled to the liquid nitrogen temperature? At what temperature will the energy transfer be effective?

19.5. Usually the absorption and luminescence spectra of trivalent ions of elements belonging to the iron group are broad, often overlapping, bands whose position and shape depend greatly on the matrix into which the ions are implanted. On the other hand, the spectra of trivalent ions of rare-earth elements depend little on the type of matrix and are narrow, well-resolved lines. Explain the difference.

19.6. In the four-level scheme of a solid-state laser, pumping is carried into the absorption band centered at $\lambda = 600$ nm, with a lasing wavelength of 1.06 μm. Find the fraction of the energy that goes to Stokes' losses in the crystal.

19.7. Condition (19.19), which requires the greatest possible gap between the lower laser level and the ground level in the four-level scheme, is favorable for lowering the lasing threshold. But what will it affect negatively?

LECTURE 20

The Ruby and Neodymium Lasers

The intracrystalline field. The energy levels of chromium ions in corundum. The ruby laser. The energy levels of neodymium ions. The neodymium laser. Laser glass. Optical homogeneity and resistance to radiation

When impurity ions are introduced isomorphically into the lattice of the crystalline matrix, they are exposed to the intracrystalline field. Occupying the places of lattice ions proper, all impurity ions in the ideal crystal are in like conditions. Theoretically, the intracrystalline field is the same for all these ions, has the same strength, orientation, and symmetry pattern and perturbs all the energy levels of all the ions in the same way. The more perfect the crystal, the smaller the spread of the values of the field strength at the sites where the ions are implanted and the smaller the level broadening caused by the intracrystalline field. Obviously, spatial inhomogeneity of the field leads to inhomogeneous broadening of the lines of the corresponding transitions (see Lecture 2) for the specimen as a whole. Inhomogeneous line broadening is especially great in glass matrices.

In some crystals there are several crystallographically nonequivalent sites into which impurity ions may be implanted. In such cases there emerge subsystems of nonequivalent ions each of which has its own characteristic spectrum and for each the above reasoning holds true.

Now let us return to the example of ruby, that is, corundum with chromium.

The structure of corundum incorporates tightly packed layers of O^{2-} ions. Between the spherical ions of oxygen, O^{2-}, there are periodically located voids of two types, tetrahedral and octahedral. The Al^{3+} ions occupy only octahedral voids, filling only 2/3 of the overall number for reasons of charge compensation. An octahedral void is a cavity in which a regular octahedron can be inscribed, with the six vertices touching six spherical oxygen ions in the case considered here. In such a void the six oxygen ions generate a strong field with cubic symmetry. But there is also a small trigonal field added to the cubic field, because the fact that every third octahedral void is empty leads to a situation in which the Al^{3+} ions in the regular crystal structure are arranged in pairs in such a way that each ion has a neighboring similar ion on one side and a void on the other.

As a result the repulsive electrostatic forces acting between like ions are not compensated for and the ions move somewhat apart along a certain direction, which is the threefold axis for the tight hexagonal packing of the oxygen ions. The symmetry of the intracrystalline field acting on the aluminum ions is altered as the additional weak field of trigonal symmetry is added to the strong field of cubic symmetry. The impurity Cr^{3+} ions replace the Al^{3+} ions and are exposed to the combined field.

In view of the Stark effect the intracrystalline field lifts the degeneracy and shifts and splits the states of a free ion. Perturbation and group-theoretic methods are employed to analyze the behavior of the states of ions in fields of various strength and symmetry. A calculation conducted in the strong-field approximation for the $^4F_{3/2}$ state describing the free Cr^{3+} ion has led to results that agree with the ruby spectra observed in experiments.

For a free ion the spin angular momentum S is added to the orbital angular momentum L. This yields the total angular momentum $J = L + S$, which for a free ion (atom) is a conserved quantity. Each J-state is $(2J + 1)$-fold degenerate in energy. In the electric field of a crystal the situation changes, since in view of the Stark effect the energy levels of the ions depend on the strength and symmetry of the field.

Two limiting cases are characteristic of strong and weak fields. A weak does not break the coupling between the orbital and spin angular momenta of the ion electrons. (In other words, a weak field cannot break the LS-coupling scheme.) This means that the Stark splitting of the ion's energy levels is small compared to the fine-structure splitting (spin-orbit multiplets), and vector J is conserved. The different projections of J on the electric field's direction correspond to the different energy sublevels ($2J + 1$ in toto). The weak influence of the field can be seen as a perturbation acting on the multiplet structure, which exists without a field. This situation is realized when the impurity ions are ions of transition elements with deep-lying and well-screened subshells being filled. These may be elements of the rare-earth group, in which the $4f$-subshell is being filled, and actinides, in which the $5f$-subshell is being filled (see Lecture 19).

A strong field disrupts the LS-coupling scheme. This case is realized when the electrons of an unfilled subshell of an ion are not deep-lying, no screening subshells are present, and the interaction of electrons with the crystalline field proves to be stronger than the spin-orbit coupling. When the LS coupling breaks down, the L and S vectors precess in an external field independently of each other, without forming a unique vector of total angular momentum $\mathbf{J}$. The energies of separate sublevels are determined by the projections M_L and M_S on the field's direction. Spin-orbit coupling can be allowed for as a perturbation after we have found the splitting of levels in the crystalline field, which lifts orbital and spin degeneracy. The case of a strong field corresponds to ions of transition elements of the iron group, whose $3d$-subshell is not screened by other subshells and is exposed to the crystalline field.

What has been said illustrates once again the earlier mentioned difference between the spectra of impurity ions of rare-earth elements and impurity ions of elements of the iron group.

Let us return to ruby crystals. The ground state $^4F_{3/2}$ of the Cr^{3+} ion corresponds to sevenfold orbital ($L = 3$) and fourfold spin degeneracy. The resulting degeneracy multiplicity is $7 \times 4 = 28$. The strong cubic field splits the 28-fold degenerate state into two orbital triplets and one orbital singlet (Figure 20.1). The relatively weak trigonal component of the intracrystalline field lifts the degeneracy still further, splitting each 12-fold degenerate triplet into six doublets. The orbital singlet also splits, producing a doublet with components whose energies differ by 0.38 cm^{-1}. In this doublet the twofold spin degeneracy left after all electric fields are accounted for is lifted by an external magnetic field (Kramers doublets). This creates a system of energy levels used in electron paramagnetic resonance studies and in building masers.

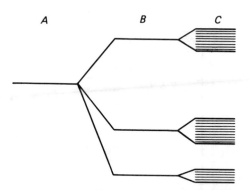

Figure 20.1. Step-by-step lifting of degeneracy by intracrystalline fields for a chromium ion; A, a free ion; B, cubic field; C, cubic and trigonal fields.

As Figure 20.1 shows, the states caused by the trigonal splitting of the orbital triplets lie considerably higher than the singlet. The lifetimes of the high-lying doublets are short (less than 100 ns), with the result that the doublets overlap, forming broad absorption bands at $\lambda = 410$ and 550 nm.

Such are the energy levels of the ground state $^4F_{3/2}$ of the Cr^{3+} ion in the cubic (and trigonal) field generated by the nearest neighbors in the corundum (Al_2O_3) lattice. The state of the free Cr^{3+} ion closest to the ground state ^4F is the ^2G state ($S = 1/2$, $L = 4$). Its 18-fold degeneracy is partially lifted in the cubic field. The ^2G level splits into four sublevels. One of these occupies a position just below the green line at $\lambda = 550$ nm of the ground state of the Cr^{3+} ion. The trigonal field and spin-orbit coupling split this sublevel into two, separated by 29 cm^{-1}. Electrodipole transitions from this doublet into the ground state are forbidden, but this restriction is partially lifted by the fact that the levels are not pure, which leads to metastability of the doublet.

As a result we have a level diagram for a Cr^{3+} ion in α-Al_2O_3 (the level diagram for ruby), which is depicted in Figure 20.2 in notation most commonly used. The transition lines are denoted in the order of increasing wave number by letters R, U, B, and Y, forming "ruby." Note that the B-,

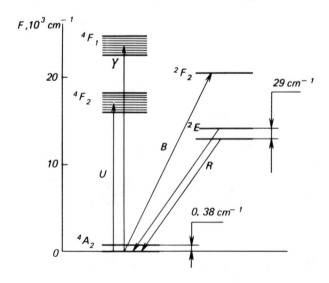

Figure 20.2. The energy level diagram of ruby.

U-, and Y-lines are observed in absorption and the R-line is the luminescence line responsible for the color of a ruby crystal. The 29-cm^{-1} splitting
of the luminescence R-line is allowed for by the notations R_1 and R_2 for
the spectral components of this luminescence doublet. At room temperature
the wavelength of the R_1-line is 694.3 nm and that of the R_2-line 692.8 nm.

According to the level diagram depicted in Figure 20.2, the ruby laser
operates by the. three-level scheme, in which level 1 corresponds to the
ground state $^4A_2$, level 2 to bands $^4F_2$ and $^4F_1$, and level 3 to doublet ^2E.
The time of nonradiative energy transfer from bands $^4F_2$ and $^4F_1$ to doublet
^2E amounts to 100 ns, and there is a rapid (100 ns) relaxation energy exchange between the doublet sublevels $\bar{E}$ and $2\bar{A}$.

If the pumping pulse is longer than 100 ns, thermal equilibrium has
time to establish itself between the sublevels of level 3. Hence, lasing is
usually observed on the R_1-line (the respective sublevel is more heavily
populated and the matrix element of the $\bar{E} \rightarrow {}^4A_2$ transition is somewhat
larger). At room temperature the amplification linewidth is about 11 cm^{-1};
at liquid nitrogen temperature the line narrows so strongly (down to 0.3
cm^{-1}) that the isotopic structure, caused by the difference in the atomic
weights of the ^{50}Cr, ^{52}Cr, ^{53}Cr, and ^{54}Cr atoms, manifests itself in the spectrum. By varying the temperature of the crystal one can change the lattice
parameters and, hence, the lasing frequency. In the -200 to $+200$ °C range
the lasing wavelength changes by $\Delta\lambda = 2$ nm.

In the pumping bands the U- and Y-absorption amounts to 2-3 cm^{-1}.
This means that the diameter of the ruby rods or the thickness of the ruby
plates must not exceed 2-2.5 cm for two-sided illumination by pumping
radiation. Otherwise, the pumping radiation cannot effectively enough penetrate the active crystal and the active medium becomes highly inhomogeneous in the cross section, which results in degradation of the parameters
of the laser light.

Spectrally the pumping bands are smooth structureless formations with
a width $\Delta\lambda \approx 100$ nm and centered at $\lambda = 410$ and 560 nm. The threshold
value of the pumping radiation density in the green band is about 3 J/cm^3.
When the threshold is exceeded considerably, the lasing energy density is
0.2-0.25 J/cm^3 in long radiation pulses (about 1 ms). The linear gain in
ruby lasers reaches 0.2-0.25 cm^{-1}.

Pumping is usually achieved by using the radiation from high-power
flashlamps. Xenon lamps are the most economical. Their efficiency reaches
50% and the effective temperature in the green region is 6500 K and that
in the violet 10 000 K. The proper design of the lamp is obviously very

important. The presence of wide U- and Y-absorption bands in the level diagram of ruby (Figure 20.2) makes it possible to employ nonmonochromatic radiation for optical pumping in the three-level scheme of obtaining population inversion. Essentially, the ruby laser transforms the luminous energy of a high-temperature source of radiation with a continuous spectrum into the energy of monochromatic radiation. This fact is of prime importance.

High-power ruby lasers are cylindrical ruby rods 2-3 cm in diameter and 20-30 cm in length. In ordinary growth conditions the c axis of the ruby crystal forms an angle of 90-60° with the rod's axis. The radiation emitted by such a crystal is linearly polarized. The field vector **E** of the radiation is perpendicular to the plane formed by the c axis and the rod's axis. Growing a crystal with zero orientation, that is, with a c axis parallel to the axis of the crystal, is possible but technologically very complicated. The radiation emitted by such zero crystals is unpolarized.

The quality of the crystal greatly determines the laser's parameters, the divergence of its radiation, the self-excitation threshold, the mode of operation, and other factors. The high thermal conductivity of ruby allows the laser to operate in the pulse periodic mode with a high repetition frequency. High-quality crystals operate at room temperature in the continuous-wave mode (pumped by the radiation emitted by mercury-vapor lamps).

All the methods of quantum electronics, such as Q-switching (giant pulses 10-100 ns long), mode locking, and power amplification, can be applied in ruby lasers.

As has been repeatedly noted earlier, the main disadvantage of the ruby laser is the three-level mechanism of its operation. This disadvantage is of a fundamental nature because it is caused by the level diagram of the trivalent chromium ion.

One of the first four-level schemes realized used the trivalent uranium ion U^{3+} in fluorite CaF_2 at 77 K. This laser ($\lambda = 2.51$ μm) gained no acceptance. More promising were found to be ions of rare-earth elements, whose unfilled 4f-subshell occupies a position closer to the nucleus than the 3d-subshell of the iron-group elements and the 5f-subshell of the actinides. As has been said, the 4f-subshell is well-screened by the 5s- and 5p-subshells. Hence, the position of levels depends little on the type of matrix. The presence of narrow luminescence lines corresponding to transitions between states of the 4f-subshell is characteristic of the spectra of rare-earth ions. Ions of rare-earth elements are also characterized by strong absorption bands associated with $4f \rightarrow 5d$ transitions suitable for pumping active media.

The bivalent ions Dy^{2+}, Tm^{2+}, and Sm^{2+}, primarily in fluorite, are active in the near IR range and the visible range. Many trivalent ions (Er, Ho, Pr, Tm, and Nd, for instance) also span these ranges.

The Nd^{3+} ion is of the greatest importance in this connection. The trivalent ion of neodymium easily activates matrices. The most promising have been found to be crystals of yttrium-aluminum garnet $Y_3Al_5O_{12}$ (YAG) and glasses. Pumping transfers Nd^{3+} ions from the ground state $^4I_{9/2}$ into several fairly narrow bands that act as the upper level. These bands are formed by a system of overlapping excited states, and their positions and widths vary somewhat from matrix to matrix. The excitation energy is rapidly transferred from the pumping bands to the metastable level $^4F_{3/2}$ (Figure 20.3). The lifetime of this level is about 0.2 ms in YAG and about 0.7 ms in glasses. The laser transition $^4F_{3/2} \rightarrow {}^4I_{11/2}$ ($\lambda = 1.06$ μm) has the highest probability. The energy gap between states $^4I_{11/2}$ and $^4I_{9/2}$, approximately 2000 cm^{-1}, ensures that the lasing of the neodymium laser proceeds according to the four-level scheme.

The closer the absorption bands to the $^4F_{3/2}$ level, the higher the lasing efficiency. The presence of a strong red absorption line is a merit of YAG crystals.

In view of the inhomogeneity of the local electrostatic fields generated by the nearest neighbors, the 1.06-μm luminescence line in glasses is strongly broadened inhomogeneously ($\Delta\lambda \approx 30$ nm). The homogeneous broadening in YAG crystals is about 0.7 nm.

The strong inhomogeneous broadening leads to a situation in which neodymium glass has a lower gain, and the respective lasers a richer mode structure, than neodymium-doped YAG. At the same time, glass allows for a larger injection (up to 6%) of active centers[4]. YAG crystals can be activated to a concentration of 1.5% in the stoichiometric substitution of Nd^{3+} ions for Y^{3+}.

Usually Nd : YAG lasers are used in a quite different way than Nd-glass lasers. In view of the high thermal conductivity and homogeneity, Nd : YAG lasers easily operate in the pulsed-periodic and continuous-wave modes.

---

[4] In lithium-lanthanum-phosphate glasses almost complete substitution of neodymium for lithium is possible. This can lead to a concentration of Nd^{3+} ions higher than (2-3) $\times 10^{21}$ cm^{-3}. The possibility of increasing the active-center concentration in crystals and glasses is discussed in Lecture 21.

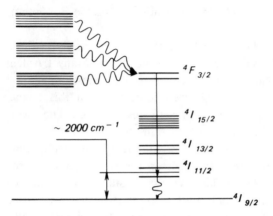

Figure 20.3. The energy level diagram of a neodymium ion.

Average power outputs of several hundred watts have been reached. On the other hand, neodymium-doped glass stores energy well because of the large volumes and higher concentration of the activator. For this reason neodymium-doped glass serves as the active media for high-energy pulsed lasers. Pulse energies of several dozen kilojoules have been reached with such lasers.

If the quality of the radiation must be high, a circuit combining a master oscillator and a power amplifier is used (MOPA technique). The master oscillator is usually an Nd : YAG laser and the power amplifier (or the final stage of the power amplifier) an Nd-glass laser.

Neodymium lasers operate in a broad range of lasing modes, from the continuous-wave mode to a substantially pulsed with pulses as short as 0.5 ps. Such short pulses are achieved by using mode locking in the broad amplification line characteristic of laser glasses.

In creating neodymium lasers, and ruby lasers, too, for that matter, all characteristic methods of controlling the parameters of laser radiation developed in quantum electronics have been used. In addition to the so-called free-running lasing, which continues practically for the entire length of the pumping pulse, wide acceptance has been gained by the Q-switching and mode-locking (self-mode-locking) methods.

In the free-running lasing mode the length of the radiation pulses is 0.1-10 ms and the energy released in the form of radiation in power amplification circuits may reach many kilojoules. The characteristic length of the Q-switching pulses is about 10 ns when optoelectronic devices are used for Q-switching. Figure 20.4 illustrates the idea of a Q-switched neodymium laser. The characteristic energy emitted by a laser of this type is 1-2 J.

Further shortening of the lasing pulses is achieved by using bleachable filters for both Q-switching (0.1-10 ns) and mode-locking (1-10 ps). Figure 20.5 gives an idea of the configuration of a self-locking laser used to generate picosecond pulses by employing a saturation filter. For the laser cavity to have only one clearly expressed period of intermode beats, the faces of the optical elements in this configuration are somewhat deflected from the normal to the cavity's optical axis and the input and output ends of the active element are positioned at the Brewster angle to this axis. Such a configuration either excludes additional reflections which the radiation may undergo in its propagation from mirror to mirror inside the laser cavity, or deflects the reflected rays from the cavity's optical axis and thus excludes the formation inside the main cavity of additional cavities with their own periods of intermode beats. The characteristic energy of a train of locked-mode pulses is about 1-2 J in lasers of this type.

In conclusion, more details about laser glass.

Glass is an excellent optical material, and its manufacture has been developed to perfection. Objects of any shape or size can be made of glass, from fibers several micrometers in diameter to disks several meters in diameter. But the main merit of glass is its high optical homogeneity. The gradient of the refractive index of good glass is $\pm(0.5\text{-}2) \times 10^{-8}$ cm^{-1}. Unfortunately, glass has a low thermal conductivity, so that pumping radiation heats it nonuniformly. Owing to the low thermal conductivity, an

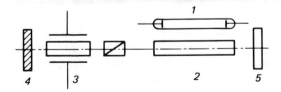

Figure 20.4. The schematic of a Q-switched laser: *1*, pumping lamp; *2*, active rod; *3*, Q-switch consisting of a Glan prism and a Poskels cell; *4*, dead-end mirror; *5*, partially transparent output mirror.

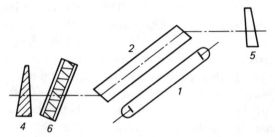

Figure 20.5. The schematic of a self-mode-locking laser (the notations are the same as in Figure 20.4). The saturation filter *6* is positioned near the left (dead-end) mirror *4*.

inhomogeneous temperature field establishes itself in glass. This leads to thermoelastic stresses, which produce optical distortions.

Since in order to maintain a high directivity of radiation the optical homogeneity of glass must be kept unaltered in the course of lasing, the thermo-optical constant

$$W = \beta + \alpha(n - 1), \tag{20.1}$$

where $\beta = dn/dT$ is the refractive-index temperature coefficient, and α is the coefficient of thermal expansion, must be as small as possible. Glass has an advantage over crystals in that there is the possibility of choosing the composition of glass and thus reducing W to values lower than 10^{-7} K^{-1} in a fairly broad temperature range (so-called athermal glass). For YAG crystals the values of W are higher by a factor of 100-1000, which, however, is partially compensated for by the substantially higher thermal conductivity of such crystals.

The resistance of the active substance of a laser to radiation is exceptionally important. Unless sufficiently resistant transparent materials are created that tolerate large fluxes of laser radiation without destruction and without violation of the optical homogeneity, it is impossible to build reliable lasers with high power outputs. In contrast to gas lasers, and primarily carbon dioxide lasers, where the weak point in the sense of optical destruction is usually the windows of the optical cells and the mirrors, the important factor for solid-state lasers is the resistance to radiation of the material of the active media.

Neodymium-doped glasses not containing metallic inclusions (platinum, iron, and like impurities of technological origin) have the highest resistance to radiation. In free-running lasing (the radiation pulse is 0.1-1 ms long) the radiation damage threshold amounts to 10^3-10^4 J/cm^2. For Q-switched pulses (1-10 ns) and the best of glasses, the threshold value of the density of radiation damaging the ends of active elements amounts to several hundred joules per square centimeter. This is true, however, for small spots (focused practically to a point). For large cross sections (about 1 cm^2) of the irradiated region, the damage threshold lowers by a factor of ten or greater primarily because of the likelihood that a defect in the material which easily initiates avalanche-like rise in damage may be within the irradiated region. For pulses shorter than 1 ns (locked modes), the damage thresholds are several joules per square centimeter.

Rubies and garnets in the Q-switched mode are destroyed at 10-30 J/cm^2.

The physical processes underlying the destruction mechanisms of solids and glasses by laser radiation are extremely diversified. Their study enters into the problem of the interaction of laser radiation with matter and is discussed in appropriate courses. Here we note only the role of the so-called self-focusing effect, the idea of which is that a strong laser field changes the refractive index of the transparent material in such a manner that a kind of effective lens is formed that increases the field density in the medium. This increase, in turn, focuses the radiation and drives its density up, which finally leads to destruction of the material.

The tendency of optical materials toward self-focusing is characterized by the nonlinear refractive index of the material, n_2. Allowing for this effect of the field, we can write the refractive index of the medium in the first approximation as

$$n = n_0 + n_2 E^2, \tag{20.2}$$

where n_0 is the linear part of the refractive index independent of the amplitude of the electric field in the laser wave, E.

The refracitve index n_2, just as the thermo-optical constant W, is an important parameter characterizing the optical properties of the active element of solid-state lasers. For the best laser glasses, $n_2 \leq 10^{-13}$ esu, and for rubies and garnets this quantity is two to four times higher.

We note once more that the technology of manufacturing high-quality optical materials is well-developed. Today there are phosphate-, borate-, beryllate-, and germanium-based neodymium-doped glasses. The choice between these must be made with an eye to the specific conditions in which the laser being designed is expected to operate.

Problems to Lecture 20

20.1. Calculate the peak output power in a pulse emitted by a ruby laser operating in the Q-switched mode if the population inversion is 5×10^{17} cm^{-3}, the cavity length 15 cm, and the losses to refraction and scattering per pass 0.06. What is the minimum rate of pumping of the upper laser level in these conditions if the level's lifetime is 2 ms?

20.2. Using the solution to Problem 19.1, find the threshold population difference, the threshold pumping power, and output lasing power in the continuous-wave mode per 1 cm^3 of an Nd^{3+}: YAG laser if the lasing wavelength λ is 1.06 μm, the lifetime of the upper laser level $^4F_{3/2}$ of neodymium 230 μs, the lasing-transition cross section σ is 4.6×10^{-19} cm^2, the lifetime of a photon in the cavity τ_{ph} is 4 ns, the cavity length l is 25 cm, and the refractive index of an Nd^{3+}: YAG crystal is 1.83.

20.3. Estimate the length of a giant pulse emitted by the ruby laser described in Problem 20.1.

20.4. The luminescence linewidth for Nd^{3+} in silicate glass is 34 nm, which is 50 times greater than the corresponding quantity for the YAG crystal. What are the negative and positive consequences of this? What causes this inhomogeneous broadening?

20.5. Estimate the required power of a lamp used to initiate lasing of an Nd-glass continuous-wave laser. The lifetime of the upper laser level is 0.5 ms and the threshold population inversion 10^{16} cm^{-3}, and the active rod has a volume of 10 cm^3. Set the efficiency of transformation of the electric energy of the lamp into the energy absorbed by the absorption bands of Nd^{3+} proper at 4%.

20.6. Suppose that in an Nd-glass laser the pumping power is considerably higher than the threshold value and that the lasing linewidth $\Delta\nu = 3$ GHz. Can lasing with a single longitudinal mode be realized in such a laser without using additional devices? What pulse length can be obtained by introducing mode locking?

LECTURE 21

Nonradiative Relaxation in Solids

The electron-phonon interaction. Weak vibronic coupling. The multiphonon relaxation probability. The vibrational spectrum of the matrix. The maximum wavelength. Examples of neodymium and erbium ions. Ion-ion interaction. Effective damping in a system of weakly coupled oscillators. The probability of energy transfer from donor to acceptor. Energy migration. Selection of optimal concentrations. Sensibilization

In earlier lectures we have repeatedly noted the important role of nonradiative relaxation in creating population inversion in lasers and masers. The concrete manifestations and mechanisms of nonradiative relaxation differ in different spectral ranges and for different active media of quantum electronics. An especially important and in many respects decisive role is played by nonradiative transitions in creating the active media of solid-state lasers (see Lecture 19). Hence, the need for a more detailed study of excitation-energy relaxation in the active media of solid-state lasers within the framework of a course of lectures on the basics of quantum electronics.[5]

Nonradiative transitions in solids, which in our course are understood to be dielectric impurity crystals and glasses, are determined by interactions of two types. The first is the electron-phonon interaction, which transfers energy between different states of a single particle (a single impurity center), and the second is the interaction between active particles (in our case the ion-ion interaction), which transfers energy between different impurity centers.

We start with the electron-phonon interaction. The general theory of electron-vibrational interaction is far from complete. Quantitative results can usually be obtained for two limiting cases, strong vibronic (i.e. vibrational-electronic) coupling and weak vibronic coupling. Coupling is considered weak (strong) if the energy of the vibrational motion of an impurity ion is small (not small) compared to that of the vibrational quantum, or phonon, which is equal to $\hbar\omega$. If the reduced ion mass is M, displacement from equilibrium position is x, and the vibration frequency is ω, an estimate of the vibration

---

[5] The author expresses his sincere gratitude to I. A. Shcherbakov for his collaboration on this lecture.

energy yields $Mx^2\omega^2$, as is known. In this case $Mx^2\omega^2/\hbar\omega$ is the dimensionless parameter whose smallness or nonsmallness determines which limiting case of the two is chosen. However, the dipole moment of the vibrating ion is very important. Its value is proportional to the displacement from the position of equilibrium, x. For this reason the square root of the energy ratio is usually extracted, and the criterion of strong vibronic coupling is written as

$$x\sqrt{M\omega/\hbar} \gg 1, \tag{21.1}$$

while the opposite case

$$x\sqrt{M\omega/\hbar} \ll 1 \tag{21.2}$$

represents the criterion of weak vibronic coupling.

Spectroscopically cases of strong and weak vibronic coupling manifest themselves differently. Strong coupling manifests a large Stokes shift between absorption and emission spectra, while for weak coupling this shift is practically extinct. For the trivalent ions of rare-earth elements, TR^{3+}, which are of principal interest to us, the optical electrons are those belonging to the $4f$-subshell. Being screened, this subshell, as we know, interacts weakly with the intracrystalline fields of the matrix and, hence, with the matrix vibrations. The vibrations of the matrix lattice have only a slight effect on the potential surfaces of the electronic states of the $4f$-subshell of an impurity ion. For this reason ion excitation in the transitions between the electronic states of the $4f$-subshell does not noticeably excite lattice vibrations. No energy is used to drive the lattice vibrations, and there is no Stokes shift between the electronic transitions in the absorption spectrum $(1 \rightarrow 2)$ and the luminescence spectrum $(2 \rightarrow 1)$ for TR^{3+} ions. Hence, nonradiative transitions between the electronic states of TR^{3+} ions should be described fairly well by weak vibronic coupling approximation, and we will restrict our discussion to this approximation.

The most general approach is based on quantum mechanics. This approach must take into account high-order perturbation theory in our case, too, in order to describe the interaction of TR^{3+} ions with the lattice caused by modulation of the static intracrystalline field by lattice vibrations. Since the energy gap between the ground and excited states of the ion, ΔE, is large compared to the energy of the phonon with the highest frequency, $\hbar\omega$ (see Lecture 19), the nonradiative transitions considered here are multiphonon. For the probability of nonradiative relaxation accompanied by the creation of p phonons in the transition between the upper (initial) state and the lower (final) state different modifications of perturbation theory produce very cumbersome ex-

pressions that incorporate the wave functions of the initial, final, and all intermediate virtual states, phonon occupation numbers, and the densities of
phonon states.

As a rule, vital information necessary for quantitative calculations is lack-
ing. A phenomenological theory of multiphonon relaxation is therefore used.
Its assumptions are:

(a) the probability of a multiphonon transition is independent of the
characteristics of the initial and final electronic states of the impurity center;

(b) the probability of a transition accompanied by the emission of p phonons is much lower than that of a transition accompanied by the emission
of $p - 1$ phonons, or

$$W^{(p)}/W^{(p-1)} = \varepsilon \ll 1; \tag{21.3}$$

(c) the small quantity ε is a characteristic of the matrix and not of an
impurity center and, therefore, is independent of p; and

(d) a nonradiative transition is accompanied by the emission of phonons
of the same frequency, so that the energy conservation law can be written
in the form

$$p\hbar\omega = \Delta E. \tag{21.4}$$

These assumptions suggest that the most energetic phonons, that is, the
phonons of the vibrational spectrum of the matrix with the highest frequencies, participate in multiphonon transitions. Remaining within the framework
of the above assumptions, we arrive at a simple formula for the dependence
of the multiphonon transition probability on the matrix temperature T and
the energy gap ΔE in the active-impurity spectrum:

$$W(\Delta E, \ T) = W_E(p)W_T(p), \tag{21.5}$$

where

$$W_T(p) = (1 - e^{-\hbar\omega/kT})^{-p} \tag{21.6}$$

gives the temperature dependence of the rate of nonradiative transitions accompanied by the emission of p phonons, and

$$W_E(p) = C^{(p)}e^{\alpha\Delta E}, \tag{21.7}$$

equal to $W(\Delta E, \ T)$ at absolute zero, gives the probability considered as a function of the energy gap ΔE. Here $C^{(p)}$ and α are the phenomenological cons-

tants characterizing the host matrix. Usually, a real situation is described fairly well qualitatively within the framework of the discussed model, although attempts to obtain quantitative information about the number and energy of the phonons participating in the relaxation process based on this approach have led to nothing.

Let us first discuss the temperature dependence of relaxation (see Eq. (21.6)).

In the presence of a certain excess of energy, the probability of a phonon of frequency ω being emitted is proportional to the number of phonons of frequency ω in the phonon mode, $n(\omega)$ (stimulated emission) plus one phonon (spontaneous emission):

$$W \propto n(\omega) + 1.$$

The phonon mode population number at temperature T is given by the Bose-Einstein distribution function

$$n(\omega, \ T) = (e^{\hbar\omega/kT} - 1)^{-1}. \tag{21.8}$$

Since in our model it is assumed that p identical phonons are emitted in a single relaxation act, the probability of the entire process is proportional to $[n(\omega, \ T) + 1]^p$, which yields formula (21.6). We note once more that $W_T(p) \to 1$ and $W(\Delta E, \ T) \to W_E(p)$ as $T \to 0$.

A weak spot in this model of multiphonon nonradiative relaxation is the single-frequency assumption (21.4). Actually, many sets of phonons with different frequencies and different multiphonon factors "cover" the gap energy ΔE, in accordance with the conditions in which vibrations are excited in one or another matrix. An important role here is played by the spectrum of possible vibrations, information on which is provided by Raman scattering.

The Raman spectra of crystals and glasses exhibit marked differences in the structure and extent of the phonon spectra of these materials. The high-frequency limit of vibrations usually lies lower in crystals than in glasses. The energy of phonons with the highest frequencies usually exceeds that of Debye phonons, $\hbar\omega_D = kT_D$ (the Debye temperature T_D for laser materials is in the vicinity of $300\,K$).

No rigorous substantiation of Eqs. (21.5)-(21.7) exists. Hence, the importance of experimental data. Usually data are obtained by observing luminescence kinetics in the selective excitation of a chosen state of the TR^{3+} impurity ion introduced into a matrix whose phonon spectrum must be known. Since the spectrum of states of TR^{3+} ions changes little from base to base, experiments of this type conducted in different matrices with different states of various ions make it possible to carry out a thorough parametric study of the dependence of the relaxation rate of an impurity center on ΔE, $\hbar\omega$, and kT.

It has been found that although different phonons may contribute to the probability of a nonradiative transition, for different ions and in different matrices excellent correspondence of relaxation rates and the high-frequency vibration limit is observed. The correspondence exists not only for the absolute value of the relaxation rate, determined by the multiphonon factor $p = \Delta E/\hbar\omega$ (where ω is the frequency corresponding to the high-frequency vibration limit), but also for the rate's temperature dependence, determined by the ratio of $\hbar\omega$ to kT and the value of p.

Let us now turn to the energy dependence (21.7). The experiments mentioned earlier have shown that in a broad range of gap energies ΔE (from 1000 to 5000 cm^{-1}) the dependence of the multiphonon relaxation rate on ΔE can be approximated fairly well by an exponential function. The absolute values of the rate are determined by the matrix and are independent of the concrete electronic states of the TR^{3+} ion, between which state ΔE is measured. As an illustration, Figure 21.1 shows the dependence of the probability of the multiphonon relaxation of Nd^{3+}, Er^{3+}, and Tm^{3+} ions in tellurate,

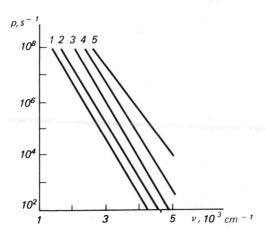

Figure 21.1. Probability of multiphonon relaxation of TR^{3+} ions vs. frequency in tellurate (*1*), germanate (*2*), silicate (*3*), phosphate (*4*), and borate (*5*) glasses.

germanate, silicate, phosphate, and borate glasses on the size of the energy gap separating the excited level and the closest lower level. Figure 21.2 shows the Raman spectra characterizing the extent of the phonon spectrum in the same glasses. There is a close resemblance between Figures 21.1 and 21.2. The reader can see that the absolute values of the multiphonon relaxation probability are the lowest in tellurium doped glass, which has the shortest vibrational spectrum. An increase in the length of the phonon spectrum leads to an increase in the rate of multiphonon relaxation because of a decrease in the multiphonon factor p, as follows from the essence of the model discussed.

The temperature-dependent part $W_T(p)$ depends weakly on p for $\hbar\omega/kT > 1$. A strong dependence on p satisfying assumption (21.3) is determined by the fact that the constant $C^{(p)}$ in (21.7) satisfies the inequality

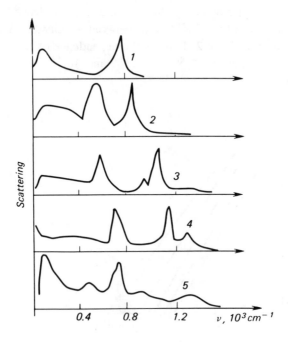

Figure 21.2. Raman spectra of tellurate (*1*), germanate (*2*), silicate (*3*), phosphate (*4*), and borate (*5*) glasses.

$C^{(p)} \ll C^{(p-1)}$. Usually $C^{(p)}/C^{(p-1)} = 0.01\text{-}0.03$, as can be determined from experimental curves similar to those depicted in Figure 21.1.

On the whole, one must conclude that Eqs. (21.5)-(21.7) correctly reflect the nature of intracenter nonradiative electron-phonon relaxation of TR^{3+} impurity ions in dielectric crystals and glasses.

It is proper now to examine the important conclusions that follow from the discussed mechanism of intracenter relaxation, formulas (21.5)-(21.7) and Figures 21.1 and 21.2.

We start with the question of the greatest possible wavelength emitted by a solid-state laser. To create population inversion effectively we must ensure that the rate of the nonradiative relaxation of the upper laser level does not exceed the rate of radiative decay of the state (see Lecture 19). As is known, the characteristic radiative lifetime of the upper laser level of TR^{3+} ions approximates 1 ms. Figure 21.1 then clearly demonstrates the impossibility of building a laser with a wavelength longer than 2-2.5 μm employing the widely used silicate and phosphate glasses, since at energy gaps smaller than 4000-5000 cm^{-1} the rate of nonradiative relaxation in these matrices begins to exceed the rate of radiative decay of the corresponding upper level. On the other hand, lasing of light with a longer wavelength is possible using crystals, which usually have a shorter phonon spectrum than glasses. One example is the lasing on the $^4I_{11/2} \to {}^4I_{13/2}$ transition in the Er^{3+} ion obtained by using CaF_2, $LiYF_4$, and $Y_3Al_5O_{12}$ in the 3-μm range. But moving to wavelengths longer than 4 μm is highly improbable.

Let us now examine the level diagram for the neodymium ion from the standpoint of the multiphonon relaxation process (see Figure 20.3).

The upper laser level $^4F_{3/2}$, of the Nd^{3+} ion is separated from the closest state (from below) $^4I_{15/2}$ by an energy gap of 6000 cm^{-1}. In accordance with the data depicted in Figure 21.1, the lifetime of multiphonon relaxation for the $^4F_{3/2}$ state proves to be much longer than the lifetime of its radiative decay. As a result the luminescence quantum yield of the upper laser level is close to 100% (in the absence of concentration quenching).

The terminal laser level $^4I_{11/2}$ of the neodymium ion is positioned above the ground state by 2000 cm^{-1}. Such an energy gap in silicate and phosphate glasses ensures a level lifetime of the order of 1 ns. Consequently, self-terminating lasing (see Lecture 14) is possible only for pulses no longer than 1 ns.

The energy gap between the upper laser level $^4F_{3/2}$ of neodymium and the main pumping bands does not exceed 2000 cm^{-1} either. Hence, the time of nonradiative relaxation from these bands to the metastable $^4F_{3/2}$ state lies in the range of 1-10 ns, with the natural lifetime of this level being 0.1-1 ms. As a result, the entire excitation energy is stored in the metastable state of neodymium.

All these conclusions have been experimentally substantiated and constitute the physical base for broad employment of the Nd^{3+} ion as the active center in laser crystals and glasses.

We now turn to the above-mentioned example of a laser that utilizes the $^4I_{11/2} \rightarrow {}^4I_{13/2}$ transition in the Er^{3+} ion. A simplified version of the respective level diagram is depicted in Figure 21.3 together with the level diagram of a neodymium ion, given for the sake of comparison. The gap between the terminal laser level $^4I_{13/2}$ and the ground state $^4I_{15/2}$ is about $6000\ cm^{-1}$, and the energy of the $^4I_{11/2} \rightarrow {}^4I_{13/2}$ laser transition lies in the $3500\text{-}cm^{-1}$ range. The multiphonon factor of nonradiative relaxation in the $^4I_{11/2} \rightarrow {}^4I_{13/2}$ transition differs considerably from that in the $^4I_{13/2} \rightarrow {}^4I_{15/2}$. In view of this the lifetime of the initial laser level $^4I_{11/2}$ is much smaller than that of the terminal level $^4I_{13/2}$, and the laser operates in the self-terminating mode, in contrast to the neodymium laser discussed earlier. The absorption bands of the pumping radiation of the erbium laser lie within the $15\ 000\text{-}25\ 000\ cm^{-1}$ range. Nonradiative transitions across the fairly broad energy gap (compared to that in neodymium) between the pumping bands and the initial laser level $^4I_{11/2}$ are considerably facilitated by the presence in the gap of many levels, which split the gap into separate regions with low multiphonon factors.

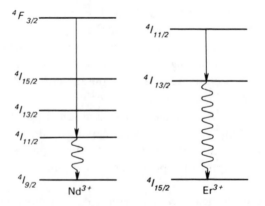

Figure 21.3. Comparison of laser energy level diagrams of Nd^{3+} and Er^{3+} ions.

The above examples demonstrate how the characteristic features of intracenter nonradiative electron-phonon relaxation of TR^{3+} impurity ions in dielectric crystals and glasses manifest themselves in the laser properties of corresponding active media.

We now turn to the ion-ion interaction. The interest in energy transfer between various impurity centers arises from a number of factors. First, as the concentration of impurity ions increases, which is always desirable because this elevates the gain and power output of the active medium, the role that energy transfer plays in energy-relaxation mechanisms that become collective increases. This, in turn, may determine the highest possible concentrations of impurities in laser crystals and glasses. Second, controllable implantation of additional impurities into the matrix may, in the process of successive acts of energy transfer and relaxation, speed up energy transfer from absorption bands to the metastable states of the upper laser levels of the active impurity centers and/or speed up the decay of their lower laser levels. Third, the presence in the matrix of additional absorbing ions capable of effectively transferring the received energy to the active ions markedly boosts the energy efficiency of active media.

The probability of nonradiative energy transfer from a donor (D) to an acceptor (A), that is, the probability of processes of the type

$$D^* + A \rightarrow D + A^*, \tag{21.9}$$

in which the excited donor returns to the ground state and the unexcited acceptor moves into the excited state, when the interaction between particles D and A separated by a distance R in a medium with a refractive index n is dipole-dipole, can be obtained by employing the first-order perturbation theory of quantum mechanics. However, a classical interpretation of the transfer process and a classical derivation of the corresponding formula are possible, too.

Within the framework of the idea of a classical oscillating electric dipole, which has been used repeatedly in these lectures, we now consider two electron dipole oscillators coupled through their electromagnetic fields. The system of equations that describes the behavior of coupled oscillators is well known:

$$\ddot{x}_1 + 2\gamma_1 \dot{x}_1 + \omega_{01}^2 x_1 = \varkappa_{21} x_2, \quad \ddot{x}_2 + 2\gamma_2 \dot{x}_2 + \omega_{02}^2 x_2 = \varkappa_{12} x_1. \tag{21.10}$$

Let the first oscillator correspond to particle D and the second to A. We also assume that the frequencies ω_{01} and ω_{02} are of the same order of magnitude and the coupling between the oscillators is weak ($\varkappa^2 = \varkappa_{12}\varkappa_{21} \ll \omega^4$). The most interesting case here is the one where the excited state of particle A rapidly relaxes, which means that the reverse transfer $A \to D$ is not present. In the classical model this situation corresponds to much heavier damping of the second oscillator ($\gamma_2 \gg \gamma_1$). The solution to Eqs. (21.10) shows then that, in accordance with the common ideas of oscillation theory, coupling between oscillators increases the damping of a high-Q oscillator.

Let us now dwell on Eqs. (21.10), which have been well-studied in oscillation theory. The solution to this system of equations was found by the German scientist Wilheim Wien at the end of the 19th century and led to such an important concept as normal oscillation frequencies (modes), which differ from the partial frequencies ω_{0i} in coupled systems with many vibrational degrees of freedom. The fractional amplitudes of oscillations at normal frequencies ω_1 and ω_2 of the first and second oscillators of system (21.10) and the energy transfer from oscillator to oscillator are determined not only by the magnitude of the coupling constant $\varkappa^2 = \varkappa_{12}\varkappa_{21}$ but also by a quantity introduced by Leonid I. Mandel'shtam (Russia) when he created the general theory of oscillations, $\chi = 2\varkappa/(\omega_{01}^2 - \omega_{02}^2)$, and depend on how close the partial frequencies are to each other.

The complete solution of system (21.10) is reduced to solving fourth-order equation and is unwieldy. However, in the weak-coupling approximation and low χ's this system can easily be solved by introducing an equivalent circuit (Figure 21.4) whose Kirchhoff equations for harmonic currents of frequency ω and amplitudes I_1 and I_2,

$$I_1(R_1 + j\omega L_1 + 1/j\omega C_1) = -j\omega M I_2,$$

$$I_2(R_2 + j\omega L_2 + 1/j\omega C_2) = -j\omega M I_1,$$

(21.11)

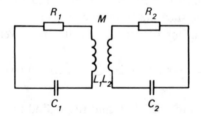

Figure 21.4. Inductively coupled oscillatory circuits.

correspond to the following differential equations:

$$q_1/C_1 + R_1\dot{q}_1 + L_1\ddot{q}_1 = -M\ddot{q}_2,$$
$$q_2/C_2 + R_2\dot{q}_2 + L_2\ddot{q}_2 = -M\ddot{q}_1. \tag{21.12}$$

Here M is the mutual inductance, q_1 and q_2 are the charges on the capacitors with capacitances C_1 and C_2, and the meaning of the other quantities is clear from Figure 21.4 (see also Eqs. (6.48)-(6.56) and Figure 6.2).

From system (21.11) we can easily proceed to an expression for the effective resistance R_{eff} of an effective single circuit formed by two inductively coupled LCR circuits:

$$R_{\text{eff}} = R_1 + R_2 \frac{\omega^2 M^2}{R_2^2 + (\omega L_2 - 1/\omega C_2)^2}. \tag{21.13}$$

Then, assuming that the χ for these circuits is low and, hence, that the first oscillator performs only a single harmonic oscillation, we can introduce, via the relation $2\gamma_{\text{eff}} = R_{\text{eff}}/L_1$, the effective damping parameter of the first circuit in the presence of coupling with the second by analogy with partial damping parameters, $2\gamma_1 = R_1/L_1$ and $2\gamma_2 = R_2/L_2$, of the corresponding circuits:

$$\gamma_{\text{eff}} = \gamma_1 + \gamma_2 \frac{\omega^2 M^2/L_1 L_2}{4\gamma_2^2 + (\omega^2 - \omega_{02}^2)^2/\omega^2}. \tag{21.14}$$

In the circuit of Figure 21.4 the coefficients of coupling of the second circuit with the first and that of the first circuit with the second are $\mu_{21} = M/L_1$ and $\mu_{12} = M/L_2$, respectively. Equations (21.10) differ from Eqs. (21.12) in the type of coupling: Eqs. (21.10) are coupled via the coordinates x_1 and x_2, and Eqs. (21.12) via the second time derivatives of the coordinates q_1 and q_2. The coupling corresponding to Eqs. (21.10) is equivalent to the one corresponding to Eqs. (21.12) if $\varkappa_{ik}x_i = \varkappa_{ik}q_i = -\mu_{ik}\ddot{q}_i$. For the harmonic oscillation $q_i = q_{0i} \exp j\omega t$ the equivalence condition assumes the form $\varkappa_{ik} = \omega^2\mu_{ik}$, which yields $M^2/L_1 L_2 = \varkappa^2/\omega^4$. Allowing for the fact that ω is close to ω_{01} and assuming that the difference $\omega_{01} - \omega_{02}$ is small, we obtain from (21.14) the following formula:

$$\gamma_{\text{eff}} = \gamma_1 + \frac{\varkappa^2}{4\omega^2\gamma_2} \frac{1}{1 + [(\omega_{01} - \omega_{02})/\gamma_2]^2}, \tag{21.15}$$

where the following trivial chain of approximate equalities is taken into account: $\omega^2 - \omega_{02}^2 \approx \omega_{01}^2 - \omega_{02}^2 \approx 2\omega(\omega_{01} - \omega_{02})$.

Let us now consider the coupling constant x^2. The energy of the electrostatic interaction of two dipoles separated by a distance R can easily be determined by referring to Figure 21.5. The dipole moments of dipoles 1 and 2 are, respectively, $p_1 = f_1 e x_1$ and $p_2 = f_2 e x_2$, where the dimensionless factors f_1 and f_2, known as oscillator strengths, characterize the effectiveness of polarization of the oscillators each formed by an optical electron (charge $-e$) and the respective ion core (charge $+e$), which are separated by a distance of x_1 and x_2, respectively. Then, in a medium with a dielectric constant ε, the interaction energy of these dipoles is

$$U_{12} = \frac{\sqrt{f_1 f_2}}{\varepsilon} \times$$

$$\left(-\frac{e^2}{R - x_1} - \frac{e^2}{R + x_2} + \frac{e^2}{R + x_2 - x_1} + \frac{e^2}{R} \right), \qquad (21.16)$$

and for $R \gg x_1, x_2$ coincides, to within the term of the order of $(x_{1,2}/R)^2$, with

$$U_{12} = -\frac{2e^2}{R^3} x_1 x_2 \frac{\sqrt{f_1 f_2}}{\varepsilon}. \qquad (21.17)$$

Hence, the force with which the second oscillator acts on the first is

$$F_{21} = -\frac{\partial U_{12}}{\partial x_1} = \frac{2e^2}{R^3} \frac{\sqrt{f_1 f_2}}{\varepsilon} x_2,$$

and the force acting from the first oscillator on the second is

$$F_{12} = \frac{2e^2}{R^3} \frac{\sqrt{f_1 f_2}}{\varepsilon} x_1.$$

Figure 21.5. Determining the strength of electrostatic interaction between two dipoles.

Since the right-hand sides of the oscillator equations (21.10) must contain the driving forces F_{12}/m and F_{21}/m, with m the electron mass, it is obvious that

$$x^2 = \varkappa_{12}\varkappa_{21} = \frac{4e^4}{m^2 R^6} \frac{f_1 f_2}{\varepsilon^2}. \tag{21.18}$$

Note that for oscillating dipoles the above calculations are valid in the near zone, that is, for $R < \lambda$, with λ the wavelength of the radiation of frequency ω in the medium considered.

The oscillator strength f is interpreted in classical dispersion theory as the fraction of elementary oscillators participating in the macroscopic polarization of the medium at frequency ω. In the quantum-mechanical approach the oscillator strength is defined not as the number of electrons of a certain type but as the number of an electron's virtual oscillators characterizing the effectiveness of dipole formation or, so to say, the extent to which a system consisting of an optical electron and the ion core can be considered a dipole at the given frequency. Quantum mechanics makes it possible to calculate the oscillator strength of the given electron system, which proves to be

$$f_k = \frac{3mc^2}{2e^2\omega^2} A_k, \tag{21.19}$$

where A_k is Einstein's coefficient for a spontaneous transition at frequency ω. Then in our case ($n^2 = \varepsilon$)

$$x^2 = \frac{9c^2}{n^4\omega^4 R^6} A_1 A_2, \tag{21.20}$$

and we get

$$\gamma_{\text{eff}} = \gamma_1 + \gamma_2, \tag{21.21}$$

with

$$\gamma_{12} = \frac{9}{4} \frac{c^6}{n^4\omega^6 R^6} A_1 A_2 \frac{1/\gamma_2}{1 + [(\omega_{01} - \omega_{02})/\gamma_2]^2}.$$

The factors $2\gamma_1$ and $2\gamma_2$ in Eqs. (21.10) have the meaning of relaxation rates for the intensity of the oscillators' vibrations. Hence, the quantity $2\gamma_{12}$, which describes the additive increase in the relaxation rate of one oscillator at the expense of a loss of energy by the other oscillator (coupled with the first), has the meaning of the probability of energy transfer from donor to acceptor:

$$W_{DA}^0 = 2\gamma_{12}.   \tag{21.22}$$

This formula was obtained for oscillators operating at fixed frequencies ω_{01} and ω_{02}. However, there are many frequencies within the spectral lines of the donor and acceptor. It is natural, therefore, to assume that the form factors of the lines, $q_D(\omega)$ and $q_A(\omega)$, give the frequency distribution of the oscillators or, in other words, the probability of finding the oscillator D or A on one or another frequency. Then the probability of the donor operating on frequency ω_{01} and the acceptor operating simultaneously on frequency $\omega_{01} + \Delta$ is equal to the product $q_D(\omega_{01})q_A(\omega_{01} + \Delta)$. The probability of energy transfer between donor and acceptor is then $q_D(\omega_{01})q_A(\omega_{01} + \Delta) W_{DA}(\Delta)$ and is determined both by the values of the partial frequencies of donor and acceptor and by the detuning between the two. Hence, the total probability of energy transfer in the $D^* \rightarrow A$ process is given by the following integral over all the frequencies ω_{01} and all the detuning values $\Delta = \omega_{02} - \omega_{01}$:

$$W_{DA} = \int\limits_0^\infty d\omega_{01} \int\limits_{-\infty}^\infty q_D(\omega_{01})q_A(\omega_{01} + \Delta) W_{DA}^{(0)}(\Delta)d\Delta.   \tag{21.23}$$

Since the width of the resonance curve of the acceptor's absorption (emission) line is considerably greater than the resonance factor $(1 + \Delta^2/\gamma^2)^{-1}$ in $W_{DA}^{(0)}$, we may assume that at all detuning values the following chain of approximate equalities holds true:

$$q_A(\omega_{01} + \Delta) \approx q_A(\omega_{01}) + \Delta dq_A/d\omega \approx q_A(\omega_{01}).$$

Then the product $q_D(\omega_{01})q_A(\omega_{01})$ does not participate in integration with respect to Δ and, using the identity

$$\int\limits_{-\infty}^\infty [1 + x^2/a^2]^{-1}dx = \pi a,$$

we arrive at the final formula

$$W_{DA} = 3\pi \frac{c^6}{n^4\omega^6 R^6} A_1 A_2 \int\limits_0^\infty q_D(\omega)q_A(\omega)d\omega,   \tag{21.24}$$

where we have allowed for the necessity of averaging over dipole orientations, which yields a factor of 2/3, and for the resonant nature of the frequency dependence of $q_D(\omega)q_A(\omega)$, which enabled taking ω^{-6} outside the integral. Note that

$$\int_0^\infty q_A(\omega)q_D(\omega)d\omega$$

is known as the overlap integral.

Formula (21.24) can be obtained, as noted earlier, within first-order perturbation theory. The fact that the final expression obtained by quantum mechanical analysis does not contain Planck's constant h makes it possible to derive and analyze this formula in a classical manner. However, writing W_{DA} in the form (21.24) makes it possible to generalize the results obtained in the dipole approximation to the case of an arbitrary multipole order, allowed for by the values of Einstein's coefficients A_1 and A_2. All quantities in (21.24) can be measured in experiments. Since $c/n\omega = \lambda/2\pi$, where λ is the wavelength in a medium with a refractive index n, and Einstein's coefficient A determines the resonant absorption cross section $\sigma = A\lambda^2/4\pi^2\Delta\nu$ (see Eq. (3.19)), we can rewrite Eq. (21.24) in the form

$$W_{DA} = 3n^2\lambda^2\sigma_D\sigma_A R^{-6}\Delta\nu_D\Delta\nu_A \int_0^\infty q_D(\omega)q_A(\omega)d\omega, \qquad (21.25)$$

where $\Delta\nu_D$ and $\Delta\nu_A$ are resonant absorption linewidths for the donor and acceptor, respectively.

We see that the rate at which energy is transferred from donor to acceptor is determined by the overlap of the spectra of these particles and the matrix elements of the respective resonant transition operators. Einstein's coefficients characterize the capacity of donor and acceptor to interact, while the overlap of the spectra ensures that such interaction is possible. The electrostatic nature of the interaction of oscillating dipoles in the near zone is expressed in the strong ($\propto R^{-6}$) dependence of the transfer rate on the distance between the dipoles. To separate the spatial dependence from the dependence determined by the properties of the interacting particles, a constant C_{DA} known as the energy transfer rate constant is often introduced into (21.25) so that the expression for W_{DA} assumes the form

$$W_{DA} = C_{DA}/R^6. \qquad (21.26)$$

A remark is in order. Equation (21.24) describes the interaction in the system consisting of a donor and an acceptor separated by a fixed distance R. Actually there is no reason to believe that distances between donor and acceptor in such systems are either fixed or equal. The distribution of such particles in a crystal, especially at low active-center concentrations in the matrix, is random and the particles lie at different distances from each other. For this reason, a macroscopic description of energy transfer through channel $D^* \to A$ requires averaging over the $W_{DA} - \varphi(W_{DA})$ distribution corresponding to the distribution $\psi(R)$ of donor-acceptor separations, the average calculated according to well-known rules of probability theory. Note that the average distance $\langle R \rangle$ between acceptors is $N_A^{-1/3}$, with N_A the acceptor number density. In accordance with what has been said, the decay of donors in time proves to be proportional to

$$\int_0^\infty \varphi(W_{DA}) \exp(-W_{DA}t) \, dW_{DA}.$$

For a Gaussian distribution of the acceptors this leads to a dependence of the following type:

$$N_D(t) \propto \exp(-\gamma/\sqrt{t}), \tag{21.27}$$

where $\gamma = (4.3)\pi^{3/2} C_{DA}^{1/2} N_A$. As a result the time dependence of the donor luminescence intensity has the form

$$I(t) = I_0 \exp[-(t/\tau_0 + \gamma\sqrt{t})], \tag{21.28}$$

with τ_0 the natural donor lifetime. Formula (21.27) has a simple physical meaning, namely, the excitation of each donor decays according to the exponential law $\exp(-W_{DA}t)$, but the W_{DA} are different for different interacting D-A pairs in view of fluctuations of R, and the sum of exponential functions is not an exponential.

In addition to energy transfer from donor to acceptor ($D^* \to A$-interaction), in real crystals the migration of excitation energy in the donor subsystem ($D^* \to D$-interaction) may be important. Since the nature of an elementary $D^* \to D$-interaction act is the same as that in the case of $D^* \to A$-interaction, the probability of the corresponding energy transfer process is characterized by a quantity C_{DD} introduced in the same manner as C_{DA} in (21.26):

$$W_{DD} = C_{DD}/R^6. \tag{21.29}$$

This quantity can be obtained from (21.24) or (21.25) by replacing the subscript A with subscript D.

By itself the $D^* \to D$-interaction does not change the population of the excited state of the donor subsystem. However, spatial displacements of the excitation change the effective interaction distances between the donor subsystem as a whole and the acceptor subsystem. Hence, migration of the excitation in the donor subsystem affects the total excitation-decay rate. This effect can be taken account of to a first approximation by introducing an additional term proportional to the migration rate W_{mig} into the exponent of the exponential function. Then, in contrast to (21.28), the time dependence of donor luminescence intensity is

$$I(t) = I_0 \exp\left[-(t/\tau_0 + \gamma\sqrt{t} + W_{mig}t)\right], \qquad (21.30)$$

where $W_{mig} = W_{mig}(C_{DD}, C_{DA}, N_D, N_A)$, with N_D the donor number density. The concrete shape of the $W_{mig}(C_{DD}, \ldots)$ function is determined by the nature of the migration processes, and further discussion of this matter would take us outside of the scope of these lectures. Suffice it to note that W_{mig} grows with C_{DD}, C_{DA}, N_D, and N_A.

Thus, the quantities C_{DD} and C_{DA}, the donor number density N_D, and the acceptor number density N_A determine the rate of decay of the excitation energy in the aggregate of interacting impurity centers. The above discussion makes it possible to resolve the questions concerning ion-ion interaction (posed at the beginning of the discussion) and to determine the optimal number densities for impurity centers of the donor and acceptor types.

Indeed, let us take, for example, a continuous-wave laser. To avoid overheating the active media, it is necessary to ensure a luminescence quantum yield close to 100%. This means that in the exponent of Eq. (21.30) the first term must be predominant:

$$1/\tau_0 \gg \gamma/\sqrt{t} + W_{mig}. \qquad (21.31)$$

Knowing $\gamma = \gamma(C_{DA}, N_A)$ and $W_{mig} = W_{mig}(C_{DD}, \ldots)$, we can select the optimum impurity-center number densities that ensure that (21.31) is valid. It is inexpedient to achieve this only by lowering N_D since this would reduce the power output. The useful (in this case) decrease in the values of C_{DD} and C_{DA} can be achieved by selecting a matrix for which the overlap integrals and corresponding oscillator strengths are small. New laser systems have been constructed on the basis of these ideas, for instance, lasers operating on neodymium pentaphosphate (NdP_5O_{14}), in which a rise in the neodymium ion concentration has been achieved up to values exceeding 10^{21} cm^{-3}, that is, more than by a factor of ten compared to values in ordinary laser crystals and glasses.

In the pulsed mode the requirements are different. Here the total excitation energy must be stored by the upper laser level of the donor subsystem in the course of the entire excitation pulse. Hence, a certain effective lifetime in relation to the decay process specified by Eq. (21.30) must exceed the pumping pulse length τ_{pump}. This means that in contrast to condition (21.31) we must have

$$1/\tau_{pump} > 1/\tau_0 + \gamma/\sqrt{\tau_{pump}} + W_{mig}. \qquad (21.32)$$

An example is the lithium-lanthanum-neodymium phosphate glass, in which the neodymium ion concentration considerably exceeds the value for ordinary laser glasses. The proper choice of the glass composition ensures that condition (21.32) is met.

Now let us go back to the properties of the 3-μm erbium laser. When studying electron-phonon relaxation we noted that the lifetime of the terminal laser level $^4I_{13/2}$ is considerably longer than the lifetime of the initial laser level $^4I_{11/2}$, which results in self-terminating lasing. Introducing appropriate acceptors, say, holmium or thulium ions, which have states in resonance with the $^4I_{13/2}$ level of erbium, may lead to a marked reduction in the lifetime of this level. If in addition one is able to reverse inequality (21.31) for the $^4I_{13/2}$ level with respect to the natural lifetime of the $^4I_{11/2}$ level, that is,

$$(\gamma/\sqrt{t} + W_{mig})_{^4I_{13/2}} \gg (1/\tau_0)_{^4I_{11/2}}, \qquad (21.33)$$

self-terminating lasing does not emerge.

Finally, energy transfer can be used to increase the population of the upper laser level. What we have in mind is the well-known phenomenon of sensibilization of luminescence, in which the energy of the exciting radiation is absorbed by one type of particles (donors) and is emitted by another type (acceptors). In this situation, naturally, acceptors are the active laser impurity. The rate of upper-level population due to the sensibilization is determined by γ and W_{mig}. Since in the sensibilization scheme donors and acceptors trade places with respect to the laser effect, inequality (21.31) reversed serves as the condition for effective continuous-wave lasing. In the pulsed mode the need for the energy to be stored by acceptors requires that the energy transfer rate be higher not only of the natural donor decay rate $1/\times \tau_D$ but also of the acceptor decay rate $1/\tau_A$. If these conditions are met, the "gain" in the population of the upper laser level is given by the ratio

$$Q = N_D \int_{\Delta\nu_D} \frac{I(\nu)}{h\nu} \sigma_D(\nu) \, d\nu \left\{ N_A \int_{\Delta\nu_A} \frac{I(\nu)}{h\nu} \sigma_A(\nu) \, d\nu \right\}^{-1}, \qquad (21.34)$$

where $I(\nu)$ is the pumping radiation spectrum, $\sigma_D(\nu)$ and $\sigma_A(\nu)$ are the absorption cross sections in donor and acceptor pumping bands, respectively, and integration is carried out in the pumping ranges $\Delta\nu_D$ and $\Delta\nu_A$. Such use of energy transfer is a promising field of research in solid-state quantum electronics.

The above discussion illustrates the important role of nonradiative relaxation and excitation-energy transfer in active media of solid-state lasers.

I would like to stress once more the profound analogy between relaxation processes in the system of impurity centers of a solid (electron-phonon and dipole-dipole interactions) and the collisional relaxation in gases (vibrational-translational, rotational-translational, and vibrational-vibrational relaxations).

The ideas discussed in this lecture have been successfully developed in the works of V. V. Osiko, A. M. Prokhorov, and I. A. Shcherbakov devoted to the creation of a new class of active media of solid-state lasers based on rare-earth gallium garnets. These crystals possess excellent optomechanical and thermal properties and can be manufactured relatively easily. Also, the rate of excitation energy transfer from donor to acceptor is higher than in YAG crystals. For the Cr^{3+}-Nd^{3+} ion pair, the rate constant C_{DA} in crystals of yttrium-aluminum, yttrium-scandium-gallium, gadolinium-scandium-gallium, and gadolinium-scandium-aluminum garnets is 9×10^{-40}, 1.4×10^{-38}, 2.2×10^{-38}, and 1.1×10^{-38} $cm^6 \cdot s^{-1}$, respectively. This has made it possible not only to create a whole new class of neodymium lasers of the pulsed, pulsed-periodic, and continuous-wave types with efficiencies severalfold higher than in the case of YAG lasers but to generate, by adding holmium and thulium ions in addition to chromium ions, 2.088-μm radiation (the $^5I_7$-$^5I_8$ transition in the Ho^{3+} ion) and for chromium and erbium ions 2.88-μm radiation. Also, crystals of these garnets have been used in lasers with continuously tunable radiation wavelengths, a topic that will be discussed in Lecture 23.

Problems to Lecture 21

21.1. Estimate the number of phonons necessary for nonradiative relaxation of the upper laser level of the Nd^{3+} ion in the gadolinium-scandium-gallium garnet (GSGG) crystal ($\omega_{max} \approx 750 \, cm^{-1}$).

21.2. Estimate the temperature at which the factor $W_T(p)$ in the expression for the probability of nonradiative relaxation, (21.5), becomes predominant. Use the data of the previous problem.

21.3. What form does $W(\Delta E, T)$ take on at high temperatures?

21.4. Use the book by Curie[6] to acquaint yourself with the quantum mechanical derivation of the formula for the probability of energy transfer from donor to acceptor. Formulate six basic conditions of the applicability of this formula. Investigate the analogy between these conditions in quantum and classical theories.

21.5. The theory of energy transfer within an aggregate of interacting particles has been successfully applied to create a new, more effective class of active media for solid-state lasers based on crystals of chromium-containing rare-earth gallium garnets activated by several types of active ions simultaneously. The effectiveness of energy transfer between particles of the same type, which is characterized by parameter C_{DA}, depends strongly on the type of matrix. Calculate and compare the values of the C_{DA} microparameters of energy transfer from chromium ions to neodymium ions in GSGG and YAG crystals with a neodymium ion concentration of $10^{20} \, cm^{-3}$ if it is known that $\gamma = 20 \, s^{-1/2}$ for YAG crystals and $\gamma = 100 \, s^{-1/2}$ for GSGG crystals.

21.6. Compare and analyze the similarities in the methods used to create population inversion in a carbon dioxide laser with He and N_2 impurities and in the active medium of a solid-state laser with several types of impurity ions.

---

[6] D. Curie, *Luminescence crystalline*, Paris: 1960.

LECTURE 22

Dye Lasers

Spectral-luminescent properties of dyes. The level diagram. The optical pumping cycle. Radiation frequency tuning. Parasitic processes. Gain. The continuous-wave and pulsed modes. Threshold pumping. Flashlamp and laser pumping. Continuous mode

A special place among condensed phase lasers is occupied by dye lasers, more precisely, lasers using solutions of organic dyes in organic solvents or in water as the active media. Such lasers have been known since 1966. Dye lasers are remarkable in that, operating in a broad range from the near IR region to the near UV inclusive, they allow continuous tuning of the lasing wavelength within a range several nanometers wide with a monochromaticity reaching 1-1.5 MHz. Dye lasers operate in the continuous-wave, pulsed, and pulsed-periodic modes. The pulse power reaches hundreds of joules, the continuous-lasing power several dozen watts, the repetition frequency several hundred hertz, and the efficiency several dozen percent (in laser pumping). In the pulsed mode the duration of lasing is determined by the length of the pumping pulses. With mode locking the length lies in the picosecond and subpicosecond ranges.

The above data refer to lasers that differ in design of cavity, pumping source, active media, etc. But the wavelength of the radiation emitted by every laser of this type can be tuned continuously, which together with any other set of laser parameters makes these sources of monochromatic radiation unique.

The properties of dye lasers are determined by the properties of their active media, organic dyes. Historically the name "dye" was attributed to organic compounds having strong colors, that is, intense and wide-band absorption in the visible part of the spectrum, that could be imparted to other materials. At present we call dyes all complex organic compounds characterized by a diversified system of conjugate chemical bonds and having strong absorption bands in the visible or near UV range. The requirement that color (or the ability to color other materials) be present is absent from this definition, so that actually the term *dye* stands for an organic compound with a certain structure and spectral-luminescent properties. Colored organic compounds contain saturated chromophor groups of the $-NO_2$, $-N=N-$, or $=CO$ type responsible for coloration. The presence of the so-called auxochrome groups of the —NH$_2$ or —OH type impart coloring properties to the compound.

What is important here is that dyes can absorb and emit radiation in the near UV and near IR ranges, just as they can in the visible spectrum. Of the many thousands of known dyes only a relatively few fluoresce in a solution. From two hundred to three hundred of these are capable of lasing, and their spectral-luminescent properties are determined by the properties of the dye and its interaction with the solvent. Although the wavelengths, width, structure, and strength of the spectra differ for different dyes and even for the same dye in different solvents, a number of general properties can be singled out that make laser dyes quite similar.

First, the absorption and emission bandwidths are about 1000 cm^{-1}. One or several additional absorption bands may lie in the spectral region corresponding to shorter wavelengths than that of the principal absorption band. Second, the fluorescence peak lies in the range of longer wavelengths than the wavelengths of the principal absorption peak (the Stokes fluorescence shift). The Stokes shift and the width of fluorescence and absorption spectra may be such that the short-wave edge of the fluorescence spectrum will overlap the long-wave edge of the absorption spectrum (Figure 22.1). Third, the fluorescence spectrum is usually a "mirror reflection" of the absorption spectrum, and the fluorescence time is usually 1 ns by the order of magnitude. Also, induced absorption spectra from excited states (excited-excited transitions) and new absorption bands of photochemical origin may (and usually do) appear.

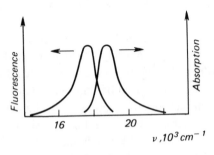

Figure 22.1. The absorption and fluorescence spectra of the rhodamine 6G dye in alcohol.

In contrast to gas lasers, whose active media consist of atoms or (fairly simple) molecules, and also to solid-state lasers operating on impurity ions, for dye lasers that employ molecules of organic dyes with molecular weight ranging from 200 to 600 and with 20 to 60 atoms to a molecule it is impossible to depict the level diagram. The wave functions of the various configurations of molecules with so many atoms and with such an exceptionally rich combination of allowed electronic, vibrational, and rotational states are unknown. The typical number of vibrational degrees of freedom of these molecules (see Lecture 15) exceeds 10^2. Hence, building the potential wells of the electronic states and the vibrational and rotational levels in them requires a space of exceptionally high dimensionality, which even if possible would be definitely unintelligible. The configuration (and, therefore, the potential energy) of a definite electronic-vibrational-rotational state of a dye molecule depends on many coordinates. For descriptive purposes it has proved useful, however, to employ a simplified representation in an arbitrary configurational space for which the entire set of configurational coordinates is replaced with a single arbitrary coordinate.

Figure 22.2 shows the energy level diagram built in this way for an organic-dye molecule. The vibrational-rotational states in this diagram are grouped near the electronic states, with a typical separation of $(1\text{-}2) \times 10^4 \, \text{cm}^{-1}$. The separation of vibrational states is $10^3 \, \text{cm}^{-1}$ by order of magnitude, and that of rotational states $1\text{-}10 \, \text{cm}^{-1}$. Depending on the orientation of the electron spins, electronic states may form either a singlet state (S) or a triplet state (T). Actually, because of spin-orbit coupling the pure triplet and singlet states mix. To characterize the arbitrariness of the level diagram of Figure 22.2, we note that the coordinates corresponding to the energy minima are different for each electronic state.

At room temperature and in equilibrium conditions the lower ($E \leqslant 200\text{-}250 \, \text{cm}^{-1}$) vibrational-rotational states of the ground state S_0 are occupied. Under optical excitation by monochromatic radiation on the $S_0 \to S_1$ transition, a certain vibrational-rotational state of S_1 becomes occupied in accordance with the Franck-Condon principle. Inside S_1 there occurs a rapid (with a relaxation time of 1-10 ps) maxwellianization of the excess energy. The radiative lifetime of the excited states of S_1 is usually 1-5 ns, whereby nonradiative relaxation within S_1 occurs faster than radiative relaxation from state S_1 on the $S_1 \to S_0$ transition.

For shifted equilibrium configurations (see Lecture 18), the uppermost vibrational levels of S_1 become excited, in accordance with the Franck-Condon principle. In the process of nonradiative relaxation the excitation energy trans-

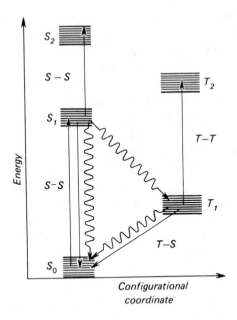

Figure 22.2. The level diagram of a dye molecule. The wavy arrows correspond to nonradiative conversion processes and the straight arrows to radiative transitions.

fers basically to the lower vibrational levels of this state. The molecule can return from the lower states of S_1 to S_0 by emitting a photon. The radiative relaxation is known as fluorescence. In view of what has been said, the energy of the emitted photon is less than that of the absorbed photon, which explains the Stokes shift in the fluorescence spectrum in relation to the absorption spectrum. The emission of fluorescence photons also takes place in accordance with the Franck-Condon principle (see Figure 22.2). The excess energy $h\nu_{pump} - h\nu_{em}$ heats the dye and solvent in the process of nonradiative vibrational relaxation.

A relaxation time of 1-10 pc is characteristic of the relaxation of vibrational energy in such a condensed medium as a liquid. The presence of a Stokes shift in the fluorescence of molecular vapor of the same magnitude as in solutions suggests an intramolecular mechanism for redistributing the excess vibrational energy. In complex molecules of the type considered here the maxwellianization of the vibrational energy may proceed through the anharmonic interactions inside vibrational modes of the molecules and between these modes.

At this point we are not interested in the mechanism of vibrational relaxation within S_1. What is important is the conclusion that for fluorescence on the $S_1 \rightarrow S_0$ transition the lower vibrational levels of state S_1 are the initial levels. No less important is where the radiative transitions $S_1 \rightarrow S_0$ end up. As has been repeatedly noted, if the equilibrium configurations of S_1 and S_0 are different, the high-lying levels of the ground state S_0 are, in accordance with the Franck-Condon principle, the lower levels in the $S_1 \rightarrow S_0$ transitions. These levels are not thermally populated. Hence, the optical pumping cycle, which incorporates nonradiative relaxation transitions in the S_1 and S_2 states in addition to S-S (singlet-singlet) absorption and S-S fluorescence, takes place according to the four-level scheme, as in the case of solid-state lasers mentioned earlier. We already know that this greatly facilitates the conditions for population inversion.

The above reasoning was carried out for the case where an $S_0 \rightarrow S_1$ transition is pumped by monochromatic radiation. From Figure 22.2 we readily see that pumping by nonmonochromatic radiation in the entire absorption of an $S_0 \rightarrow S_1$ transition changes nothing and that all previous reasoning remains valid.

In the event of fluorescence of an excited dye not placed in a cavity, that is, spontaneous emission in free space, the spectral properties of the emitted radiation are determined only by the relative position of the S_1 and S_0 states and the requirement that the Franck-Condon principle be satisfied for the equilibrium-populated vibrational levels of these states. The situation changes drastically when the excited dye is placed in a high-Q selective cavity (commonly known as a dispersive cavity). For simplicity we assume that the cavity is single-mode and single-frequency. When this cavity is tuned to a definite frequency within the fluorescence line, radiative depopulation of the appropriate upper level in S_1 occurs exactly at this frequency because of the well-known positive feedback effect. The radiatively depopulated level becomes populated in the process of intrastate maxwellianization. In view of the high rate of this process (1-10 ps), almost always the entire energy stored by the S_1 state participates in single-frequency emission. Stimulated emission in the cavity and the vibrational relaxation of excitation in the S_1 state form the radiative-relaxation channel for transferring the energy of pumping of the S_1 state into laser radiation on $S_1 \rightarrow S_0$ transitions.

Obviously, tuning the cavity within the spectral width of the fluorescence line of the dye results in tuning the frequency of the radiation. Here, up to a pulsed mode with a pulse length shorter than the relaxation lifetime in S_1 (mode locking), the energy stored by all the vibrational levels of the S_1 state is pumped into the single-frequency radiation of the tuned frequency. In a nondispersive cavity the emission takes place at the frequency corresponding to maximum gain, that is, at the peak of the fluorescence line. The similarity with rotational competition in the case of carbon dioxide lasers (see Lecture 16) is obvious.

The radiative transitions $S_1 \rightarrow S_0$ are not the only channel by which excited molecules can leave S_1. Radiative transitions between S_1 and other excited singlet states are possible. These singlet-singlet transitions may lead to absorption at the fluorescence frequency and cause losses in the dye laser that are induced by pumping and depend on the pumping intensity. The spectra of S-S absorption in dyes have been practically ignored by researchers and the extent of their effect on the lasing process is generally unknown. The obvious requirement here is that the fluorescence spectrum must not coincide with the S-S absorption spectrum, which is usually met by the dyes that have gained wide acceptance.

More dangerous in this respect is the triplet-triplet (T-T) absorption, whose spectrum usually overlaps somewhat with the fluorescence spectrum. The point is that between states of different multiplicity, say S_1 and T_1 (see Figure 22.2), nonradiative transitions (intercombination singlet-triplet conversion) may occur in the complex dye molecules. Atoms, molecules, or molecular complexes possessing strong spin-orbit coupling and entering as constituents into both dye and solvent drive up the rate of singlet-triplet conversion since they increase the strength of the interaction between singlet and triplet states of the dye and the overlap of wave functions of these states.

Singlet-triplet conversion decreases the number of molecules in the S_1 state that are capable of making the transition to state S_0 and in this way reduces the fluorescence quantum yield. Moreover, by populating the metastable state T_1 the singlet-triplet conversion makes possible triplet-triplet absorption on $T_1 \rightarrow T_2$ transitions, which increases with the dye excitation and may hinder lasing.

The triplet state T_1 is a long-lived metastable state. Radiative transitions into state S_0 are possible but have a fairly low probability. The emission on $T_1 \rightarrow S_0$ transitions is known as phosphorescence. The lifetime of state T_1 with respect to phosphorescence is 1 ms. Since the radiative lifetime cannot be influenced and more rapid depopulation of state T_1 is necessary to diminish T-T absorption, it is important to search for triplet state quenchers, which when added to the solution increase the probability of nonradiative T_1-S_0 conversion. The most effective of the known triplet state quenchers in lasers using dye solutions are oxygen and unsaturated hydrocarbons of the C_8H_8 or $C_{10}H_{12}$ type. The concentration of the quenchers must not be too high so as not to reduce the lifetime of the upper laser level.

Concluding this discussion of parasitic processes of the depopulation of state S_1, we also note the possibility of nonradiative transitions (internal conversion) between states of like multiplicity, S_1 and S_0, that lower the fluorescence quantum yield. The probability of internal conversion is determined by the structure of the dye molecule and is usually moderate if compared to the probability of fluorescent decay of state S_1.

The fluorescence quantum yield of dyes used in lasers lies between 0.01 and 1.00.

If we now proceed from the above reasoning, it is convenient to start by considering a simple two-level model containing the ground singlet state S_0 and the first excited singlet state S_1. Denoting by N_0 and N_S the total population number densities of states S_0 and S_1, respectively, and by $\sigma_{SS}(\nu)$ the cross section of the radiative transition $S_1 \rightarrow S_0$, we can write the following formula for the dye gain at frequency ν:

$$\alpha(\nu) = \left[N_S \exp \left(- \frac{E_1}{kT} \right) - N_0 \exp \left(- \frac{E_0}{kT} \right) \right] \frac{\delta E}{kT} \sigma_{SS}(\nu), \qquad (22.1)$$

where E_1 and E_0 are the energies of the initial and final groups of levels in states S_1 and S_0, respectively. The energies E_1 and E_0 are measured in relation to the minimum energy in the respective state. The radiative transition at frequency ν occurs between states S_1 and S_0. The energy interval $\delta E \ll kT$ isolates in a practically continuous energy spectrum of the electronic states a group of closely packed (actually merging) vibrational-translational states that serve as the initial and final levels in the $S_1 \rightarrow S_0$ transition (Figure 22.3).

Indeed, the continuous spectrum of an electronic state formed by the juxtaposition of many close-lying vibrational states corresponds to the case of inhomogeneous spectral-line broadening. Participating in the stimulated transition at frequency ν is that fraction of the total number of molecules excited to the state considered whose energy falls into the interval corresponding to homogeneous broadening. The number of molecules inside a certain interval

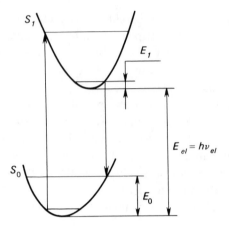

Figure 22.3. Determining the conditions for positive gain.

δE is equal to the product $\varrho(E)\delta E$, where $\varrho(E) = dN(E)/dE$ is the level density in the vicinity of energy E. In turn, the level density obeys the Boltzmann distribution, $\varrho = C \exp(-E/kT)$, where the normalization constant C is determined from the requirement that $\int_0^\infty \varrho \, dE$ be equal to the total number of molecules in the state. Applying this reasoning to states S_1 and S_0 and assuming δE and temperature T constant, we can easily find Eq. (22.1).

Equation (22.1) implies that gain is positive if

$$N_S > N_0 \exp\left(-\frac{E_0 - E_1}{kT}\right). \tag{22.2}$$

The frequency at which gain is observed, as Figure 22.3 shows, is

$$\nu = \nu_{el} - (E_0 - E_1)/h, \tag{22.3}$$

where $h\nu_{el}$ is equal to the difference in minimum energies in S_1 and S_0, and ν_{el} is known as the frequency of a purely electronic transition. If $E_0 > E_1$, gain may be positive for $N_S < N_0$, that is, in the absence of complete inversion, with ν always being lower than ν_{el}. As we see, electronic-vibrational-translational transitions realize the conditions for partial inversion in dyes, just as vibrational-rotational transitions do in carbon monoxide lasers. Note that under complete maxwellianization inside an excited singlet state, E_1 is close to zero and N_S may be notably smaller than N_0, which greatly simplifies the requirements on the pumping source.

Amplification can be achieved at frequencies greater than ν_{el} when $E_1 > E_0$, but this requires complete inversion ($N_S > N_0$), which is difficult to have because of the short lifetime of the excited singlet state S_1.

At $E_1 \approx 0$ formula (22.1) can be written in the form

$$\alpha(\nu) = \left[N_S - N_0 \exp \left(- \frac{h(\nu_{el} - \nu)}{kT} \right) \right] \frac{\delta E}{kT} \alpha_{SS}(\nu), \qquad (22.4)$$

which is convenient for analyzing the frequency dependence of $\alpha(\nu)$. The frequency spectrum of $\sigma_{SS}(\nu)$ coincides with the fluorescence spectrum, but the factor inside the square brackets in (22.4) is spectrally dependent and with N_S and N_0 fixed grows as ν decreases. Hence, the maximum in the amplification band $\alpha(\nu)$ is shifted toward the long-wave region in relation to the peak in the fluorescence line. As N_S increases, the role of the second term inside the square brackets (the frequency-dependent term) diminishes, the amplification band shifts toward the short-wave region as we move toward complete inversion ($N_S > N_0$), approaching in both position and form the fluorescence spectrum.

The two-level approximation (22.1) does not allow for singlet-singlet and triplet-triplet transitions, which lead to re-absorption of emitted light in the system of excited dye levels and, hence, a drop in gain. As noted earlier, the most dangerous is the triplet-triplet absorption. Denoting the cross section of T-T absorption by $\sigma_{TT}(\nu)$ and the population number density of the first excited triplet state T_1 by N_T, we can easily write the condition for positive gain for the case of complete intrastate maxwellianization of states S_1 and T_1:

$$N_S \sigma_{SS}(\nu) > N_T \sigma_{TT}(\nu). \qquad (22.5)$$

The triplet state becomes populated in the process of intercombination S-T conversion, which proceeds with a characteristic time τ_{ST}. In transitions to the ground singlet state the nonradiative depopulation of the triplet state (triplet state quenching) proceeds with a characteristic time τ_T. (Relaxation in the phosphorescence process can be ignored because of its low rate.) In steady-state conditions $N_S/\tau_{ST} = N_T/\tau_T$. As a result we arrive at a purely spectroscopic condition for positive gain in the continuous-wave mode:

$$\sigma_{SS}(\nu) > \sigma_{TT}(\nu)\tau_T/\tau_{ST}. \qquad (22.6)$$

This clearly shows the desirability of triplet state quenching, which lowers τ_T, and of lowering the intercombination S-T conversion probability, which increases τ_{ST}.

If condition (22.6) is not met, lasing in the cw mode is impossible. However, in the pulsed mode, under intense pumping with a steep pulse, condition (22.5) may be met for a short time in view of the difference in the rate of variation of population number densities N_S and N_T, where N_S and N_T are time-dependent if condition (22.6) is not met. A simple estimate can be done in the three-level approximation (ground singlet level and excited singlet and triplet levels) if we ignore the stimulated emission at the amplification frequency and use the following rate equations:

$$\frac{dN_S}{dt} = -(1/\tau_S + 1/\tau_{ST})N_S + P(t)N_0,$$

$$\frac{dN_T}{dt} = -N_T/\tau_T + N_S/\tau_{ST}, \qquad (22.7)$$

$$N_0 + N_S + N_T = \text{const},$$

where $P(t)$ is the rate at which particles are pumped from the ground state to the excited singlet state (this rate is proportional to the optical pumping power).

The exact solution of these equations can be obtained for any pumping pulse by using a computer. If we assume that $P(t)$ is abruptly switched on at $t = 0$ and remains constant (equal to P) and that the pumping intensity is so high that $PN_0 \gg N_S (1/\tau_S + 1/\tau_{ST})$ and yet N_0 can be assumed constant, then after switch-on N_S grows linearly in time: $N_S = PN_0 t$. In the least favorable case, where there is no triplet state quenching ($\tau_T \to \infty$), we find that after the switch-on, N_T grows in proportion to the square of time:

$$N_T = PN_0 t^2/2\tau_{ST}.$$

The integrating nature of the process of population the triplet level when the triplet state lifetime is great ($\tau_T \gg \tau_{ST}$) leads to a time interval during which triplet-triplet absorption is nonessential. If, in accordance with (22.5), we equate the product $N_S(t)\sigma_{SS}(\nu)$ with the product $N_T(t)\sigma_{TT}(\nu)$, we arrive at the following condition:

$$t \lesssim 2[\sigma_{SS}(\nu)/\sigma_{TT}(\nu)]\tau_{ST}, \qquad (22.8)$$

which provides the upper bound on the time in which amplification exists in the presence of noticeable T-T absorption. Notwithstanding its approximate nature, (22.8) describes the real situation fairly well. For a characteristic inter-combination conversion lifetime of 10-100 ns this estimate predicts a nano-second duration of lasing in dyes unable to operate in the continuous-wave mode.

Now let us turn to the question of how to estimate the pumping intensity needed to achieve positive gain in the above cases. We start with the continuous mode.

If we allow for T-T absorption and intrastate maxwellianization, formula (22.4) assumes the form

$$
\alpha(\nu) = \left\{ \left[N_S - N_0 \exp \left(-\frac{h(\nu_{\text{el}} - \nu)}{kT} \right) \right] \right.
$$
$$
\left. \times \sigma_{SS}(\nu) - N_T \sigma_{TT}(\nu) \right\} \frac{\delta E}{kT}, \tag{22.9}
$$

where we have assumed that the spectral interval of homogeneous broadening δE is the same in the singlet and triplet states. The positive-gain condition takes on the simple form

$$
N_S > N_0 \exp \left(-\frac{h(\nu_{\text{el}} - \nu)}{kT} \right) + N_T \frac{\sigma_{TT}(\nu)}{\sigma_{SS}(\nu)}. \tag{22.10}
$$

In steady-state conditions $N_S = N_0 P (1/\tau_S + 1/\tau_{ST})^{-1}$ and $N_T = N_S \tau_T / \tau_{ST}$ (see Eqs. (22.7)), and, as is known, the pumping rate is given by the formula

$$
P = I_{\text{pump}} \frac{\sigma_{SS}(\nu_{\text{pump}})}{h \nu_{\text{pump}}}, \tag{22.11}
$$

where I_{pump} is the pumping intensity, ν_{pump} the pumping frequency, and $\sigma_{SS}(\nu_{\text{pump}})$ the cross section of S-S absorption in the $S_0 \rightarrow S_1$ transition at ν_{pump}. Hence, we arrive at the following condition for positive gain at frequency ν:

$$
I_{\text{pump}} > \frac{h \nu_{\text{pump}}}{\tau_S} \left(1 + \frac{\tau_S}{\tau_{ST}} \right) \frac{\sigma_{SS}(\nu) / \sigma_{SS}(\nu_{\text{pump}})}{\sigma_{SS}(\nu) - \sigma_{TT}(\nu) \tau_T / \tau_{ST}}
$$
$$
\times \exp \left(-\frac{h(\nu_{\text{el}} - \nu)}{kT} \right). \tag{22.12}
$$

The physical meaning of this condition is clear and agrees completely with the above discussion. Usually $\tau_S \ll \tau_{ST}$. In this case, when the spectroscopic condition (22.6) is well satisfied, the above expression simplifies considerably,

$$I_{pump} > \frac{h\nu_{pump}}{\tau_S \sigma_{SS}(\nu_{pump})} \exp\left(-\frac{h(\nu_{el} - \nu)}{kT}\right), \tag{22.13}$$

and shows even more clearly that the threshold pumping intensity must exceed the intensity of saturation of the singlet-singlet $0 \to 1$ transition, greatly weakened by the Boltzmann factor owing to the four-level nature of population inversion in dye lasers. A numerical estimate strongly depends on the size of $\nu_{el} - \nu$, which usually varies between 200 and 2000 cm^{-1}. At $\tau_S = 1$ ns, $\sigma_{SS}(\nu_{pump}) = 5 \times 10^{-18}$ cm^2, $\lambda_{pump} = 0.4$ μm, and at room temperature the threshold value of I_{pump} varies between 10^3 and 10^7 W/cm^2.

In the pulsed mode with pulse lengths greater than the intrastate maxwellianization but shorter in accordance with (22.8) than the intercombination conversion time τ_{ST}, the condition for a positive gain (22.4) assumes a form different from (22.10):

$$N_S > N_0 \exp\left(-\frac{h(\nu_{el} - \nu)}{kT}\right). \tag{22.14}$$

Determining the threshold pumping energy requires solving Eqs. (22.7) and subsequent integration of the pumping rate with respect to time until condition (22.14) begins to be met owing to the energy used to pump the molecules in the $S_0 \to S_1$ transition. In the general case this is not possible, but a simple estimate can be made in the following manner.

In time intervals shorter than the lifetime of the singlet state S_1 the state becomes populated primarily by the pulsed mode in the effective two-level system $S_0 \to S_1$. According to Eq. (3.43) the saturation energy density in pulsed saturation amounts to $F_{sat} \geq h\nu/2\sigma$. The four-level nature of population inversion in dye lasers for $\nu < \nu_{el}$ and a corresponding Stokes shift leads to a Boltzmann weakening of the requirements on the number of particles transferred to the upper state. As a result we arrive at the following estimate of the pumping energy density (pump fluence) that leads to a positive gain in a time interval shorter than the lifetime of the excited singlet state:

$$F_{pump} > \frac{h\nu_{pump}}{\sigma_{SS}(\nu_{pump})} \exp\left(-\frac{h(\nu_{el} - \nu)}{kT}\right). \tag{22.15}$$

But it we are interested in time intervals greater than τ_S, then $F_{\text{pump}} = I_{\text{pump}}\Delta t$, where $\tau_S < \Delta t < t$, I_{pump} is given by (22.12) or (22.13), and in the case of dyes not obeying the spectroscopic condition (22.6) time t is limited by condition (22.8).

Note that the above discussion of the threshold values of pumping intensity (fluence) was carried out without allowing for parasitic energy losses in the laser cavity and energy losses due to useful emission from the laser cavity and had in view the so-called electronic amplification characteristic, in principle, of the active substance and the pumping mode considered here.

Without dwelling on technical details, we note that the pumping of dyes by the radiation of flashlamps has required constructing special flashlamps with high intensities and short switch-on fronts not exceeding 0.1-1 μs. The best results are achieved, however, by laser pumping. The pulsed mode uses the second harmonic of the ruby laser (0.347 μm), the second (0.53 μm), third (0.353 μm), and fourth (0.265 μm) harmonics of the neodymium-doped glass laser, and the radiation emitted by the copper-vapor, nitrogen, and excimer lasers. The latter make it possible to obtain a high repetition frequency of pulses.

What is interesting here is the possibility of pumping dyes by radiation emitted by mode-locked lasers. If the optical paths of the cavities of the pumping laser and the dye laser are equal or multiples of each other, the dye laser emits ultrashort pulses (in the picosecond range). Such short pulses of light combined with the possibility of continuously tuning the wavelength offer a unique tool for studies in the spectroscopy of fast processes, photochemistry, and photobiology.

In the continuous mode the pumping source is an argon laser. By focusing its radiation into a small spot 10-100 μm in diameter it is possible to create the required high pumping intensity. To eliminate thermo-optical distortions in the active medium of the laser and the consequences of possible photolysis of the dye, the active liquid must be pumped through at a rate so high that it is completely replaced in the active region in the course of several microseconds. The best of the known designs is depicted schematically in Figure 22.4. A thin plane-parallel jet of the solution is directed to the cavity's optical axis at the Brewster angle. Such lasers make it possible to attain exceptionally high monochromaticity and good stability.

To conclude: dye lasers with continuous pumping can successfully operate under active and passive mode locking (spike length 1-2 ps). The more common types of laser dyes are oxazine dyes, carbon-bridged dyes, xanthene dyes and their derivatives, and coumarin derivatives. The best is the xanthene

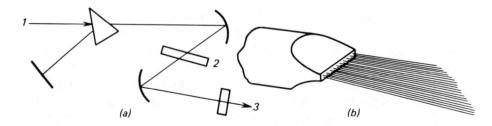

Figure 22.4. Diagram of a jet dye laser (a) and the schematic of the jet (b): *1*, pumping radiation; *2*, plane of the jet; and *3*, laser radiation.

derivative rhodamine 6G, which covers the 550-560 nm region. The best solvents are ethyl alcohol and distilled water. The dye concentration must not exceed one hundredth of the molar concentration to avoid dimer formation and concentration quenching.

Solid solutions of dyes in polymers of the polystyrene and polymethylmethacrylate types are also possible. For some applications it has proved expedient to soak in a dye solution a fine-pore monodispersed spongy glass matrix, transparent for laser radiation and pumping radiation.

The wavelength of dye-laser radiation is tuned by employing the dispersive cavities briefly discussed at the end of Lecture 10. Distributed feedback cavities whose feedback grating is formed by the coherent (laser) pumping radiation are used for dynamic control. If this radiation is split into two beams that interfere in the dye volume, gain undergoes spatial modulation. The period of the induced lattice is determined by the angle of convergence of the interfering beam which, in turn, fixes the tuning of the distributed feedback cavity. For instance, when a thin layer of the rhodamine 6G dye dissolved in a polymer film was pumped by the radiation of a helium-cadmium laser, the wavelength of the radiation produced by the resulting distributed feedback laser was found to vary between 640 and 560 nm as the angle between the interfering pumping beams was changed from 47 to 57°.

Problems to Lecture 22

22.1. What is the origin of the term $\exp\left[-h(\nu_{el} - \nu)/kT\right]$ in the formulas for the threshold pumping intensity?

22.2. Estimate the "gain" that results from using threshold intensity pumping of a dye laser owing to the choice of the four-level scheme at room temperature with $\nu_{el} - \nu \approx 200\text{-}2000 \text{ cm}^{-1}$.

22.3. By how many times will the efficiency of a laser operating on the four-level scheme be higher than that of a laser operating on the three-level scheme? Consider two cases: (a) pumping is close to the threshold value, and (b) pumping is close to saturation.

22.4. Estimate the rate of pumping of the upper laser level of a dye needed to reach complete population inversion in the continuous-wave mode for a singlet state lifetime τ_S of 1 ns.

22.5. Estimate the threshold pumping energy density per pulse for a dye restricted by intercombination conversion and excited by the radiation from a copper-vapor laser ($\lambda = 510.5 \text{ nm}$). Put $\nu_{el} - \nu$ equal to 210 cm^{-1} and σ_{SS} to $5 \times 10^{-18} \text{ cm}^2$ and assume that all processes occur at room temperature.

22.6. What must be the size of the spot created by focusing the pumping radiation from a copper-vapor laser onto the cell with the dye solution if the excess over the threshold is taken equal to 10, the average pumping power is 2 W, and the repetition frequency f_{rep} is 10 kHz?

22.7. Calculate the length of a quasiconfocal cavity that would be optimal for the laser of the previous problem.

22.8. A tunable dispersive element placed in the cavity of a dye laser (see Problem 24.6) for spectroscopic purposes selects a lasing band 5-cm^{-1} wide. (a) Estimate the length and repetition frequency of the pulses in passive mode locking. (b) The time that population inversion exists in the dye is fairly short, say 6 ns. What measures must be taken for mode locking to be stable?

LECTURE 23

Color-Center Lasers

F-centers. Methods of coloring crystals. Anion vacancies in AHC. F-, F_2-, F_2^+-, and F_2^--centers. Absorption and luminescence spectra. The optical pumping cycle. Laser parameters. Vibronic and non-phonon transitions. The chromium ion. The alexandrite laser. Gallium garnets with chromium

The broad luminescence line inherent in organic dyes ensures the possibility of continuous and broad-band tuning of the wavelength of the radiation emitted by the corresponding lasers. The simplicity of implementing spectral tuning of the laser radiation is the main merit of dye solutions as the active laser substances. Unfortunately, there are no dyes that luminesce in the IR region at wavelengths exceeding 1.0-1.5 μm. Hence the great importance of the possibility of employing ionic crystals with color centers as the active media of tunable lasers.

The luminescence of color centers (*F*-centers) is in many respects similar to that of dyes. This explains the similarity of terminology used to describe objects so different in nature as *F*-centers in crystals and organic dyes.

The luminescence spectrum of color centers is very often shifted toward the long-wave region of the spectrum in contrast to the case of organic dyes. This suggests the possibility of broadening the region of tunable lasing up to 3.0-3.5 μm.

Let us consider the structure and optical properties of *F*-centers in ionic crystals in more detail. In many cases the forbidden band of ionic crystals is wide enough for these crystals to be insulators transparent in the entire optical region of the spectrum. Examples are well known. These are the alkali halide crystals (AHC's) such as LiF, NaCl, and KCl, the alkaline earth fluorides with the fluorite structure such as CaF_2, BaF_2, and SrF_2, and the metal oxides such as corundum Al_2O_3. Defects in the crystal lattice may lead to the appearance of additional levels in the forbidden band and, hence, to the appearance in the crystal's absorption spectrum of additional absorption bands characteristic of both the crystal and the type of defect. If an absorption band lands in the optical region of the spectrum, a colorless crystal acquires a distinctive color.

The spectral features of the characteristic absorption bands do not depend on how they were obtained since the corresponding color centers are formed on the basis of the intrinsic point defects of the crystal proper. Aggregates of point defects possessing an intrinsic absorption frequency form a vast class of objects with the generic name "*F*-centers". One must bear in mind, however, that *F*-centers or, in other words, intrinsic color centers in ionic crystals represent only one type of crystal-structure defects.

Crystallography knows of many types of crystal-structure imperfections, that is, deviations from the ideal in the structure of real crystals. Point defects are imperfections in the crystal structure whose sizes in all three dimensions are commensurable with the characteristic interatomic distance in the structure. One of the simplest point defects is a vacancy, a lattice site in a crystal unoccupied by an atom or ion. In ionic crystals an anion vacancy, that is, a point defect representing the absence of a negative ion in the respective site of the crystal lattice, acts as a local effective positive charge. A free electron that happens to be near a vacancy is fixed by the vacancy's field in the place of the absent anion. In an AHC an electron localized in the vicinity of a halide (anion) vacancy of the crystal lattice forms the simplest color center possible, an intrinsic *F*-center. Figure 23.1 shows the schematic and model of an *F*-center in an AHC. This structure of an *F*-center suggests viewing the *F*-center as a hydrogenlike atom, which model describes fairly well the spectral properties of this color center.

The main methods used to create color centers in AHC's are the photochemical, the additive, and the electron-beam coloration.

In photochemical (radiative) coloration the *F*-centers form during irradiation of the crystal by UV, X-ray, or gamma radiation. A halogen ion residing in a site of the crystal's lattice absorbs a photon, loses an electron, transforms

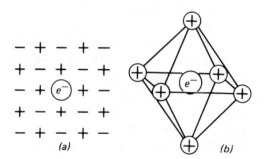

Figure 23.1. The schematic (a) and model (b) of an *F*-center in an AHC.

into a neutral atom, and transfers to the interstice, creating an anion vacancy. The liberated electron is captured by a vacancy and forms an *F*-center. The *F*-centers created in this way are characterized by low thermal and optical stability. The colored crystals are easily decolorized as a result of absorption of light in the characteristic *F*-band even at low temperatures and also when heated. The reason for this instability is that when the crystal is heated or the *F*-centers in the principal absorption band are excited, the electrons leave the vacancies and recombine with the halogen atoms displaced to the interstice. (These atoms are hole centers with strong electron affinity or, in other words, constitute a deeper electron trap.)

In additive coloration the *F*-centers form in the process of heating the specimen in an alkali metal vapor. For example, colorless NaCl and KCl crystals become bright blue in a sodium vapor. When heated, the alkali metal easily ionizes, and the liberated electrons diffuse through the crystal. When the crystal is cooled, the electrons wandering in it are localized at anion vacancies, which results in the formation of *F*-centers.

Vacancies exist in crystals with any structure and at any temperature. In crystals of stoichiometric composition in equilibrium conditions, the relative number of point defects is estimated by the formula $n/N \approx \exp(-E/kT)$, with E the activation energy for defect formation. Usually the activation energy of a vacancy is of the order of 1 eV. Hence up to the melting points of crystals the relative number density of equilibrium vacancies does not exceed 10^{-5}-10^{-4}. In the case of ionic crystals formation of point defects requires that the crystal maintain electroneutrality in its entirety. For this reason vacancies are created in pairs: either a vacancy and an appropriate interstitial ion (Frenkel pair or defect) or two vacancies of opposite signs (Schottky defect). To estimate the equilibrium concentration of such pairs the same formula can be used as the one given above, with E interpreted as the pair formation energy.

When the crystals are heated in an alkali metal vapor, the interstitial anions become bound by the metal's ions. For this reason the recombination partners leave the crystal and there is no irreversible recombination of the *F*-center electrons with the hole centers in additive-colored crystals, the corresponding *F*-centers being stable.

In electron-beam coloration the *F*-centers form in the process of bombardment of the crystal by high-energy electrons, which penetrate the crystal quite deeply and are trapped by anion vacancies.

Elementary electron-vacancy formations tend to aggregate and form pair centers and color centers of a higher order, which encompass several crystal lattice sites. Color centers of this type are denoted by F_2, F_3, F_4, etc. The structure of an F_2-center, that is, an aggregate of two F-centers in neighboring anion sites, is similar to that of a quasimolecule of hydrogen. The respective schematic drawing can be seen in Figure 23.2.

In an AHC, aggregates of F-centers are formed in moderate quantities together with single F-centers under strong additive and radiative coloration, especially under coloration by ionizing radiation. Raising the temperature of the crystal and thereby increasing the rate of migration of vacancies in the crystal, facilitates formation of associated F-centers. Note that the destruction of F-centers by light in the intrinsic absorption region of the F-band results in photochemical transformation of F-centers into F_2-, F_3-, F_4-, and higher-order centers. Generally speaking, a set of several types of color centers is usually formed during coloration.

The production of centers of a specified type requires additional, and sometimes rather complicated, photochemical and thermal treatment. Of considerable interest are such varieties of F_2-centers as F_2^+ and F_2^--centers. An F_2-center that has captured an additional electron forms an F_2^- color center, while an F_2^+-center is an ionized color center, that is, two neighboring anion vacancies holding on to a single electron. The production of F_2^+ and F_2^--centers requires step-by-step treatment of the crystals so as to create preferable conditions for the formation of F_2-centers and their subsequent ionization or, on the contrary, conditions for electron capture. The methods employed for this include X-ray or electron-beam irradiation at liquid helium temperatures, ultraviolet irradiation in the near and vacuum regions of the spectrum, and irradiation in the F_2-center absorption band. In this way it has proved possible to achieve high concentrations of centers of the required type (up to 10^{17}-10^{18} cm^{-3}).

The excited states of color centers have an energy of 1.5-3.0 eV (the optical

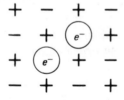

Figure 23.2. The schematic of an F_2-center in an AHC.

spectrum). As noted earlier, the absorption bands in transitions to these states lead to coloration of crystals specific for each type of center. When excited in the absorption band region, the color centers exhibit a strong luminescence in the near IR range, characterized by broad spectral bands and a lifetime of 1-1000 ns. Figure 23.3 shows (a) the absorption spectrum of electron color-centers in KCl, (b) the absorption and luminescence spectrum of F-centers

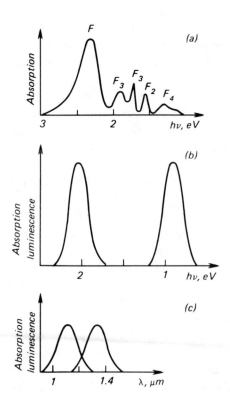

Figure 23.3. (a) The absorption spectrum of color centers in KCl, (b) the absorption and luminescence spectrum of F-centers in KBr, and (c) the absorption and luminescence spectrum of F_2^+-centers in KF.

in KBr, and (c) the absorption and luminescence spectrum of F_2^+-centers in KF. We see that the adsorption-luminescence spectra of color centers strongly resemble the spectra of organic dyes. This resemblance is especially noticeable in the case of F_2^+ and F_2^--centers.

It is proper now to discuss the question of the width of color-center absorption and luminescence spectra greatly exceeding 10^3 cm^{-1}. Residing at the sites of the crystal lattice, the color centers are strongly affected by the intracrystalline field. In contrast to impurity rare-earth ions, whose $4f$-subshell is screened and the effect of the crystalline field on which is weakened (see Lectures 20 and 21), the color centers are directly bound to the lattice. The result is that from the standpoint of relaxation characteristics and the absorption and emission bandwidth even the simplest of these, the F-center proper, cannot be considered a free quasiatom of hydrogen (and, respectively, the F_2-center cannot be considered a quasimolecule of hydrogen). Hence, an approximation in which the selection rules known for the atom (molecule) of hydrogen can be applied to the F-center (F_2-center) and the corresponding states can be classified makes it possible only to estimate fairly well the oscillator strengths of the possible transitions.

At the same time one must not forget that an electron in the neighborhood of an anion vacancy is under the influence of quasielastic forces generated by the nearest-neighbor ions (see Figures 23.1 and 23.2). The presence of these forces results in the F-center having vibrational degrees of freedom. Consequently, the structure of electron energy levels of a color center becomes more complicated owing to the presence of a system of vibrational sublevels, which makes the spectrum bandlike. The vibrational motion of an F-center in a solid is complicated. The great number of vibrational degrees of freedom results in the potential energy of an electronic-vibrational state being a function of many coordinates. Hence, as in the case of the large molecules of organic dyes, it has proved expedient for purposes of schematic description to use a simplified representation in an arbitrary configurational space in which the entire set of configurational coordinates is replaced with a single arbitrary coordinate.

Figure 23.4 shows a simplified configurational diagram of an F-center. The vibrational states that fill the potential wells of the ground and excited electronic states have a lifetime of the order of 1 ps with respect to phonon relaxation and overlap. The result is the emergence of broad continuous bands corresponding to the electronic absorption and luminescence transitions (see Figure 23.3). This explains the resemblance of the spectra discussed here to the absorption and luminescence spectra of organic dye molecules in solutions

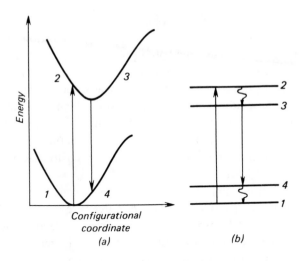

Figure 23.4. (a) Configurational diagram and (b) energy level diagram of an *F*-center.

(see Lecture 22). Note that a vibrational relaxation time of 1-10 ps is characteristic of relaxation processes in condensed media when there is strong coupling between the relaxing state and the surroundings.

The configurational diagram depicted in Figure 23.4 illustrates the principle of operation of a color-center laser. The optical pumping cycle fully resembles the cycle used in dye lasers. The same is true for the possibility of tuning the laser radiation.

The pumping radiation initiates a transition from state *1* to state *2*. For an *F*-center this is the $1S \rightarrow 1P$ transition between the states of a hydrogenlike atom with the highest probability and with which the appearance of an absorption *F*-band is associated. This is followed by a nonradiative transition in the system of the vibrational levels of the excited electronic state to the so-called relaxed excited state *3*. The transition of the color center from the excited electronic state to the ground state, $3 \rightarrow 4$, is accompanied by luminescence. The pumping cycle is completed by the rapid nonradiative transition $4 \rightarrow 1$. We see that the process is four-level. Population inversion is achieved on the $3 \rightarrow 4$ transition owing to the difference in the lifetimes of states *2*, *3*, and *4*. The vibrational relaxation of states *2* and *4* takes place in 1 ps, while the lifetime of state *3* with respect to luminescence on the $3 \rightarrow 4$ transition amounts to

10 ns. Hence, for not very high pumping intensities on the $1 \to 2$ transition, population inversion can be achieved on the $3 \to 4$ transition.

We see that color centers in ionic crystals represent a promising active medium for tunable lasers. This stems from the possibility of maintaining a four-level excitation mode, the broad character of the absorption and luminescence spectra, a considerable Stokes shift in the radiation, high oscillator strengths, and a high number density of centers (10^{17}-10^{18} cm^{-3}). The theoretical analysis of the conditions for population inversion, the operational modes, and the parasitic effects in tunable color-center lasers is similar to that developed for studying organic-dye lasers.

Strictly speaking, F-centers are not the best active centers in the lasers considered here. First, the losses inherent in F-centers are considerable, in fact, equivalent to S-T and T-T absorption in organic dyes. Second, owing to an unnecessarily high Stokes shift the wave function of the relaxed excited state 3 overlaps slightly with the wave function of the unrelaxed ground state 4 and, therefore, the oscillation strength of the $3 \to 4$ transition is low (of the order of 10^{-2}). Finally, F-centers are not sufficiently stable either thermally or with respect to irradiation in the F-band.

The best results are achieved with centers F_2, F_2^+, and F_2^- and with F_A and F_B. The last two types of color centers are F-centers in which one (in the case of an F_A-center) or two (in the case of an F_B-center) of the nearest-neighbor cations from an anion vacancy are replaced with cations of the homologous series of alkali metals (Li$^+$, K$^+$, Na$^+$, etc.) but in such a way that the ionic radius of the impurity is smaller than the ionic radius of the cation of the AHC base. These centers, characterized by high thermal and optical stability, form in the process of additive coloration of AHC's and additional irradiation in the absorption F-band of the initially obtained F-centers at an appropriately chosen temperature. Centers of these two types cannot form in LiF crystals. The best results obtained with such crystals involve lasers in the continuous-wave and pulsed modes operating on F_2^+-centers in the 0.8-1.1 μm tuning range and on F_2^--centers in the 1.1-1.3 μm. These lasers primarily use laser pumping and the gain values reach several dozen percent. The output power in the continuous-wave mode amounts to 0.1-1.0 W, and the energy of pulsed lasing reaches several dozen millijoules per pulse. As in the case of dye lasers, introduction of mode locking results in picosecond lasing with tuning in the entire amplification band. The most effective in KCl and RbCl crystals are centers F_A and F_B, which ensure tuning in the 2.25-2.90 and 2.25-3.30 μm ranges.

The luminescence spectrum of the known color centers in AHC's ranges from 0.6 to 4.0 μm. Lasers using these color centers have been built, and the tuning ranges utilized are 0.63-0.73, 0.8-1.5, and 2.25-3.30 μm. There are many types of color centers. The choice, as in the case of dyes, must be guided by the lasing and working characteristics in the required wavelength band. We also note the possibility of building color-center lasers that use not only AHC's but alkali-earth metal fluorides as well, say CaF_2.

A common feature of condensed-medium lasers is the impurity nature of their active centers immersed in a kind of matrix, a crystal, glass, or liquid. What is important in all these lasers is the four-level optical pumping cycle. Also important are the vibrational (phonon) spectrum of the matrix and the degree of screening of the electron motion in the active center. (Quantization of this motion produces the energy states used in the lasers.) The latter has a strong effect on the laser characteristics, which differ considerably in the cases of weak and strong coupling of the electron motion with the vibrational.

Indeed, in the case of TR^{3+} ions, the $4f$-subshell containing an optical electron is strongly screened and the coupling with the crystalline field is weak, with the result that the coupling with lattice vibrations is weak, too. Owing to the weak vibronic coupling, the lifetimes of the excited states are great, the lines are narrow, and the Stokes shift is absent from the luminescence spectrum (see Lecture 21). The four-level nature of the optical pumping cycle is provided by the system of levels of a free ion whose position changes little from matrix to matrix.

In the case of color centers and dye molecules, the field of the nearest-neighbor surroundings directly affects the electron motion, which means that the coupling with the vibrations of the lattice, glass, or solvent is strong. Owing to this strong coupling with the vibrations of the condensed medium, the lifetimes of the excited states are small, the lines are broad, and the Stokes shift in the luminescence spectrum is considerable. The four-level nature of the optical pumping cycle is provided by the rapid relaxation of the electronic states over the vibrational sublevels (these states are strongly coupled with the vibrations) and by the fact that the transitions between electronic states obey the Franck-Condon principle.

The ions of transition metals belonging to the iron group, the first in the order of the increasing atomic number of elements in the Periodic Table, contain unscreened $3d$-electrons in the unfilled subshells (see Lecture 19). Hence, when the ions are introduced into the crystalline matrix, they may be noticeably influenced by the crystalline field and, hence, the lattice vibrations. As a result there appears a vibrational structure in the electron transition spectrum. The transitions become electronic-vibrational, and not only the electron impurity-ion energies change simultaneously but the vibrational mode energy of the crystal does as well. The luminescence spectra of such crystals at moderate vibronic coupling contain both narrow lines corresponding to purely electronic transitions and broad bands corresponding to the vibronic structure of the impurity-center spectrum of states and originating in the electron-phonon coupling.

The intensity and shape of the vibronic bands are determined by the nature and strength of the electron-phonon coupling or, in other words, by the individual properties of the electronic state of the impurity center, the strength and symmetry of the intracrystalline field, and the phonon spectrum of the matrix. The broad bands of electron-phonon transitions demonstrate a noticeable Stokes shift in relation to the narrow lines of purely electronic transitions. In this situation it becomes possible to build a four-level laser operating on vibronic transitions in a broad frequency band.

Such lasers with a sufficiently broad tuning range in the near IR have indeed been realized on the basis of V^{2+}, Ni^{2+}, and Co^{2+} ions, which belong to the iron group, in MgF_2 and ZnF_2 matrices. Unfortunately, these lasers operate only at low (liquid nitrogen) temperatures, and reference to such lasers would be inappropriate in this course if not for the discovery of the alexandrite laser, that is, a laser operating on chromic chrysoberyl $BeAl_2O_4:Cr^{3+}$ at room and higher temperatures.

Chromium ions, which belong to the same iron group, are well known as the impurity centers of the active media of solid-state lasers. Suffice it to recall that the first laser radiation was generated by a laser that used as the active medium a ruby crystal, which is chromic corundum (sapphire) $Al_2O_3:Cr^{3+}$. Lasing in the ruby laser takes place, as is known, at room temperature and follows the three-level scheme on a nonphonon line corresponding to the transition between the excited metastable state ^2E and the ground state $^4A_2$ (see Figure 20.2). At 77 K lasing has been realized on the same transition with crystals of the $Y_3Al_5O_{12}$ garnet. Thus, alexandrite is the third crystal, after ruby and garnet, in which chromium ions act in the lasing process, but, in contrast to the first two examples, according to the four-level scheme. Below we give the main features of the alexandrite laser.

The chrysoberyl crystal $BeAl_2O_4$ is orthorhombic. The Al^{3+} ions in it occupy two nonequivalent octahedral positions, one with mirror symmetry and the other with inversion symmetry. The total number density of octahedral positions of Al^{3+} is 3.5×10^{22} cm^{-3}. Just as ruby is α-corundum Al_2O_3 in which a fraction of the Al^{3+} ions is replaced isomorphically with Cr^{3+} ions, alexandrite is chrysoberyl $BeAl_2O_4$ in which a fraction of the Al^{3+} ions is replaced isomorphically with Cr^{3+} ions.

Alexandrite is a high-temperature (with a melting point of 1870 °C), dense (3.69 g/cm^3), and durable crystal. Its thermal conductivity is twice as great as that of garnet and only twice as low as that of sapphire. It expands moderately and almost isotropically as the temperature grows. Both the mechanical and thermal properties of the crystal satisfy the requirements imposed on laser materials fairly well, and its optical properties are also satisfactory. Alexandrite crystals possess fine optical properties up to chromium ion concentration numbers of 0.4-0.5 in relation to the concentration of Al^{3+}. Optical losses within the range of lasing frequencies are about 0.003 cm^{-1}. The optical strength of alexandrite is the same as that of ruby. Alexandrite is a biaxial crystal with a refractive index of about 1.75. Crystals several centimeters long are grown by the Czochralski process (crystal pulling from a melt). A serious shortcoming of alexandrite is the complexity of the technological process for its preparation owing to the presence of beryllium and its oxide, beryllia, which are toxic substances.

As in ruby and garnet, chromium ions in alexandrite are in the octahedral crystalline surroundings, and their absorption and luminescence spectral lines have been clearly identified. In the luminescence spectrum there are two well-resolved R-lines and broad ^2T and ^4T bands. The strength and relative position of these transitions change from matrix to matrix in view of the repeatedly mentioned unscreened nature of the $3d$-electrons of the chromium ion. Figure 23.5 shows the luminescence spectrum of alexandrite. An interesting feature of this spectrum is the presence, alongside a broad band, of a narrow line that sets this spectrum and similar spectra of ions of transition metals apart

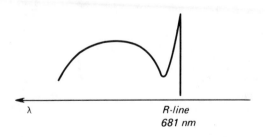

Figure 23.5. The luminescence spectrum of alexandrite.

from the luminescence spectra of color centers and organic dyes. The broad luminescence band corresponds to the vibronic transitions. The luminescence radiation is linearly polarized, with vector $\mathbf{E}$ parallel to the b axis of the crystal. The absorption spectrum is the same for both polarizations and spans the entire visual range. Of the two nonequivalent positions of Cr^{3+} in the $BeAl_2O_4$ crystal lattice the one with mirror symmetry is responsible for emission in the vibronic band.

Figure 23.6 depicts the level diagram or, more precisely, the configurational diagram of the chromium ions in alexandrite. The electronic-vibrational $^4T_2 \rightarrow {}^4A_2$ transition is responsible for the wide-band luminescence and tunable lasing. We see that the four-level nature of the optical pumping cycle arises from the presence of a Stokes shift of the $^4T_2$ state. A distinctive feature of the alexandrite crystal is the essentially smaller, compared to ruby and garnet, energy gap between states ^2E and $^4T_2$ (about 800 cm^{-1}). The maxwellianization time inside these states is very short (about 0.1 ps) and their natural lifetimes are 1.5 ms and 6 μs, respectively. The long lifetime of the ^2E state contributes to particle build-up in the system of the upper laser levels, while

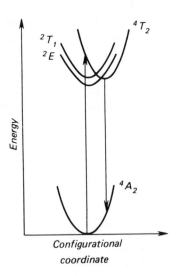

Figure 23.6. The level diagram for a chromium ion in alexandrite.

the short lifetime of the $^4T_2$ state and, hence, the large value of the $^4T_2 \rightarrow {}^4A_2$ transition cross section guarantee the high value of gain in the vibronic transitions. The mutual position of the $^4T_2$ and ^2E states characteristic of alexandrite leads to an increase in gain in the vibronic spectrum with temperature because heating leads to an increase in the population of the $^4T_2$ state with respect to the lower lying metastable state ^2E, which accumulates particles in the process of optical pumping.

Thus, the electron-phonon alexandrite laser can be described by the model depicted in Figure 23.7. Here the initial laser level is the $^4T_2$ multiplet fed by particles from the storage level ^2E. The final level is one of the levels of the equivalent vibrational band corresponding to the tuning range. The terminal laser level within this band is chosen by a selective device for laser tuning as in the case with dye lasers. The phonon relaxation of the terminal level to the ground state $^4A_2$ state takes 0.1 ps, as pointed out earlier. We note once more that in optical pumping the excited particles accumulate on the metastable level ^2E, characteristic of Cr^{3+} ions, and are transferred via thermal excitation across the energy gap of 800 cm^{-1} to the $^4T_2$ level. It must be emphasized that such a specific structure of the upper laser state and the storing of energy

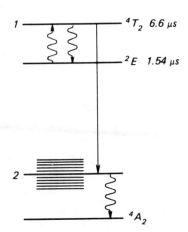

Figure 23.7. The level diagram for a mode of an alexandrite laser: *1*, the initial level, and *2*, the terminal level, in the *1* → *2* laser transition.

in this state makes it possible to realize the Q-switched mode and to generate giant pulses, which cannot be done with dyes in view of the small lifetime.

The tuning range of the alexandrite laser lies within 700 to 820 nm. The red limit is fairly sharp and determined by absorption from excited states, similar to the singlet-singlet and triplet-triplet absorption in organic dyes. The blue edge of the tuning range is limited by the R-line (681 nm) and the dip in the luminescence spectrum that emerges when we approach the R-line from the long-wave side. Within the tuning range the amplification cross section amounts to 7×10^{-21} cm^2 at 300 K and grows to 2×10^{-20} cm^2 at 475 K.

The energy characteristics of the alexandrite laser are very attractive. When a xenon flashlamp is used to pump an alexandrite rod 6 mm in diameter and 75 mm long, in the single-pulse mode the rod emits 5-10 J per pulse, and in the pulsed-periodic mode with a repetition frequency of 5-100 Hz the average power output is 35-70 W. In the Q-switched mode giant pulses 30-200 ns long have been obtained, with the radiation tunable in wavelength in the entire 700-820 mn tuning band.

The alexandrite laser has demonstrated the vast possibilities of the Cr^{3+} ion, which have hardly been exhausted in the ruby laser. Since the energy of the $^4T_2$ state depends strongly on the type of matrix into which chromium ions are implanted, the spectroscopic situation similar to the one in the case of alexandrite can be expected to occur with other matrices. This, too, can lead to low-threshold lasing by the four-level scheme, with tuning the radiation frequency in a broad spectral range.

Indeed, in crystals of rare-earth gallium garnets, which were discussed at the end of Lecture 21, the position of the $^4T_2$ level of the chromium ion with respect to the position of the ^2E level proved favorable (a small energy gap!) for increasing the population of this level in optical pumping and, hence, for obtaining a high gain in a fairly broad wavelength range. Another favorable circumstance in this respect is that crystals of rare-earth gallium garnets have a lower melting point than $Y_3Al_5O_{12}$ and $BeAl_2O_4$, do not contain toxic components, and have satisfactory mechanical properties and thermal conductivity. The proximity of the ionic radii of Ga^{3+} and Cr^{3+} makes it possible to introduce chromium ions into gallium garnets in large quantities.

Continuous-wave lasing at room temperature has been obtained on the electronic-vibrational $^4T_2 \rightarrow {}^4A_2$ transition of the chromium ion. This was done by pumping radiation emitted by a krypton laser into crystals of yttrium-scandium-gallium, yttrium-gallium, lanthanum-lutecium-gallium, gadolinium-gallium, and gadolinium-scandium-gallium garnets activated by chromium ions. The overall region of continuous tuning lies within a wavelength range of 700 to 900 nm. The most promising of the above garnets, the gadolinium-scandium-gallium garnet (GSGG) with chromium, made it possible to achieve a 766-820 nm tuning range.

Further development of tunable lasers based on rare-earth garnets with chromium could be very promising.

Problems to Lecture 23

23.1. What are the main merits of color-center lasers compared to dye lasers?

23.2. By how many times will the threshold pumping power required for reaching complete population inversion increase when a KCl:Li crystal with F_A-centers is heated from the liquid nitrogen temperature to room temperature if the lifetime of the upper laser level is approximately inversely proportional to temperature?

23.3. What are the shortest pulses in mode-locked lasing that can be achieved with a tunable color-center laser if the bandpass of the tunable dispersive element is 5 cm^{-1} wide?

23.4. The threshold pumping power of color-center lasers is usually much lower than that of dye lasers. Why?

23.5. The levels $^4T_2$ and ^2E of chromium ions in alexandrite are separated by a gap of $\Delta E = 800$ cm^{-1}. The probability of radiative decay of level ^2E is A_1 and that of $^4T_2$ is A_2. It is also known that A_1/A_2 is of the order of 10^{-2} and the corresponding ratio of the degeneracy multiplicities of the levels, g_1/g_2, is 1/3. Find the intensity of steady-state luminescence from level $^4T_2$ at $T = 300$ K and at $T = 77$ K if every second Q photons land in the absorption band of Cr^{3+} ions and nonradiative relaxation can be ignored. (*Hint*: Set up the transport equations of the (2.17) or (19.1) type for the population numbers of levels $^4T_2$ and ^2E and allow for the fact that the intensity of luminescence from level $^4T_2$ is $g_2 n_2 A_2 h\nu$, with n_2 the population number of the $^4T_2$ level.)

23.6. Can an alexandrite-chromium tunable laser operate at low temperatures?

LECTURE 24

Semiconductor Lasers I

Distinctive properties of semiconductor lasers. Allowed bands. Direct- and indirect-band semiconductors. Recombination luminescence. The Fermi level. The quasi-Fermi levels. The condition for population inversion. Nonradiative recombination. Internal quantum yield

Quantum electronics deals basically with bound states corresponding to discrete energy levels and relatively narrow lines of resonance transitions. The more an electron of an active center is isolated from external influences, the greater the extent to which its lifetime in a bound state is determined by the spontaneous decay time and the closer the resonance-transition linewidth is to the smallest value possible, and vice versa the greater the external influence on the electron, the transitions between whose bound states are considered possible laser transitions, the broader, generally speaking, are the corresponding luminescence and amplification lines. Examples abound. Among them are dye lasers, which we have already considered, color-center lasers, the alexandrite laser, and high-pressure molecular lasers.

Up to this point we have studied lasers whose active centers are characterized by the presence of fairly narrow discrete energy levels. But population inversion can also be achieved when the energy spectrum of active centers contains broad allowed bands separated by a distinct forbidden band. An example of a laser operating on these principles is the semiconductor laser. The current lecture is devoted to a study of such lasers.

A distinctive feature of semiconductor lasers is the fact that population inversion is created between states in the electronic energy bands of a semiconducting crystal. These energy bands appear because of the splitting of the energy levels of valence electrons of the atoms composing the crystal lattice, electrons that are in the strong and spatially periodic intracrystalline field of the atoms of the crystal proper. In this field the individual atoms lose their valence electrons, which cease to be localized. In other words, in a strong periodic field the valence electrons of a semiconducting crystal are collectivized. The collectivized motion of the collective electrons in the allowed bands of the semiconducting crystal ensures the electrical conductivity of the crystal.

As is known, the electrons in the conduction band and the holes in the valence band are the electric current carriers in semiconductors. Suppose that some sort of an external influence (say, pumping) is used to create an excess of electrons in the conduction band and an excess of holes in the valence band (in relation to the equilibrium numbers of electrons and holes, respectively). The return to equilibrium, that is, the recombination of the excess electrons and holes, can be achieved by emission of radiation on band-to-band optical transitions (Figure 24.1a).

The probability of band-to-band radiative transitions is high if the transitions are direct, that is, can be depicted by a straight vertical line on the diagram describing the dependence of the current-carrier energy E on the

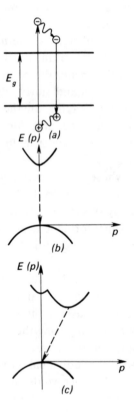

Figure 24.1. Band-to-band optical transitions in semiconductors.

quasimomentum p of the carriers. The fact is that band-to-band optical transitions result in the creation of free current carriers. Hence, the momentum conservation law must hold true in such transitions, which means, in view of the smallness of the photon momentum, that the electron quasimomentum must remain constant in such transitions. In the case of so-called direct-band semiconductors, for which the maximum on the potential curve that limits the valence band from above and the minimum on the potential curve that limits the conduction band from below occur at the same value of the quasimomentum p (Figure 24.1b), the vertical transitions connect the top of the valence band with the bottom of the conduction band, that is, the parts of both allowed bands that get populated primarily with excess current carriers. A transition is allowed in this case, and the creation of nonequilibrium population inversion is maximally facilitated. Thus, direct-band semiconductors are most suited for creating lasers. In the case of indirect semiconductors (Figure 24.1c), the current carriers build up in states with different quasimomenta, vertical transitions are impossible because of the absence of recombining particles, and nonvertical (indirect) transitions are strictly forbidden.

The high probability of radiative transitions in direct-band semiconductors and the high density of states in the bands allow for exceptionally high values of gain, exceeding 10^4 cm^{-1} in some especially favorable situations. In lasers based on solid dielectrics with impurity active centers, which we discussed in previous lectures and which are usually called solid-state lasers, the transitions between discrete energy levels of isolated ions are employed, in contrast to the case of semiconductor lasers. The active particle density and the transition rates in solid-state lasers are considerably lower, with the result that the gain is considerably lower, too, and, hence, the active elements have much bigger linear dimensions. The closest to semiconductor lasers are organic-dye and color-center lasers, characterized by wide energy bands because of the intense interaction of the active centers with the ambient and by the high gain because of high center density numbers, though this gain is not as high as in semiconductor lasers.

An important feature that puts semiconductor lasers apart from other lasers based on condensed media is the electrical conductivity of the semiconducting crystal which makes it possible to pump them by applying electric current and thus to directly transform electric energy into laser light.

Thus, semiconductor lasers occupy a special place in quantum electronics, differing from all other lasers in a number of important features. From the practical standpoint, the following merits of semiconductor lasers are the most important.

1. Compactness due to the exceptionally high gain.

2. High efficiency due to the high efficiency in transforming the input energy into laser light when fairly perfect semiconducting single crystals are pumped via electric current.

3. Broad range of lasing wavelengths due to the possibility of selecting a semiconductor material with a band gap that corresponds to emission of the radiation resulting from band-to-band transitions that occur practically anywhere in the spectral interval from 0.3 to 30 μm.

4. Continuous tuning of radiation wavelength due to the dependence of the spectral-optical properties of semiconductors and primarily the band gap on temperature, pressure, magnetic field strength, and other factors.

5. Small time lag due to the smallness of the relaxation times and to practically no time lag in creating a nonequilibrium excess of electrons and holes in pumping via electric current, which leads to the possibility of modulating the radiation by varying the pumping current with frequencies reaching 10 GHz.

6. Simplicity of design due to the possibility of pumping via direct electric current. This ensures high compatibility of semiconductor lasers with IC's of semiconductor electronics, devices of integral optics, and optical fiber communication networks.

The drawbacks of semiconductor lasers are, as is often the case, a continuation of their merits. The small dimensions lead to low values of the power output. In addition, semiconductor lasers, like any device of semiconductor electronics, are sensitive to overloads (they disintegrate under irradiances of several megawatts per square centimeter) and to overheating, which causes a jump in the self-excitation threshold and even to irreversible disintegration when the temperature is raised above a temperature specific for each type of laser.

Lasing has been achieved through the use of several dozen different semiconducting materials. These materials, aside from the usual requirements of purity and single-crystal nature, must possess a high optical homogeneity and a low probability of the nonradiative recombination of electrons and holes.

Let us examine the basic luminescence mechanism in semiconductors, the radiative recombination of electrons and holes. Recombination of electrons and holes in semiconductors is a process that makes an electron to transfer from the conduction band to the valence band with the result that a conduction electron-hole pair ceases to exist. Recombination always means that a current carrier has moved to a lower energy level, either to the valence band or to an impurity level in the forbidden band.

In conditions of thermodynamic equilibrium recombination balances the process of thermal carrier generation, and the rates of these mutually opposite processes are such that the net effect is the setting in of the Fermi distribution for electron and hole energies.

Many recombination mechanisms are known. They differ by the direction in which the energy liberated as a result of recombination is transferred. If the excess energy is liberated in the form of a photon, radiative recombination occurs. This elementary act of light generation in semiconductors is similar to radiative decay of an excited state in systems with a discrete spectrum. Nonradiative recombination is also possible. In this case the liberated energy is used to excite the crystal lattice, that is, to heat the crystal. Obviously, nonradiative recombination is similar to the nonradiative relaxation of excitation energy in systems with a discrete spectrum.

In radiative recombination the total number of emission acts is proportional to the product np of the electron and hole densities, n and p. At moderate carrier densities this recombination channel is inefficient. At high carrier densities, that is, exceeding 10^{16}-10^{17} cm^{-3}, semiconductors become effective sources of light through recombination emission in a fairly narrow range of wavelengths near the intrinsic-absorption edge of the semiconductor.

Recombination emission involves band-to-band transitions (see Figure 24.1a). Recombination luminescence has a band-to-band nature not only in intrinsic semiconductors but also in heavily doped semiconductors. In the latter case the energy spectrum of the semiconductor is strongly distorted near the edges of the forbidden band because as a result of heavy doping the impurity levels broaden into the impurity band, which partially or completely merges with the intrinsic band. The energy of the photons emitted in the course of recombination emission may somewhat differ from the band gap of the doped semiconductor. Practically, however, all processes of radiative recombination employed in semiconductor lasers have a characteristic common feature, namely, the transition energy $\hbar\omega$ is close to the value of the band gap, E_g.

The presence of spontaneous radiative recombination indicates the possibility of creating a laser. For gain to appear in the emission spectrum of spontaneous recombination, stimulated emission of photons must prevail over absorption. A necessary condition for this is population inversion. In semiconductor lasers, therefore, population inversion must manifest itself on levels involved in radiative recombination transitions. Let us consider the conditions necessary for population inversion to take place.

Analysis of such conditions requires knowing the energy levels, their excitation cross sections, relaxation times, and other factors. But since the semiconductors that form the active media in such lasers constitute a very broad class, this avenue of research is not rational even if possible. Fortunately, sufficiently general thermodynamic considerations that also allow for the special features of the statistics of electrons in semiconductors can provide the general conditions for population inversion to establish itself in these materials.

Irrespective of the concrete mechanism of radiative recombination, the emerging photons obey the general laws of radiation. The rate at which photons of a definite frequency ω fill up a radiation mode of volume V is

$$dN_\omega/dt = (A + BN_\omega)/V, \tag{24.1}$$

where N_ω is the number of photons in the mode. The first term on the right-hand side of Eq. (24.1) is responsible for spontaneous emission, and the second corresponds to the difference in rates of stimulated emission and absorption of photons. In the case of radiative recombination, a single electron-hole pair ceases to exist every time a photon is emitted, and a new pair is created in each absorption act.

We can find the relation between the coefficients of spontaneous (A) and stimulated (B) emissions by using thermodynamic considerations in a way similar to that introduced in Lecture 1. In accordance with the Bose-Einstein statistics, the equilibrium occupation number per mode (for two polarizations) at a temperature T is

$$\overline{N}_\omega = \frac{2}{\exp{(\hbar\omega/kT)} - 1}. \tag{24.2}$$

At equilibrium $d\overline{N}_\omega/dt = 0$; hence, in view of (24.1) we have

$$A/B = -\overline{N}_\omega. \tag{24.3}$$

Further analysis requires taking account of the special features of semiconductors. In the spectrum of electronic states let us isolate two levels with energies $E_2 > E_1$. The radiative recombination rate on the $E_2 \to E_1$ transition is proportional to the product of the electron density number on level E_2 and the hole density number on level E_1.

Electrons obey the Fermi-Dirac statistics, as we know. The probability of an electron occupying a state with energy E is given by the Fermi distribution formula

$$f(E) = \left(\exp \frac{E - F}{kT} + 1\right)^{-1},$$ (24.4)

where F is the Fermi energy or level. The probability of finding a hole on level E is simply the probability of not finding an electron on this level and is, hence, equal to

$$1 - f(E) = \left(\exp \frac{F - E}{kT} + 1\right)^{-1}.$$ (24.5)

Then the spontaneous recombination rate, proportional to the number of electrons on level E_2 and the number of holes on level E_1, can be represented in the form

$$A = A_0 f(E_2)[1 - f(E_1)],$$ (24.6)

with A_0 the proportionality factor.

Similarly, B, which in (24.1) determines the difference in rates of stimulated emission and absorption, is given by the formula

$$B = B_2 f(E_2)[1 - f(E_1)] - B_1 f(E_1)[1 - f(E_2)],$$ (24.7)

where B_1 and B_2 are proportionality factors. Substituting these formulas for A and B into (24.3) and allowing for the equilibrium distribution functions (24.2) and (24.4), we arrive at the equation

$$\exp \frac{\hbar\omega}{kT} - 1 = 2 \frac{B_2}{A_0} \left(\frac{B_1}{B_2} \exp \frac{E_2 - E_1}{kT} - 1\right).$$ (24.8)

By hypothesis, photons with an energy $\hbar\omega$ appear as a result of direct-band radiative recombination occurring between levels E_2 and E_1; hence, $\hbar\omega = E_2 - E_1$. Then Eq. (24.8) is always satisfied if $B_2 = B_1 = A_0/2$. This means that

$$B = \frac{A_0}{2} [f(E_2) - f(E_1)].$$ (24.9)

Hence, the difference between the rate of stimulated emission of photons on a band-to-band transition under radiative recombination and the rate of photon absorption on the same transition is positive if

$$f(E_2) > f(E_1). \tag{24.10}$$

If $f(E)$ is the Fermi distribution function (24.4), represented in Figure 24.2, then for the pair of levels $E_2 > E_1$ this condition is not met in thermodynamic equilibrium.

We now recall that the levels E_2 and E_1 are separated by the forbidden band and located, respectively, in the conduction band and valence band. Nonequilibrium carriers, that is, nonequilibrium electrons in the conduction band and holes in the valence band created by some pumping source, have finite lifetimes in the bands. During these lifetimes thermodynamic equilibrium is violated and, hence, there is no unique Fermi level for the system as a whole. However, if in the course of a time interval shorter than the carrier lifetimes in the bands there sets in a quasi-equilibrium state of the Fermi type in the electron gas and a different quasi-equilibrium state of the same type in the hole gas such that there is a unique temperature for the entire system but there is no equilibrium between the gases, we can introduce the so-called quasi-Fermi levels separately for electrons in the conduction band, F_n, and for the holes in the valence band, F_p.

It is not clear a priori whether it is justified to introduce separate quasi-equilibrium distribution functions for holes and electrons in the valence band and in the conduction band, respectively. Justification lies in the fact that at least in some semiconductors the maxwellianization time for electrons within

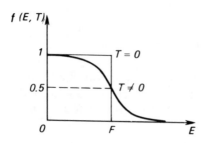

Figure 24.2. The Fermi distribution function (F is the Fermi energy or level).

a band (0.1 ps) is shorter by a factor of 1000 to 10 000 than the characteristic time of band-to-band maxwellianization (1 to 10 ns).

Let us return to the inversion condition (24.10). If in accordance with the above discussion we represent $f(F_2)$ and $f(E_1)$ in the form

$$f(E_2) = \left(1 + \exp \frac{E_2 - F_n}{kT}\right)^{-1},$$

$$f(E_1) = \left(1 + \exp \frac{E_1 - E_p}{kT}\right)^{-1}, \tag{24.11}$$

where F_n is the electron quasi-Fermi level, and F_p the hole quasi-Fermi level, then using (24.10) we can find an equivalent but more graphic condition:

$$F_n - F_p > E_2 - E_1. \tag{24.12}$$

Since the minimum value, $E_2 - E_1$, is the band gap E_g, the condition for population inversion on band-to-band transitions assumes the simple form

$$F_n - F_p > E_g. \tag{24.13}$$

Hence, pumping, which creates the nonequilibrium condition, must be strong enough so that the Fermi-quasi level appears inside the respective allowed energy bands. This means that the electron and hole gases are degenerate, and all the levels in the valence band with energies $E_1 > F_p$ are practically vacant and all the levels in the conduction band with energies $E_2 < F_n$ are almost completely occupied by electrons (Figure 24.3). Then the photons whose energies obey the inequalities

$$E_g < \hbar\omega < F_n - F_p \tag{24.14}$$

cannot generate transitions from the valence band to the conduction band and are therefore not absorbed. Reverse transitions, from the conduction band to the valence band, are possible. It is stimulated radiative recombination on these transitions that generates laser light, and the inequalities (24.14) determine the corresponding amplification bandwidth.

The intensity of radiative recombination is determined by the concrete features of the band structure of the semiconductor, the square of the matrix element of the appropriate transition, and the number density of the recombining pairs. The rate of radiative recombination, that is, the number of transi-

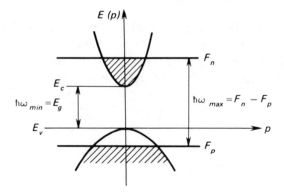

Figure 24.3. Inverse electron distribution in a one-valley semiconductor.

tions in which radiation is emitted per unit volume and per unit time, depends on the same factors, obviously. Since the radiative and nonradiative recombination channels are parallel, the resulting recombination rate is equal to the radiative recombination rate $(1/\tau_{rad})$ plus the nonradiative recombination rate $(1/\tau_{nonrad})$:

$$1/\tau = 1/\tau_{rad} + 1/\tau_{nonrad}. \tag{24.15}$$

Obviously, the fraction of radiative recombination acts in the recombination process as a whole is equal to the ratio of the respective rates: $1/\tau_{rad} \div 1/\tau$. The same ratio specifies the fraction of energy stored in the nonequilibrium electrons and holes and released through the radiative recombination channel. This gives us the definition of the internal quantum yield of radiative recombination:

$$\eta_{int} = \frac{1/\tau_{rad}}{1/\tau_{rad} + 1/\tau_{nonrad}}. \tag{24.16}$$

This quantity characterizes the quality of the semiconducting material. Proper choice of doping and the manufacture of perfect crystals make it possible to obtain values of η_{int} close to one for many semiconducting materials.

Multiphoton nonradiative recombination on a band-to-band transition has a low probability of occurring. At moderate carrier concentrations the most important mechanism is recombination through intermediate states in the forbidden band localized near impurities and defects. There is a certain similarity between the process considered and the mechanisms of nonradiative relaxation through intermediate levels in the case of impurity dielectric crystals (see Lecture 21). Many impurities and defects, most of which are poorly identified, act as recombination centers. Too high a concentration of these centers emerging because of technological errors makes the semiconducting material unsuitable for creating a laser crystal.

As the carrier concentration grows, Auger recombination becomes ever more important. Auger recombination is recombination of an electron and a hole in which no electromagnetic radiation is emitted, and the excess energy and momentum of the recombining electron and hole are given up to another electron or hole. This process becomes noticeable at high concentrations of free carriers because its realization requires the collision of three carriers. The role of Auger recombination grows in narrow-gap semiconductors.

The equations describing cascade and Auger recombination processes are nonlinear. For this reason nonradiative decay of electron-hole pairs is nonexponential and, strictly speaking, cannot be characterized by a lifetime constant in relation to this process. For a rough estimate of the order of magnitude of the rates of nonradiative recombination one can use the experimental values of the cross sections of the corresponding processes. In the cascade process the center-trapping cross section may amount to 10^{-12}-10^{-22} cm^2. The corresponding recombination coefficient varies between 5×10^{-6} and 5×10^{-16} cm^6/s. For a recombination center concentration of 10^{16} cm^{-3} this leads to an effective recombination rate varying between 5 and 5×10^{10} s^{-1}. The recombination coefficient for the Auger process usually varies between 10^{-25} and 10^{-32} cm^3/s. For a carrier concentration of 10^{19} cm^{-3} this yields an effective recombination rate of 10^6 to 10^{13} s^{-1}.

These estimates show that with real materials it is indeed possible to reach an internal quantum yield of radiative recombination close to one (see Eq. (24.16)). Properly performing the technological process of growing the semiconducting crystal may rule out the unfavorable effect of cascade recombination. On the other hand, Auger recombination, whose rate increases with concentration like n^3, is nonremovable in principle.

Thus, when the inversion condition (24.13) is met and $\tau_{rad} \ll \tau_{nonrad}$, effective emission of laser light by semiconducting crystals becomes possible.

The first to suggest using semiconducting crystals in quantum electronics was Basov (1959).

Problems to Lecture 24

24.1. Can we speak of population inversion in a semiconductor whose entire valence band except the edge is filled with electrons and the conduction band with holes?

24.2. Derive inequality (24.12).

24.3. What is the best way to raise the efficiency of a semiconductor laser: to raise the temperature or lower it?

LECTURE 25

Semiconductor Lasers II

Diode injection lasers. Distribution of carriers in a semiconducting crystal with a *p-n* junction. Injection of carriers. The band structure of a semiconductor with a *p-n* junction. A degenerate semiconductor with a *p-n* junction. The band structure. Population inversion by the injection of carriers into the *p-n* junction of a degenerate semiconductor. Efficiency. Power output. Heterostructures. Wavelength ranges of the emitted radiation. Tuning

Let us now examine the methods used to create nonequilibrium electron-hole pairs. Inversion in the electron distribution in the valence and conduction bands of a semiconducting crystal can be achieved in several ways. Highly effective is electron-beam pumping. This creates nonequilibrium carrier pairs and, hence, generates laser radiation in many materials, including wide-gap (the short-wave range). However, the most widespread method is the excitation of the semiconductor by direct electric current, which injects electrons and holes into the region of the *p-n* junction of a semiconductor diode. Injection (diode) semiconductor lasers created in this way have gained widest acceptance because of their simplicity, reliability, and high efficiency.

When two semiconductors, one of the *p*-type and the other of the *n*-type, are brought into contact, potential barriers build up inside the boundary layer, and these noticeably change the carrier concentrations inside the layer. The properties of the boundary layer, that is, the transient transition between the *p*-type semiconductor and the *n*-type semiconductor or simply the region of the electron-hole junction (*p-n* junction), depend on the applied voltages. In many cases such a dependence leads to a nonlinear volt-ampere characteristic of the semiconductor junction, which acquires the properties of a semiconductor diode.

It is well known that *p-n* junctions are widely used in semiconductor electronics to rectify electric current and transform, amplify, and generate electric oscillations. Here we will briefly discuss their laser application based on the creation of nonequilibrium carriers in the junction region.

To avoid uncontrollable surface effects, *p-n* junctions are created by forming the required distribution of the donor and acceptor impurities inside a

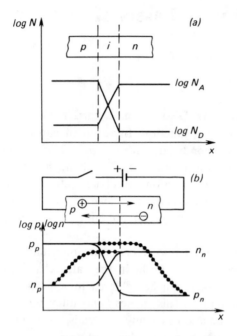

Figure 25.1. A schematic of a *p-n* junction (a), and injection of carriers into the *p-n* junction (b).

single crystal (Figure 25.1). If these impurities are completely ionized, then on the left-hand side of the crystal, so to say, where the acceptor concentration N_A is high, the *p*-type conductivity is predominant, with the majority carrier concentration p being approximately $N_A - N_D$ (N_D is the donor concentration). Correspondingly, on the right-hand side of the crystal the *n*-type conductivity is predominant, with the majority carrier concentration n being approximately $N_D - N_A$. Between the *p*- and *n*-region lies a transient layer of the technological *p-n* junction where the impurity concentrations change markedly across the layer. Within a thin region of the transient layer where the donors and acceptors compensate each other ($N_D \approx N_A$) there is intrinsic (*i*) conductivity. Strictly speaking, a *p-n* junction is a *p-i-n* junction.

The required distribution of donors and acceptors is created by various technological means: by fusing semiconductors of the *p*- and *n*-type, by adding the required impurity to the melt in the crystal growth process, by diffusing impurities from the gaseous or liquid phase into the crystal, by employing the ion implantation method, etc. Acceptors are atoms of elements belonging to the columns of the Periodic Table to the left of the group containing the main element of the semiconducting crystal. Correspondingly, donors belong to the group to the right of the main element. For instance, for silicon and germanium (group IV) the elements of group III are acceptors, while those of group V are donors (say, boron and phosphorus, respectively). For gallium arsenide, GaAs (a semiconducting compound of the III—V type), zinc and cadmium (group II elements) are acceptor impurities, and selenium and tellurium (group VI) are donor impurities.

In an equilibrium semiconductor with a p-n junction, the concentration p_p of majority carriers, holes in the p-region, is high and constant if there is no current passing through the junction. In the junction region the hole concentration reduces, and in the electron region, where holes are minority carriers, assumes the low value p_n. Similarly, the electron concentration changes from a large value n_n in the n-region (majority carriers) to a small value n_p in the p-region (minority carriers).

If an external voltage is applied to the junction so that the "plus" of the voltage source is connected to the p-region and the "minus" to the n-region, a positive current flows through the junction (the current through a diode flows in the positive direction). Holes from the p-region rush to the n-region and electrons from the n-region to the p-region. The holes that have arrived at the n-region and the electrons that have arrived at the p-region become minority carriers in these regions. They must recombine with the corresponding majority carriers in the p- and n-region adjoining the p-n junction.

The carrier lifetime with respect to recombination is finite and recombination does not take place instantly, so that along the current in a certain volume outside the junction the electron concentration in the p-region and the hole concentration in the n-region considerably exceed the respective equilibrium values n_p and p_n in these regions. Then to compensate for the space charge, the leads supply electrons to the n-region and holes to the p-region. As a result the concentration of carriers of these two types increases on both sides of the junction, that is, there appears a quasineutral region of elevated conductivity near the junction. This constitutes the essence of the phenomenon of injection of carriers into the p-n junction. The distribution of carrier concentration during injection is shown in the lower part of Figure 25.1 together with the equilibrium carrier concentration (in the absence of a current).

If we assume that the thickness of the junction region is smaller than the electron (and hole) diffusion length, then for nondegenerate semiconductors the concentrations of injected carriers at the boundaries of the p-n region are

$$n = n_p \exp \frac{eU}{kT} , \quad p = p_n \exp \frac{eU}{kT} , \qquad (25.1)$$

where e is the electron charge, and U the voltage across the junction. At $T = 300$ K we have $e/kT \approx 40$ V^{-1}. For this reason even a small voltage greatly changes the minority carrier concentration on the boundaries of the p-n junction. At $U \approx 0.25$ V the change is roughly 10^4-fold.

Let us now discuss the band structure of a semiconductor with a p-n junction. For an intrinsic semiconductor the band gap is a characteristic constant. In p-type semiconductors the acceptor impurities produce energy levels lying inside the forbidden band and adjoining the top of the valence band. At high impurity concentrations these levels merge with the valence band and in this way narrow the band gap from below. In n-type semiconductors the situation is just the opposite, that is, the donor impurities narrow the band gap from above. Hence, in the same semiconducting crystal the forbidden band in the p-region lies higher and in the n-region lower.

In the transition from one region to the other, that is, in the presence of a p-n junction, the boundaries of the bands change continuously in such a manner that the forbidden band of the p-region smoothly transforms itself into the forbidden band of the n-region. As a result the band structure of an equilibrium semiconductor with a p-n junction acquires the shape schematically shown in Figure 25.2a. Besides, in an equilibrium semiconductor, the Fermi level, common for the entire volume of the crystal, must lie below the middle of the forbidden band in the region with an excess of acceptors and above the middle of the forbidden band in the region with an excess of donors.

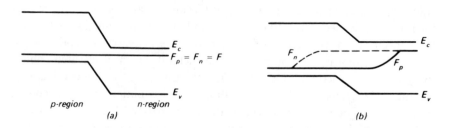

Figure 25.2. Quasi-Fermi levels in a nondegenerate p-n junction: (a) without injection, (b) with injection.

The shift of the forbidden bands of the p- and n-regions with respect to each other in a semiconductor with a p-n junction is necessary, as shown by Figure 25.2a, so that the same Fermi level is situated both below the middle of the forbidden band of the p-region and above the middle of the forbidden band of the n-region.

When carriers are injected and a voltage U is applied across the p-n junction, the equilibrium breaks down, the shift of the forbidden bands of the p- and n-region with respect to each other diminishes by eU, and, what is most important, the Fermi level F splits into a quasi-Fermi level F_p for holes and a quasi-Fermi level F_n for electrons, and these levels differ considerably in the vicinity of the junction. Far from the junction region they merge again, but near the junction

$$F_n - F_p = eU. \tag{25.2}$$

In semiconductor theory the validity of (25.1) and (25.2) follows from simple energy considerations.

The pattern of the quasi-Fermi levels in a p-n junction when minority carriers are injected into a nondegenerate semiconductor is shown in Figure 25.2b. Even under strong injection it is difficult to move the quasi-Fermi levels F_p and F_n in a nondegenerate (i.e. lightly doped) semiconductor so far apart that the population inversion conditions (24.13) are met. Figure 25.2 illustrates this case. But if the p- and n-regions in the crystal are heavily doped, the electron and hole gases can be strongly degenerate in the corresponding regions of the crystal.

The fact that the fermion distribution differs considerably from the Boltzmann distribution serves as a criterion of the degeneracy of the fermions. Distribution (24.4) and Figure 24.2 show that this difference, which is small when $\exp[(E - F)/kT] \gg 1$, becomes noticeable when $\exp[(E - F)/kT] \lesssim 1$ (weak degeneracy) and highly important when $\exp[(E - F)/kT] \ll 1$ (strong degeneracy). Since there are no electrons in the forbidden band, the lower bound on the electron energy is the bottom of the conduction band, E_c. Hence, the electron gas is strongly degenerate if

$$\exp\frac{E_c - F_n}{kT} \ll 1. \tag{25.3}$$

This means that the electron Fermi level F_n must lie inside the conduction band. Similarly, when the hole gas is strongly degenerate, the corresponding Fermi level F_p must lie inside the valence band.

The Fermi level is a kind of characteristic energy that depends on the semiconductor type, its state, and its composition. For our case it is important that the position of the Fermi level be uniquely connected with the carrier concentration. If in the event of strong doping of a semiconductor by ionizing impurities of the p- or n-type the carrier concentration exceeds the so-called effective density of states in the valence or conduction band, the Fermi level lies inside the corresponding band and the hole or electron gas degenerates. For the sake of reference we note that when the temperature is 300 K and the effective carrier mass is equal to the rest mass of a free electron, the effective density of impurity states above which the semiconductor degenerates is 2.5×10^{19} cm^{-3}.

Thus, for laser diodes it is advisable to use heavily doped semiconductors, in which the electron and hole gases in the n- and p-regions are strongly degenerate. Then even with no injection the Fermi level lies in the p-region inside the valence band and in the n-region inside the conduction band (Figure 25.3a). Distortion of the band structure near the p-n junction when carriers are injected in the positive direction leads to the population inversion condition (24.13) being met, as shown schematically in Figure 25.3b. The width of the active region, where $F_n - F_p \geq E_c - E_v = E_g$, can be considerably greater than the junction's technological width. Within this region and in the spectral interval (24.14) there are the conditions for stimulated-recombination band-to-band transitions. The active region shown in Figure 25.3a in energy vs. distance

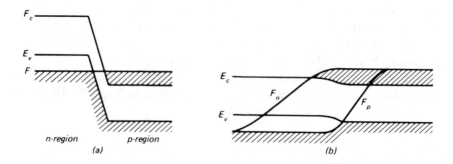

Figure 25.3. Quasi-Fermi levels in a degenerate p-n junction: (a) without injection, (b) with injection. The hatched sections depict the regions filled by electrons.

coordinates corresponds to the inversion region shown in Figure 25.3b in energy vs. quasimomentum coordinates.

The laser effect with population inversion created by carrier injection into the *p-n* junction has been realized in many one-valley direct-gap semiconductors. One of the best is the gallium arsenide laser. The diode is a thin section of a GaAs single crystal with transverse and longitudinal dimensions of about 0.1-1 mm. The sample is cut out of a heavily doped substance of the *n*-type (Te and Se acting as donors). After diffusion or implantation of a *p*-type semiconductor (Zn and Cd acting as acceptors), the upper part of the sample acquires *p*-type conductivity, and not far from the surface (at a distance of 10-100 μm) there forms a planar layer of a *p-n* junction 1-10 μm thick. The contact surfaces of the *p*- and *n*-region are covered with a gold film. The crystal is attached (by glue or by soldering) to a heat-conducting substrate usually by its *n*-region. Diamond is the best support material, but sapphire is also often used. Figure 25.4 is a schematic of such a laser. Under a large forward current (current densities of about 1000 A/cm^2), population inversion is created in the thin layer of the *p-n* junction. The gain is high. For this reason the lateral facets of the crystal ($n = 3.6$ and $R = 30\%$) may act as Fresnel mirrors and lasing occurs in crystals of small length.

Cleaving lateral facets of a crystal has proved to be highly successful in forming perfect plane-parallel mirrors of the cavity of a semiconductor laser. The radiation leaves the laser through the narrow strips formed by the intersection of the active layer with the partially reflecting facets of the crystal. The characteristic angular dimensions of the directivity diagram corresponding to

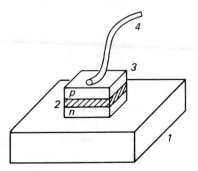

Figure 25.4. The schematic of an injection laser: *1*, substrate; *2*, plane of the *p-n* junction; *3*, semiconducting crystal; *4*, lead.

this section are $5° \times 50°$. Higher directivity can be achieved by using an external cavity. Then the facets of the semiconductor diode must be either bleached or Brewster-angle oriented in relation to the cavity axis. Usually meeting these requirements poses serious difficulties.

Efficiency, which is defined as the ratio of the power of emitted radiation to the pumping power dissipated by the diode, is directly proportional to the internal quantum yield of recombination emission, η_{int} (see formula (24.16)), and the ratio of the gap energy E_g (expressed in volts) to the diode voltage drop E_{diode}:

$$\eta = \eta_{int} E_g / E_{diode}. \tag{25.4}$$

For values of η_{int} close to one and a small voltage drop on the leads and in the material of the p- and n-region of the crystal, the value of η may be extremely high. In the case of gallium arsenide, the laser efficiency reaches 0.7-0.8 when liquid nitrogen is used as a coolant, so that injection semiconductor lasers are the most effective. However, their power output is moderate primarily because the p-n junction region is so small. In the continuous-wave mode and for an emitting surface of 10^{-4} cm^2 the emitted power reaches 10 W (GaAs, 77 K).

The power output of a laser is proportional to the quantum yield and the difference between the pumping current density J and the threshold value J_{thr}:

$$P \propto (J - J_{thr}) \eta_{int}. \tag{25.5}$$

As the temperature rises, the value of η_{int} drops because the role of nonradiative recombination grows. In addition, a rise in temperature strongly drives down the difference in the rates of stimulated emission and absorption (24.7), which determines the gain value. Allowing for condition (24.13), which determines the possibility of creating population inversion in semiconductors, and using the definition of quasi-Fermi levels (24.11), we notice that the first term on the right-hand side of Eq. (24.7) decreases as the temperature grows while the second term increases. As a result, to drive the gain up we are forced to increase the injection current. This leads to a sharp increase in the threshold value of the current density with temperature. When the current grows, the crystal heats up and at some temperature continuous-wave lasing becomes impossible. The temperature at which the semiconductor laser still emits radiation in the continuous-wave mode is determined by the design of the diode and the possibilities of heat removal. At liquid helium temperatures it is possible to remove about 30-40 watts of heat, at liquid nitrogen temperatures about 10 watts, and at room temperature about 1 W.

The way in which the threshold current density depends on temperature is largely determined by the diode's design and operating conditions. For GaAs-based lasers the threshold current density grows very rapidly in the vicinity of 77 K, approximately like T^3, which is really a very strong dependence. The characteristic value of J_{thr} at 77 K for these lasers is about $(2\text{-}3) \times 10^2$ A/cm^2. Within the range from helium to room temperatures the threshold current densities of GaAs-based lasers vary from 10^2 to 10^5 A/cm^2.

Thus, in the continuous-wave mode the restriction on the power output of a semiconductor laser is set only by overheating due to the pumping current.

The power output of a pulsed gallium arsenide laser is about 100 W at liquid nitrogen temperatures and at an injection pulse 0.5-1 μm long. As usual, restriction in power in the pulsed mode is conditional on the optical self-destruction of the crystal. At room temperature the pulse-periodic mode sets in in such lasers, with repetition frequency up to 10 kHz and a peak power output of several watts.

Considerable improvement of the characteristics of semiconductor lasers, primarily drastic lowering of the threshold current density and the related possibility of continuous-wave operation at room temperature, has been achieved by using anisotropic heterojunctions.

So far we have considered p-n junctions formed by distributing p- and n-impurities in the same single crystal. In such junctions, also called homojunctions, the properties of the crystal are the same on both sides of the interface. But if the single-crystal layer of one semiconductor is grown on the single-crystal substrate of another, a heterojunction forms. Needless to say, such growth without substantially disturbing the single-crystal nature of the entire specimen is possible only for semiconducting materials whose crystal lattices differ very little. This is usually achieved by isoperiodic substitution via the epitaxial growth method. Examples of such pairs are GaAs-$Al_xGa_{1-x}As$, GaAs-$GaAs_xP_{1-x}$, and CdTe-CdSe. By implanting acceptor and donor impurities in the proper manner semiconductor heterojunction diodes can be created. The schematic of the diode of a semiconductor laser with a double heterojunction is shown in Figure 25.5.

Creation of perfect heterostructures requires that the contacting materials have the same type of lattice and the same period, which is just the situation with GaAs and AlAs. Hence, in the solid solution of $Al_xGa_{1-x}As$ the substitution of gallium for aluminum and vice versa has practically no effect on the

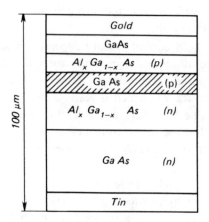

Figure 25.5. The schematic of the diode of a semiconductor laser with a double heterojunction. The active region is hatched.

lattice period, and heterostructures obtained on the basis of this material have practically no mismatch defects.

In a heterojunction the *p-n* junction is extremely abrupt and there is no diffusive spreading of the injected carriers, which concentrate as a result of injection in the well-defined narrow junction region. This lowers the threshold injection-current density. In addition, the refractive index of the three-component semiconductor AlGaAs is considerably lower than that of the binary semiconductor GaAs. Hence, the effect of an optical guide emerges, and the radiation concentrates in the active region and does not penetrate the absorption region not affected by pumping. All this combined with the fact that the diode is mounted on a substrate with high thermal conductivity has made it possible to lower the threshold current density at room temperature more than 100-fold, driving it up to several hundred amperes per square centimeter. As a result stable lasing by continuous-wave semiconductor lasers has been achieved with an output power of up to 100 mW at room temperature, which has greatly increased the practical importance of these devices. Laser communications form the main area of their application, primarily in optical fiber communication networks.

Distributed feedback has found wide application in heterostructure lasers. Periodic spatial modulation is usually carried out by the creation of a corrugate pattern on the surface of the crystal. Such patterns are produced by applying ordinary microelectronics techniques, such as electron and ion etching and selective photoetching through specially designed etching masks. There has also been acceptance of holographic methods and methods using photoetching in combination with illumination of the field by interfering beams of coherent light. Epitaxial growth methods are used to grow laser heterostructures with crystal facets periodically corrugated in composition, that is, gallium arsenide and its solid solution. For instance, gallium arsenide has a refractive index $n = 3.5$ while its isoperiodically substituted modification $Ga_{0.7}Al_{0.3}As$ has $n = 3.4$. This has made it possible for a corrugation period of 0.11 μm and a crimp height of 0.5 μm to build a corrugated waveguide with a waveguide thickness of only 3 μm. Note that dynamically distributed feedback can be achieved by exciting acoustic vibrations of appropriate frequency in the crystal.

The first to suggest using heterostructures in semiconductor injection lasers was the Russian physicist Zhores I. Alferov in 1963. The suggestion was realized under his guidance in 1968.

The wavelength range of the radiation emitted by semiconductor lasers is determined, as (24.14) implies, by the energy gap. Gallium arsenide ($E_g \approx 1.5$ eV) emits in the 0.84-μm range. Semiconductor solid solutions of varying composition make it possible to span a broad wavelength range. For instance, depending on composition the AlGaAs system emits in the 0.63-0.90 μm range, the AlGaSb system in the 1.20-1.80 μm range, the GaInAs system in the 0.9-3.4 μm range, and so on. In some solid solutions, say, in $Pb_xSn_{1-x}Se$ and $Cd_xHg_{1-x}Te$, the gap energy passes through zero as the percentage of the component varies. This enables constructing long-wave lasers with wavelengths up to 30-40 μm. In the last case to prevent the conduction band from being thermally populated deep cooling of the pumped crystal is required.

The wavelength of the radiation emitted by semiconductor diode lasers may vary within broad limits. In contrast to lasers of other types, it is determined by transitions not between discrete energy levels of atoms or molecules but between bands of allowed states in the semiconductor and depends on many factors affecting the band structure, such as pressure, temperature, and magnetic field. Discrimination of a single mode in this case is facilitated by the large spectral mode separation caused by the small dimensions of the active medium of the laser (see Eq. (10.21)). Usually, the separation of two adjacent modes is 0.5-3 cm^{-1}. The spectral width of a single mode slightly exceeds 10^{-4}-10^{-3} cm^{-1} owing to the instabilities of temperature and injection current.

Laser tuning by changing the pressure and temperature has gained the widest acceptance.

In lasers based on ternary compounds of lead, such as $Pb_{1-x}Sn_xTe$ and $Pb_{1-x}Sn_xSe$, hydrostatic compression changes the parameters of the crystal lattice and thereby makes the energy gap smaller. As a result the lasing wavelength increases considerably: in the accessible pressure range of 0-15 kbar a tuning by 1000 cm^{-1} is possible. An example is the PbSe-based diode laser, in which the lasing wavelength is tuned from 8.5 to 22 μm by a pressure of 14 kbar. The advantage of laser tuning by pressure variation is that pressure has no effect on the threshold current and the conditions necessary for laser self-excitation. The disadvantages are the notable technical complexity of the device as a whole and the considerable time lag of tuning.

Another method of tuning is based on lowering the temperature of the laser from the highest possible to that of liquid helium. The characteristic tuning in this widely used method amounts to 100-200 cm^{-1}. One drawback is the considerable nonlinearity in the temperature dependence of the lasing wavelength. In addition, temperature variation strongly affects the laser self-excitation threshold, as follows from the above discussion. Technically speaking, when there is a closed-cycle self-contained cryogenic device, the temperature method of tuning the radiation's wavelength proves to be quite convenient. Of course, a combination of these two methods, the temperature and the pressure variations, can be applied to a single laser, which considerably broadens the operational tuning range.

Fine and relatively inertialess tuning is achieved during the passage of the injection current pulse through the *p-n* junction thanks to the Joule heat liberated at the junction. Current tuning is quasi-continuous (Figure 25.6) because a shift in the amplification band is accompanied by a shift in the spectral-mode frequencies of the laser owing to the temperature variation in the refractive index of the active medium. These processes proceed at different rates, and because of this the tuning range is not continuous but consists of separate sections of curves of the continuous tuning of laser modes emerging as a result of refractive-index variations. The continuous tuning range usually spans a region from 0.5 to 3 cm^{-1}.

A convenient method for scanning the lasing frequency provides a varying current component of given shape and frequency to the constant injection current below the threshold value. The maximum rate of current tuning is determined by the thermal inertia of the laser and may reach 10^4 cm$^{-1} \cdot$s^{-1}. Fine tuning of the laser to a given fixed wavelength and maintaining this

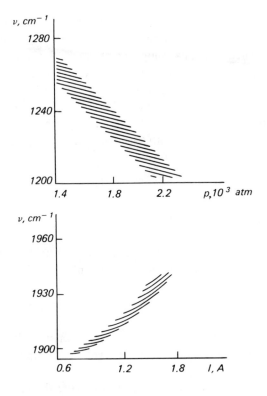

Figure 25.6. Tuning curves of narrow-gap diode lasers: above, tuning a PbSe laser by pressure variation; below, tuning a PbTe laser by current variation.

wavelength steadily require a high stability of the laser temperature and the pumping current.

The main fields of application of tunable diode lasers are high-resolution spectroscopy and detection of various impurities in gases and liquids.

Concluding the discussion of semiconductor diode lasers, it is well to note that electron-beam pumping of semiconductors makes lasing possible in a broad spectral range, most effectively in the short-wave region. Population inversion is created via avalanche formation of secondary, ternary, etc. electron-hole pairs in a semiconducting crystal irradiated by a beam of electrons whose energies range from several dozen kiloelectronvolts to several hundred. This does not require high electrical conductivity or the presence of *p-n* junctions. For this reason the method of electron-beam pumping can be applied to wide-gap semiconductors. Such lasers have high power outputs (up to 10^6 W per pulse) owing to the possibility of pumping large (compared with injection lasers) semiconductor volumes. The best examples here are lasers based on such well-known wide-gap semiconductors as CdSe, CdS, ZnSe, and ZnS with lasing wavelengths 0.69, 0.49, 0.46, and 0.33 μm, respectively.

The first semiconductor lasers with electron-beam pumping were constructed in 1964 by a group led by Basov.

Problems to Lecture 25

25.1. Derive formulas (25.1) and (25.2). (*Hint*: Use a textbook.)

25.2. What is laid off along the horizontal axis in Figures 25.2 and 25.3?

25.3. Does the degeneracy of carriers in semiconductors increase with temperature or decrease?

25.4. Depict the diode's volt-ampere characteristic. Mark the operational region of an injection laser. What determines its upper and lower boundaries?

25.5. How, in qualitative terms, does varying the temperature affect the amplification spectrum of an injection laser?

25.6. Compare the spot width of the mode of an injection homostructure laser whose cavity is formed by the facets of the semiconducting crystal proper, and the width of the amplification region (the specimen size is about 2 mm, $\lambda = 0.84$ μm, and $d_{\text{dif}} \approx 1$ μm).

25.7. Why is a heterostructure laser usually more effective than a homostructure laser? Give at least three reasons.

25.8. How are the amplification and emission spectra of a semiconductor laser related to the emission spectrum of a light-emitting diode operating on the same junction?

25.9. Estimate the angular divergence in the junction plane and in the normal plane of the radiation emitted by a gallium arsenide laser (see Problem 25.4).

LECTURE 26

Free-Electron Lasers

Generating microwaves by electron currents. Radiation wavelength
and synchronism for an ultrarelativistic electron beam. Gain in
undulator radiation emitted by relativistic electrons. The free-
electron undulator laser

When studying the physical bases of quantum electronics and the operational
principles of various lasers, we saw that the stimulated-emission effect in systems with discrete energy levels (discrete energy bands), that is, systems that
are essentially quantum, made it possible for quantum electronics to encompass by a single method a tremendous spectral range, from radio waves to
vacuum UV radiation. A significant fact must be mentioned at this point.
Notwithstanding the existence of lasers and masers, generation of electromagnetic waves in the radio (microwave) range is based primarily on the interaction
of free electron fluxes with waveguide and resonant structures, that is, is im-
plemented by classical methods.

Yet stimulated emission is not in essence a strictly quantum effect and can
be maintained in classical systems. It is, therefore, justifiable to ask whether
electron generation methods can be carried from the microwave range to the
optical. By analogy with the situation in which essentially quantum systems
can generate oscillations in a range extending from the microwave to the vacuum UV, it is resonable to expect that the stimulated emission of radiation
of such classical objects as free electrons can also be used to generate electromagnetic oscillations in a broad spectral range from radio waves to light
waves.

In accordance with the meaning of the word "laser", it is natural to call
a device that generates electromagnetic waves through the use of stimulated
emission of light by electron fluxes a free-electron laser.

In classical vacuum electronics of the microwave range, the size of the oscillator devices or their characteristic parts is comparable to the wavelength of
the emitted radiation. The region where the electrons interact with the high-
frequency radiation field, that is, where the electrons kinetic energy is transformed into the energy of the radiation field, is part of the oscillating system
of an electrodynamic resonant structure, which is usually characterized by
many natural frequencies. The time it takes the electrons to fly through the
interaction region or, more characteristically, from one element of the resonant

structure to another coincides, at least in order of magnitude, with the oscillation period of the emitted radiation.

For all oscillators of microwave vacuum electronics it is highly important to ensure synchronism between electron displacement and the electromagnetic wave into which the electron energy is being pumped. As the electrons interact with the high-frequency field of the wave, bunches of particles form in the electron beam (phase grouping). In their subsequent movements the electron bunches give off their energy to the radiation field, interacting with the field's component for which the synchronism condition is properly satisfied.

Grouping of electrons into bunches is characteristic of all microwave devices, although its concrete realization may be carried out differently in different devices. The most pictorial is the process of electron bunching in a travelling-tube, which utilizes the prolonged interaction of the field with the electron flux propagating rectilinearly along the direction of the travelling electromagnetic wave. At moderate, that is, essentially nonrelativistic, velocities of electron translational motion, the synchronism of the interaction is ensured by slowing down the propagation of the wave. The simplest slowing-down device is a single-threaded wire helix. The wave propagates along a loop of the helix at a velocity close to that of light. Along the helix's axis z the phase velocity of the wave, v_z, is approximately equal to $cd/2\pi a$, with a the radius of the helix and d the lead of the helix. The electrons moving along the axis in step with the wave gather into bunches under the action of the accelerating and decelerating sections of the wave. Figure 26.1 illustrates the simplest case when the longitudinal component of the electric field of the travelling wave can be written as $E_z = E_0 \sin(\omega t - \omega z/v_z)$. The forces acting on the electrons are depicted by arrows. The reader can see that the electrons from regions AB and BC must gather in plane B, those from regions CD and DE in plane D, etc. Correspondingly, planes A, C, E, etc. must become electron-depleted. Thus, in planes B, D, etc. separated by a distance equal to the spatial period of the $E_z(t, z)$ wave, bunches of electrons form.

In the event of exact synchronism (ideal phase-matching), when the velocity of translational electron motion along the z axis is equal to the phase velocity of electron motion ($v = v_z$), the electrons are motionless in relation to the field of the travelling wave and there is no exchange of energy between the electrons and the wave. When v is greater than v_z, the bunches overtake the wave, which corresponds to motion from left to right in Figure 26.1, that is, against the decelerating force. In the process the kinetic energy of electron motions is transformed into the energy of the microwave field, which means that the field is amplified. This amplification is coherent, and with appropriate

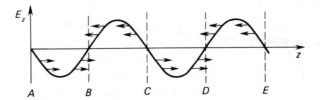

Figure 26.1. Grouping of electrons by the electric field of the longitudinal component of a slowed down travelling wave. The E_z vs. z distribution is shown at $t = 0$.

feedback radiation is emitted. As usual, this emission begins as a result of fluctuation bursts, in whose radiation field the formation of electron bunches intensifies, and the bunches in turn produce more electromagnetic radiation. In terms of quantum electronics this means stimulated emission.

The travelling-wave tube has an exceptionally wide bandpass. In the case at hand the spatial period of bunch formation, Λ, is Tv_z, with T the temporal oscillation period. For a helical decelerating system, $v_z \approx cd/2\pi a$ and, to a first approximation, is independent of the field's frequency, that is, dispersion is absent. The synchronism conditions, which determine the oscillation frequency of the amplified field, depend only on the potential difference of the electrostatic field accelerating the electrons and are not bound to any characteristic natural spatial period of the retarding structure, which can therefore be considered homogeneous. Because travelling-wave tubes have a wide bandpass, they are commonly used to amplify weak microwave signals. Devices used to construct oscillators usually have a narrower bandpass. Their characteristic feature is a spatially inhomogeneous periodic structure with its own resonance frequencies, that is, an essentially dispersive structure.

Consider an electron flying at a velocity V through a spatially periodic system characterized by a spatial period Λ. The electron is under a periodic force acting with a temporal period Λ/V. Hence, the electron's acceleration varies periodically with a frequency V/Λ . The corresponding electron trajectory is spatially-periodic, and an electron moving along such a trajectory emits radiation. When the electron velocity is essentially nonrelativistic, the radiation frequency is equal to the frequency of periodic motion, V/Λ. For macroscopic spatial periods Λ the radiation wavelength

$$\lambda = c\Lambda/V, \tag{26.1}$$

corresponds in the nonrelativistic case to the radio-frequency range.

The situation changes radically as the electron's velocity grows. The relativistic Doppler effect leads to a sharp decrease in the radiation wavelength in comparison to the characteristic dimensions of the radiating system. At electron velocities close to the speed of light ($V \lesssim c$), we have $\Lambda' = \Lambda(1 - V^2/c^2)^{1/2}$, in view of Lorentz contraction, and the frequency of the periodic action on the electron in the reference frame co-moving with the electron increases up to $\gamma V/\Lambda$, with $\gamma = (1 - V^2/c^2)^{-1/2}$ the relativistic factor. In the laboratory reference frame the inverse Lorentz transformation leads to a relativistic Doppler increase in radiation frequency in the direction of electron motion by a factor of $\gamma(1 + V/c)$. For ultrarelativistic electrons ($V \approx c$) we can put $1 + V/c \approx 2$, and as a result of this the wavelength of bremsstrahlung in the laboratory reference frame decreases sharply in comparison to the spatial inhomogeneity period Λ:

$$\lambda \approx \Lambda/2\gamma^2. \tag{26.2}$$

To estimate the value of γ it is expendient to use the relativistic relation linking the energy $W = mc^2$ of a highly energetic particle with the particle's mass $m = \gamma m_0$, with m_0 the particle's rest mass. In electron accelerator physics, electron energy is usually measured in electronvolts, using the relation $W = eU$, where e is the electron charge and U the potential difference of the electrostatic field required to accelerate the electron to a velocity V. Then

$$\gamma = eU/m_0c^2. \tag{26.3}$$

As is known, the electron rest energy m_0c^2 is 511 keV. This means that at $eU = 50$ MeV the relativistic factor γ is approximately 10^2, so that at $\Lambda = 1$ cm the radiation wavelength λ falls into the visual range.

Hence, if we wish to create free-electron lasers, we must focus on essentially relativistic cases where

$$\gamma = (1 - V^2/c^2)^{-1/2} \gg 1, \quad \text{or} \quad |V - c| \ll 1. \tag{26.4}$$

Relativistic effects manifest themselves not only in a sharp increase in the frequency of the radiation emitted by moving electrons. Obviously, when electrons move at relativistic speeds, the synchronism between the electron and light beams is provided for automatically. Electron accelerators, which generate beams of high-energy electrons, usually operate in the pulsed mode. If in the time it takes an electron bunch to fly through the region where the bunch interacts with the light wave the electron bunch misses the light wave by a distance less than the wavelength, synchronism can be assumed not violated.

True, there are also weakly relativistic devices, such as the cyclotron-resonance maser developed under the guidance of the Soviet Academician A. V. Gaponov-Grekhov. Such masers have proved to be highly promising sources of high-power short-wave microwave radiation. The common way to obtain high power outputs at a fairly low oscillation frequency (in the micro-wave and far IR ranges) is to employ heavy-current beams of moderate-energy electrons. Moving into the optical range requires the use of high-energy elec-tron beams (see Eq. (26.2)), which in view of the high energy have a fairly low number density. In the case of strong currents, collective effects in the beam's plasma play an important role. (At the beginning of this lecture we discussed the importance of collective effects in creating conditions for stimu-lated emission.) In the case of weak currents the electron-field interaction is essentially single-particle. Having in view a laser that uses ultrarelativistic elec-trons ($\gamma \gg 1$), we will carry out the subsequent discussion in the single-particle approximation.[7] A quantitative criterion for the validity of the single-particle approach will be given at the end of this discussion.

In a free-electron undulator laser the beam of relativistic electrons (usually a train of short electron packets) flies through a fairly extended region in which the magnetic field is spatially periodic (Figure 26.2). Systems ensuring that the field is spatially periodic are known as undulators (from the Late Latin word *undula* meaning "small wave") or wigglers (the spatially varying magnet-ic field causes the electron beam to "wiggle" and hence to radiate). Magnetic undulators produce near the beam's axis a time-constant transverse spatially periodic field that is either linearly or circularly polarized.

Let us consider a laser with a helical undulator whose magnetic field at the axis is circularly polarized. Under circular polarization of the wave propagating along the z axis parallel to the electron beam, the electrons are acted upon by the vector potential $\mathbf{A}_1$ of the undulator field and the vector potential $\mathbf{A}_2$ of the field of the electromagnetic wave, with

$$\mathbf{A}_1 = \frac{H}{q} (\mathbf{x} \cos qz + \mathbf{y} \sin qz),$$

(26.5)

$$\mathbf{A}_2 = \frac{Ec}{\omega} \left[\mathbf{x} \cos \left(\frac{\omega}{c} z - \omega t \right) - \mathbf{y} \sin \left(\frac{\omega}{c} z - \omega t \right) \right].$$

---

[7] The author expresses his sincere gratitude to M. V. Fyodorov for his collaboration on the material that follows.

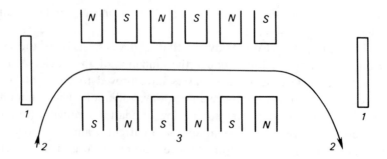

Figure 26.2. The schematic of a free-electron laser: *1*, mirrors; *2*, electron beam; *3*, magnetic undulator; N and S stand for the North and South poles of the undulator magnets.

Here **x** and **y** are the unit vectors directed along the **x** and **y** axes that are perpendicular to each other and to the z axis, E is the electric field strength, ω the frequency of the electromagnetic wave propagating along the z axis, and $q = 2\pi/\Lambda$, with Λ and H the period and strength of the undulator's magnetic field.

In the reference frame moving at the initial electron velocity $V \approx c$ the potentials (26.5) assume the form

$$\mathbf{A_1'} = \frac{H}{q}\left[\mathbf{x}\cos\frac{\Omega}{c}(z' + Vt') + \mathbf{y}\sin\frac{\Omega}{c}(z' + Vt')\right],$$

$$\mathbf{A_2'} = \frac{Ec}{\omega}\left[\mathbf{x}\cos\frac{\omega'}{c}(z' - Vt') - \mathbf{y}\sin\frac{\omega'}{c}(z' - Vt')\right], \tag{26.6}$$

where the "primed" quantities refer to the moving reference frame and, in accordance with (26.3) and (26.4),

$$\omega' = \omega(1 - V/c)^{1/2}(1 + V/c)^{-1/2} \approx \omega/2\gamma, \quad \Omega = cq\gamma.$$

The first formula in (26.6) shows that the vector potential of the undulator field in the co-moving reference frame resembles that of a plane wave of frequency Ω. In other words, a relativistic electron "perceives" the static spatially periodic magnetic field as an electromagnetic field propagating head-on and having a wavelength of Λ/γ. The resonance condition $\omega' \approx \Omega$ determines the field frequency in the vicinity of which amplification and lasing is possible in a free-electron undulator laser. In the laboratory reference frame this condition yields

$$\omega_0 = 2\gamma^2 cq,$$ (26.7)

which is equivalent to formula (26.2).

The equation of motion of an electron in the co-moving reference frame is (see § 17 in the well-known book *The Classical Theory of Fields* by L. D. Landau and E. M. Lifshitz (4th ed., Oxford: Pergamon Press, 1975))

$$\frac{d}{dt'} \left(\mathbf{p}' + \frac{e}{c} \mathbf{A}' \right) = \frac{e}{c} \mathbf{z} \left(\mathbf{V}' \frac{\partial \mathbf{A}'}{\partial z'} \right),$$ (26.8)

where $\mathbf{z}$ is the unit vector along the z axis, $\mathbf{p}'$ and $\mathbf{V}'$ are the electron momentum and velocity, and $\mathbf{A}' = \mathbf{A}_1' + \mathbf{A}_2'$ and we have allowed for the fact that $\mathbf{A}'$ is independent of transverse coordinates. In view of this independence it is easy to write the first integral of motion for Eq. (26.8), which determines the electron motion in the xOy plane:

$$\mathbf{p}_\perp' = -\frac{e}{c} \mathbf{A}'.$$ (26.9)

If we assume that the motion of the electron in the co-moving reference frame is nonrelativistic, the integral of motion (26.9) immediately yields the electron velocity in the transverse plane:

$$V_x' = -\frac{e}{m_0 c} A_x', \quad V_y' = -\frac{e}{m_0 c} A_y'.$$ (26.10)

As Eqs. (26.6) show, the vector potential $\mathbf{A}$ has no longitudinal component A_z, which corresponds to the character of the double winding of the helix of the solenoid that generates the undulator field and to the fact that the electromagnetic wave propagating in the undulator is transverse. Then, according to Eq. (26.8), the equation for the longitudinal component p_z' of the electron momentum assumes the form

$$\frac{dp_z'}{dt'} = \frac{e}{c} \left(V_x' \frac{\partial A_x'}{\partial z'} + V_y' \frac{\partial A_y'}{\partial z'} \right).$$ (26.11)

Substituting into (26.11) the velocity components V_x' and V_y' from (26.10), we easily find that

$$\frac{dp_z'}{dt'} = -\frac{e^2}{2m_0c^2}\frac{\partial}{\partial z'}(A_x'^2 + A_y'^2).$$

(26.12)

The sum of the squares of the transverse components of the total vector potential $\mathbf{A}'$ is given by the following formula:

$$A_x'^2 + A_y'^2 = \frac{H^2}{q^2} + \frac{c^2E^2}{\omega^2} + 2\frac{cEH}{q\omega}$$

$$\times \cos\left[\frac{\Omega + \omega'}{c}z' + (\Omega - \omega')t'\right].$$

(26.13)

Substituting this into (26.11) and allowing for the fact that $p_z' = m_0dz'/dt'$ for nonrelativistic motion in the co-moving reference frame, we arrive at the following equation for the electron longitudinal coordinate z' in this frame:

$$\frac{d^2z'}{dt'^2} = \frac{e^2}{m_0^2c^2}\frac{eH}{q\omega}(\Omega + \omega')\sin\left[\frac{\Omega + \omega'}{c}z' + (\Omega - \omega')t'\right].$$

(26.14)

The expression inside the square brackets determines the phase of the electron motion in the undulator field and in the field of the wave propagating through the undulator:

$$\varphi \equiv \frac{\Omega + \omega'}{c}z' + (\Omega - \omega')t'.$$

(26.15)

The relationship that links φ with the longitudinal coordinate z of the electron motion and the time t in the laboratory reference frame can be obtained via the inverse Lorentz transformation:

$$\varphi = \left(\frac{\omega}{c} + q\right)z - \omega t.$$

(26.16)

In the neighborhood of a resonance, that is, at $\omega' \approx \Omega$, we have

$$\frac{d^2\varphi}{dt^2} = 4\frac{e^2\Omega^2}{m_0^2c^3}\frac{EH}{q\omega}\sin\varphi.$$

(26.17)

Going over to the laboratory reference frame and substituting $\Omega = cq\gamma$ and $\omega \approx \omega_0 = 2\gamma^2 cq$ into Eq. (26.17), we get the equation

$$\frac{d^2\varphi}{dt^2} = 2\frac{e^2 EH}{\gamma^2 m_0^2 c^2}\sin\varphi. \tag{26.18}$$

Thus, the equation of motion of an electron in an undulator has been reduced to the equation of the classical simple pendulum for the phase of the motion. This proves the great similarity between the free-electron laser and microwave electronic devices, which in the fixed-field approximation are described by analogous equations.

Further analysis requires knowing the initial conditions. At the moment when an electron enters the undulator the phase has an arbitrary value φ_0, generally speaking. The second initial condition can easily be obtained by finding the derivation of Eq. (26.16), which serves as the definition of phase. As a result at $t = 0$ we have

$$\varphi = \varphi_0, \quad \frac{d\varphi}{dt} = \dot{\varphi}_0 = \left(\frac{\omega}{c} + q\right)V - \omega \approx cq - \frac{\omega}{2\gamma^2} = -\frac{\omega - \omega_0}{2\gamma^2}. \tag{26.19}$$

Note that the initial rate of phase variation is proportional to the detuning of the radiation frequency from the resonance value.

The combination of Eq. (26.18) and the initial conditions (26.19) fully defines the motion of an electron in the fields of the undulator and the wave and enables determining the basic characteristics of the laser.

Let us find the energy emitted by an electron in the undulator per pass. The energy emitted per unit time is defined as the work performed by the field of the wave on the electron times minus one:

$$\frac{d}{dt}F(t) = -c\mathbf{E}\cdot\mathbf{V}, \tag{26.20}$$

where by definition $\mathbf{E} = -c^{-1}\partial\mathbf{A}_2/\partial t$. This equation establishes a simple relation between the emitted energy F and phase φ.

Indeed, if we allow for (26.9), we see that the transverse electron velocity in the laboratory reference frame is given by the formula

$$\mathbf{V}_\perp = \frac{1}{\gamma}\mathbf{V}'_\perp = -\frac{e}{\gamma m_0 c}\mathbf{A}' = -\frac{e}{\gamma m_0 c}\mathbf{A} = -\frac{e}{\gamma m_0 c}(\mathbf{A}_1 + \mathbf{A}_2). \tag{26.21}$$

Substituting (26.6) into (26.21) and the result into (26.20) and performing simple transformations, we get

$$\frac{d}{dt} F = - \frac{e^2}{\gamma m_0 c^2} \frac{cEH}{q} \sin \varphi. \tag{26.22}$$

But $\sin \varphi$ is linked to $d^2\varphi/dt^2$ by the pendulum equation (26.18), which yields the sought relation in the fairly simple form

$$\frac{d}{dt} F = - \frac{\gamma m_0 c^2}{cq} \frac{d^2\varphi}{dt^2} = - \frac{W}{cq} \frac{d^2\varphi}{dt^2}. \tag{26.23}$$

Here $W = \gamma m_0 c^2$ is the total energy of a relativistic electron.

Integrating this equation with allowance for the initial conditions (26.19) for $d\varphi/dt$ and assuming, naturally, that $F(0) = 0$ yields

$$F = \left(\frac{\omega_0 - \omega}{2\Omega} - \frac{1}{cq} \dot{\varphi} \right) W. \tag{26.24}$$

Next we employ the well-known first integral of motion of a pendulum. This integral expresses the law of conservation of energy: the sum of kinetic and potential energies of a pendulum at an arbitrary moment in time is equal to the sum at time $t = 0$. Using our notation together with the initial conditions (26.19) yields

$$(\dot{\varphi})^2 + \frac{4e^2 c^2 EH}{W^2} (\cos \varphi - \cos \varphi_0) = \text{const} = \left(\frac{\omega - \omega_0}{2\gamma^2} \right)^2. \tag{26.25}$$

Note that the value of the total energy of a pendulum is determined both by the initial phase φ_0 and by the detuning $\omega - \omega_0$. In the low-signal approximation

$$2L^2 e^2 EH/W^2 \ll 1, \tag{26.26}$$

where we have allowed for the fact that the time it takes an electron to pass through an undulator of length L is L/c, that is, as $E \to 0$, our pendulum is in rotational motion about a position of equilibrium rather than in vibrational motion, performing complete rotations with a circular frequency $(\omega - \omega_0)/2\gamma^2$. This means that in the low-signal approximation Eq. (26.25) can be solved by the method of iterations with respect to the term $4e^2c^2EH(\cos\varphi - \cos\varphi_0)/W^2$.

Not wishing to clutter up our discussion with complicated formulas, we only note that in the zeroth-order approximation the emitted energy, given by Eq. (26.24), is nil. In the absence of an electromagnetic wave field there is neither emission nor absorption. In the next, first-order, approximation the amount of emitted energy proves to be proportional to $\cos\varphi_0$ or $\sin\varphi_0$. But in high-energy electron accelerators the electron beam consists, as noted earlier, of electron bunches (electron packets) of finite duration and of a longitudinal size usually no less than 1 mm, which considerably exceeds the wavelength of the light. Hence, the emitted energy must be averaged over the initial phase φ_0.

As a result of such averaging the emitted energy "vanishes" in the first order in E. Only in the second order in iterations does Eq. (26.25) yield a nonzero average rate of phase variation, which with the aid of (26.24) makes it possible to determine the average energy $\overline{F}$ emitted by an electron per pass. This quantity is linked in a natural way to the value of the radiation gain per pass in power, $1 + G$, namely,

$$G = \frac{4\pi}{E^2}\,N_e\overline{F},\tag{26.27}$$

where N_e is the electron number density. After performing rather clumsy calculations we arrive at the following expression:

$$G = \frac{4\pi^{3/2}N_e e^4 H^2 L^3}{m_0^3 c^{9/2}\omega_0^{3/2}\Lambda^{1/2}}\,\frac{d}{du}\,\frac{\sin^2 u}{u^2},\tag{26.28}$$

where we have introduced the quantity $u = 4\pi n(\omega - \omega_0)/\omega_0$, with $n = L/\Lambda$ the number of undulator periods and L the length of the undulator.

The factor $\dfrac{d}{du}\dfrac{\sin^2 u}{u^2}$ in (26.28) determines the dispersive dependence of G on ω. Amplification is possible $(G > 0)$ if $u < 0$ or $\omega < \omega_0$. Maximum gain is attained at $|u| = 1$. This condition determines the bandwidth

$$\Delta\omega = \omega/4\pi n,\tag{26.29}$$

which is a consequence of the finiteness of the undulator length $(L = n\Lambda)$ and resembles the common homogeneous linewidth. Note, however, that formula (26.28) was obtained on the assumption that the electron beam is monoenergetic. This is not always the case in reality, and if the energy spread of the electrons in the beam, $\Delta W/W$, is sufficiently large, there emerges inhomogeneous broadening, which may prove to be much greater than homogeneous.

If $\Delta W/W > 1/4\pi n$, inhomogeneous broadening exceeds homogeneous and formula (26.28) must be averaged over the electron energy distribution function $f(W)$. In the case of strong inhomogeneous broadening the factor $u^{-2}\sin^2 u$ can be approximated by the delta function: $u^{-2}\sin^2 u = \pi\delta(\omega)$. Then after averaging over W we get

$$G = \frac{e^4 H^2 N_e L\Lambda^{5/2}}{4\pi^{1/2}m_0 c^{3/2}\omega^{1/2}}\frac{df(W)}{dW}\bigg|_{W=W_0}. \tag{26.30}$$

Here we have introduced the notation $W_0 = m_0 c^2(\omega/2qc)^{1/2}$ and assumed that the distribution function $f(W)$ is normalized to unity, $\int f(W)dW = 1$, in view of which $(df/dW)_{max} \approx 1/(\Delta W)^2$. Maximum gain is attained for $W_0 - \overline{W} < 0$, $|W_0 - \overline{W}| \approx \Delta W$, with $\overline{W}$ the average electron energy in the beam. This implies that the amplification bandwidth in this case is

$$\Delta\omega = (\Delta W/W)\omega. \tag{26.31}$$

Formula (26.30), which is valid for $\Delta W/W > 1/4\pi n$, allows for a straightforward similarity between free-electron lasers and lasers whose operation is based on transitions between discrete levels of atoms or molecules. Indeed, the condition that absorption be negative $(G > 0)$ is met if $(df/dW)|_{W=W_0} > 0$. This means that amplification is carried out by the electrons belonging to the ascending wing of the electron energy distribution function, while absorption is due to the electrons belonging to the descending wing. In other words, amplification is observed if the number of electrons with higher energy in the vicinity of W_0 is greater than the number of electrons with lower energy, which is simply the condition for population inversion on levels in relation to a system with a continuous spectrum. In the event of inhomogeneous broadening, $\Delta W/W > 1/4\pi n$, negative absorption is achieved when the common condition for population inversion in the vicinity of energy W_0 is met, with the condition dependent on frequency ω and the undulator period $\Lambda = 2\pi/q$.

Formulas (26.28) and (26.30) were obtained in the single-particle approximation. Yet, as has been noted, in the case of strong electron currents collective effects in the beam's plasma may play an important role, generally speaking. However, if in the reference frame co-moving with the relativistic electrons the product of the increment of the plasma instability development and the electron transit time in the undulator is small, no instabilities develop and collective effects can be ignored. The maximum instability-development increment in the plasma is determined by the plasma frequency $\omega_p = (4\pi e^2 N_e/m_0)^{1/2}$. The condition that the interaction be single-particle comes down to the requirement that $\omega_p' t' \ll 1$, where ω_p' and t' are, respectively, the plasma frequency and the interaction time interval in the co-moving reference frame. The inverse Lorentz transformation for time and longitudinal coordinate leads to the condition that $\gamma^{-3/2} c^{-1} L (4\pi e^2 N_e/m_0)^{1/2} \ll 1$, which is always met for ultrarelativistic electrons with a great margin.

So we see that a beam of relativistic electrons propagating rectilinearly in a magnetic undulator is capable of amplifying and, hence, under appropriate feedback generating radiation at wavelengths determined by the spatial period of the undulator, Λ, and the value of γ, that is, the electron energy. The wavelength tuning of the radiation then is naturally achieved by adjusting the electron energy. Lasers of this type can operate, theoretically, on wavelengths ranging from the submillimeter to the far UV. For characteristic undulator lengths of several meters and a spatial period of 1-3 cm, the relative amplification linewidth would be 10^{-4}-10^{-3}. Usually the relative non-monoenergetic value for electron beams exceeds this figure.

By using double superconducting helixes a circularly polarized magnetic field with an induction of several kilogauss can fairly easily be generated in undulators. In these conditions, for an electron beam current of several amperes ($N_e \approx (0.5\text{-}1) \times 10^{11}$ cm^{-3}) and an electron energy of 20-30 Mev, gain per pass amounting to several percent (of the order of ten percent) and lasing with a peak power output of the order of 10^4 W have been achieved in the near IR.

Lasing has also been achieved in the medium IR range by employing linear accelerators and undulators that use permanent samarium-cobalt magnets with a gap field of about 0.3-0.5 T. Varying the electron energy from 20 to 10 MeV changes the radiation wavelength almost by two octaves, from 9 to 35 μm (see (26.2) and (26.3)). For a 1-Hz repetition frequency of the pulses of accelerated electrons, the average lasing power output was found to be 0.5-1 W. A pulse power output of the order of 10 MW corresponds to a peak current of 50 A.

The temporal (and, correspondingly, the spectral) structure of the radiation emitted by such free-electron lasers is fairly complicated. Electron accelerators used as sources of electron beams usually operate in the pulsed mode. The length of the pulses in the electron beam is of the order of 1 μs, as a rule. These pulses, however, are by no means smooth; rather, they constitute a regular train of short electron bunches (packets) of an essentially shorter duration (usually several picoseconds). In some accelerators it has proved possible to realize a continuous train of such bunches.

To complete the feedback circuit it is necessary that the temporal distance between the bunches in the regular train entering the undulator be an integral multiple of twice the transit time of the radiation travelling in the cavity. Only in this case will the radiation generated by the electron bunches and stored in the laser cavity in the course of multiple reflections from the cavity mirrors enter the undulator in the form of wave packets in step with the electron packets and become amplified in the course of the entire time during which the train of electron bunches exists. As a result the laser radiation manifests itself in the form of a train of short pulses separated by time intervals equal to, or as an integral multiple of, twice the transit time of the radiation travelling between the laser cavity mirrors, the distance between which is practically equal to the undulator length.

This mode of operation is in all respects equivalent to mode locking in dye lasers (see Lecture 22) pumped by a train of laser pulses with a time interval that is an integral multiple of the transit time of the radiation travelling through the laser.

The nonmonochromaticity of the radiation emitted by a free-electron laser is, therefore, determined by the length of the electron bunches. Bunches 3-ps long occupy a spatial region with a linear dimension of about 1 mm and result in an emission-spectrum width of about 10 cm^{-1}. Increasing the monochromaticity of the radiation of free-electron lasers requires longer electron bunches (assuming, of course, that the peak value of the electron current remains unchanged).

In accordance with what has been said, free-electron lasers realized on the basis of linear accelerators whose electron current consists of 1000-2000 spikes each 35-ps long with an overall length of 100 μs at a maximum repetition frequency of 1 Hz produce radiation in the 1000-cm^{-1} range with a peak power output of 10 MW, while the power output averaged over the train of spikes is 10 kW and the lasing power output averaged over a large time interval reaches 1 W.

The efficiency of transforming the energy of the electron beam into light is about one percent.

The spectral properties of the radiation correspond to those discussed above. For a small excess over the self-excitation threshold, the width of the emission spectrum exceeds severalfold the Fourier transform limit for an optical pulse 30-ps long, which results in a 0.3% relative width at $\lambda = 10.6\ \mu$m. At high power output levels the relative broadening noticeably increases, reaching 4% at $\lambda = 10.6\ \mu$m.

At the same time we must mention the high optical quality of the radiation emitted by free-electron lasers, which is due to the smallness of the Fresnel number of their long and thin cavities (see (7.11) and (7.12)) and the homogeneity of their active medium. The beam of light generated by a laser whose parameters are given above is focused into a spot only 5% greater in area than the spot determined by the diffraction limit.

Simpler accelerating systems, such as microtrons and Van de Graaf generators, have found wide acceptance in creating free-electron lasers for the far IR spectral region at wavelengths ranging from 100 to 400 μm.

Problems to Lecture 26

26.1. Why do free-electron lasers require accelerators?

26.2. How is tuning of the lasing frequency achieved in free-electron lasers?

26.3. How does gain depend on the current density?

26.4. What must be done to drive down the lasing wavelength when designing free-electron lasers?

26.5. Which electron beam is preferable, a narrow or a wide? What restriction must be imposed?

26.6. In what respects does the radiation emitted by a free-electron laser differ from synchrotron radiation from the practical viewpoint?

LECTURE 27

The Seventeen Best-Known Lasers

Brief descriptions. Characteristic properties. Methods of creating
conditions for population inversion

In earlier lectures we discussed the most interesting lasers that exist today.
For the sake of reference here are brief descriptions of their characteristic
features.

The helium-neon laser (Lecture 13). Lasing in the cw mode involves the
$3s \rightarrow 2p$ transition ($\lambda = 0.63$ μm), the $2s \rightarrow 2p$ transition ($\lambda = 1.15$ μm), and
the $3s \rightarrow 3p$ transition ($\lambda = 3.39$ μm) of neutral neon atoms. The upper laser
levels of neon are excited in the quasiresonance transfer of excitation energy
from helium to neon in inelastic collisions, which proceed with a small energy
gap.

The metastable helium state that transfers its energy to neon in collisions
is excited by the electrons in the glow-discharge plasma. Collisional energy
transfer effectively excites the level being populated if the energy transfer rate
noticeably exceeds the rate of level decay,

$$N\sigma v \gg 1/\tau, \tag{13.7}$$

a condition that is well-satisfied for a neon-helium mixture with the partial
pressure ratio ranging from $1 \div 5$ to $1 \div 15$. For population inversion to exist
in the cw mode the lower laser level must become rapidly depopulated, that
is, the effective lifetime of the upper level must exceed the lifetime of the lower
level:

$$\tau_2 > \tau_1. \tag{14.8}$$

In the helium-neon laser the lower levels become depopulated during collisions
with the walls of the gas-discharge tube.

A characteristic feature of the helium-neon laser is the competition of the
lasing 0.63-μm and 3.39-μm lines, which have the common initial level $3s$.

The argon laser (Lecture 14). The active medium consists of Ar^+ ions.
Lasing takes place in the cw mode and involves transitions between high-lying
levels of argon ion configurations, $3p^4 4p \rightarrow 3p^4 4s$. Emission at $\lambda = 514.5$ nm
and $\lambda = 488.0$ nm is the strongest. Population inversion is created in a low-
pressure high-current capillary discharge during a cascade process of ioniza-

tion of an atom followed by excitation of the ion in collisions with the discharge electrons. The lower laser level is depopulated radiatively.

The argon laser is characterized by low efficiency and high power output, with the result that the structural elements (electrodes and the walls of the discharge capillary) function in an extremely harsh environment.

The cadmium laser (Lecture 14). Lasing takes place in the cw mode at $\lambda = 441.6$ nm and $\lambda = 325.0$ nm and involves transitions between the states $^3D_{3/2,5/2}$ and $^2P_{3/2,1/2}$ of the Cd$^+$ ions. Population inversion is created by an electric discharge in a mixture of helium and cadmium vapor in conditions when the excitation energy is transferred from the metastable state of the helium atom to the cadmium atom through collisions, which leads to ionization of the cadmium atom and excitation of the ion (Penning ionization). The discharge electrons supply the energy to the metastable state of helium.

The helium-cadmium laser is similar to the helium-neon laser in the mechanism of excitation of the upper laser levels and to the argon laser in the mechanism of depopulation of the lower levels.

The copper-vapor laser (Lecture 14). The active media is a vapor consisting of neutral copper atoms. Lasing involves transitions between the copper-atom configurations $3d^{10}4p$ and $3d^94s^2$ at $\lambda = 578.2$ nm and, primarily, at $\lambda = 510.5$ nm, and the upper laser levels $^2P_{1/2,3/2}$ are resonance levels while the lower levels $^2D_{3/2,5/2}$ are metastable. Hence, for the active media of the copper-vapor laser we have an inequality that is the opposite of (14.8). As a result population inversion can exist for only a short time compared to the lifetime of the lower laser level. This means that the copper-vapor laser belongs to the class of self-terminating lasers. Population inversion is created by a gas discharge in a mixture of copper vapor and a buffer gas.

The discharge electrons in such a laser primarily populate the resonance level and weakly excite the metastable level, whose population in the course of radiative transitions from the upper level leads to the disappearance of population inversion. Lasing is essentially pulsed. During the rapid discharge the gain switch-on mode is maintained in which the radiation pulse may be much shorter than the lifetime of the upper level. The repetition frequency of the lasing pulses cannot exceed the inverse of the lifetime of the lower level.

Self-terminating lasers have a high limit transition efficiency

$$\eta_{\text{lim}} = \frac{g_{\text{low}}}{g_{\text{low}} + g_{\text{up}}} \frac{h\nu}{E_{\text{up}}} \tag{14.21}$$

owing to the fact that the metastable (lower laser) level occupies a fairly low position on the energy scale. The copper-vapor laser has $\eta_{lim} \approx 0.38$.

The carbon dioxide laser (Lectures 15 and 16). Lasing takes place in the pulsed and cw modes and involves vibrational-rotational transitions within the system of the lower vibrational levels of the ground electronic state of the carbon dioxide molecule. The upper laser level is the first excited state of the asymmetric stretching vibrational mode, $00°1$, while the lower laser level may either be the first excited state of the symmetric stretching vibrational mode, $10°0$ ($\lambda = 10.6$ μm), or the second excited state of the deformation vibrational mode, $02°0$ ($\lambda = 9.6$ μm). Population inversion is created primarily in an electric discharge when the energy is transferred, via collisions, from the N_2 molecules excited by discharge electrons and also when the vibration $00°1$ of the carbon dioxide molecule is directly excited in collisions with electrons.

The lower laser levels become depopulated through collisions because of the high cross section of the deactivation of the deformation vibration in collisions with the buffer gas, for which helium is often used. What is essential is the presence of a Fermi resonance between levels $10°0$ and $02°0$, which links the symmetric stretching and deformation vibrations of the carbon dioxide molecule. In the most common active mixture for the carbon dioxide laser, $CO_2 \div N_2 \div He$, carbon dioxide emits radiation, nitrogen stores the energy, and helium depopulates the lower laser levels. In addition, helium facilitates the electric discharge and cools the gas mixture. Gasdynamic excitation and energy transfer from molecules excited chemically are also possible.

Electric-discharge carbon dioxide lasers operating at low pressure (several dozen torr) use longitudinal discharge in fairly long gas-discharge tubes: at high pressure (of the order of 1 atm and higher) transverse discharge has found application. At high pressure both self-sustained and non-self-sustained discharges are possible, the latter being preferable. The discharge is initiated and sustained through the ionization of the gas in a discharge gas by UV radiation, by high-energy electron beams, etc.

Tuning of the carbon dioxide laser in the frequency range from 900 to 1100 cm^{-1} is carried out at pressures lower than a few atmospheres in steps of 1-2 cm^{-1} following the lines that correspond to the vibrational-rotational transitions in the P- and R-branches belonging to the vibrational $00°1 \rightarrow 10°0$ and $00°1 \rightarrow 02°0$ bands, and continuously at pressures higher than 5-6 atm. Owing to the effect of rotational competition, practically all the energy stored in the nonequilibrium distribution of particles over the levels can be emitted at a single frequency.

The carbon dioxide laser is characterized by high efficiency and high power and energy outputs in the cw and pulsed modes.

The gasdynamic laser (Lecture 16). The source of radiant energy is the thermal energy of an equilibrium-heated molecular gas. Population inversion is created as a result of the sudden cooling due to vibrational relaxation processes that proceed at different rates for different vibrational modes of the polyatomic molecule or for different components of the gas mixture. The cooling rate must be sufficiently high so that the relaxation processes have no time to depopulate the upper laser level and, hence, the level's population corresponds to the high initial temperature of the gas. In the case of the carbon dioxide laser this refers to the relaxation of the energy stored on the $00^\circ 1$ level of the carbon dioxide molecule and in the vibrationally excited nitrogen. Cooling is also accompanied by relaxation processes that depopulate the lower laser level. In the case of the carbon dioxide laser this means the vibrational relaxation of the deformation vibrations of the carbon dioxide molecule. The required rapid cooling of great masses of flowing gas is achieved gasdynamically in the supersonic flow of the compressed heated gas into a vacuum.

Combustion of fossil fuel can be used for the heating process and for formation of carbon dioxide. Then instead of costly helium it becomes expedient to use the water vapor formed in the combustion process as the buffer gas, in collisions with whose atoms (or molecules) the levels of the carbon-dioxide deformation mode become depopulated.

Gasdynamic lasers operate in the pulsed mode (explosion, shock waves) and in the cw mode (combustion, electric heating). Actually, gasdynamic lasers are heat engines that directly transform thermal energy into the energy of coherent electromagnetic radiation.

The chemical laser (Lecture 17). Population inversion is created during the nonequilibrium distribution of energy among the internal degrees of freedom of the products of exothermic chemical reactions at the expense of the energy liberated in the course of the reaction, which proceeds, as a rule, on transitions between the vibrational levels of molecules in the gaseous phase. Lasing takes place on the vibrational-rotational transitions of the diatomic molecules of hydrogen-halide compounds formed primarily in a substitution reaction of the type

$$A + BC \rightarrow AB^* + C. \tag{17.1}$$

To reduce the share of energy used for the initiation of such a chemical reaction, chain self-sustaining reactions are employed. The longer the reaction chain, the greater the extent to which the radiant energy is determined by the energy stored in the initial reactants and the smaller the role played by the energy of the initiating effect, which triggers the reaction.

Both pulsed and cw modes of operation are possible. In the pulsed mode, UV photodissociation or electron-beam radiolysis initiates the chain reaction, which proceeds fairly rapidly so that relaxation processes have no time to destruct population inversion. The cw mode can be realized by pumping through the gases and removing the reaction products from the reaction cell. Bringing mutually unstable reactants into chemical contact and rapidly removing reaction products make it possible to realize a purely chemical laser without initiation.

The following wavelengths are characteristic of chemical lasers: 2.7 μm for HF, 3.7 μm for HCl, 4.2 μm for HBr, and 4.3 μm for DF. Also resonant transfer of excitation energy, $DF^* \rightarrow CO_2$, is widely used to create a chemical carbon dioxide laser. Among the merits of chemical lasers are the possibility of obtaining population inversion in large volumes and at high mass flow rates of the active substance and the absence, theoretically, of any sizable expenditure of energy for creating population inversion at the moment and place where it was obtained.

A characteristic feature of these lasers is a chemical mechanism for creating population inversion in which the radiant energy exceeds the energy of initiation of the chemical reaction.

The photodissociation laser (Lecture 17). Population inversion is achieved on the transitions between the electron energy levels of atoms that are the products of pulsed photodissociation of stable molecules. An example is the iodine laser, in which excited iodine atoms are obtained by pulsed photolysis, say, according to the scheme

$$CF_3I + h\nu \rightarrow I^* + CF_3. \tag{17.33}$$

Radiation is emitted in the $^2P_{1/2} \rightarrow {}^2P_{3/2}$ transition, with both states belonging to the same electron configuration $5p^5$ of iodine. The radiation wavelength is 1.315 μm. The UV radiation emitted by a nonmonochromatic source of dissociating light is absorbed in a broad molecular band, and the excited atoms (the dissociation products) emit within a narrow line characteristic of atomic spectra. Large volumes of the initial molecular gas can be subjected to photodissociation. Hence, the photodissociation laser has the potential for generating high-energy pulses.

A characteristic feature of the photodissociation laser, which makes it similar to the chemical laser, is that the active substance and the lasing medium are created simultaneously.

The carbon monoxide laser (Lecture 18). Lasing takes place on vibrational-rotational transitions in the ground electronic state of the carbon monoxide molecule. The radiation wavelength ranges from 5 to 6.5 μm. Vibrational excitation occurs via direct population of the upper vibrational levels of the carbon monoxide molecule in collisions with the electrons generated in a gas discharge, in energy transfer from the vibrationally excited nitrogen molecules, etc. In this respect the carbon monoxide laser is similar to the carbon dioxide laser.

In view of the specific character of the relaxation of vibrational energy in the single-mode and essentially anharmonic oscillator that is the carbon monoxide molecule, complete vibrational population inversion is absent. The population distribution over the vibrational levels exhibits a plateau. In the presence of a plateau, that is, when the vibrational population numbers are equal for at least two levels, partial rotational inversion manifests itself. Lasing, here essentially cascade by nature (see Figure 18.3), is observed only in the P-branch. The plateau's size, which fixes the spectrum of lasing frequencies, depends on the ratio of the rates of vibrational relaxation of different vibrational levels, and is determined by the temperature and composition of the gas mixture. At $T \approx 100$ K the plateau may extend from $V = 5$ to $V = 35$. The difference in the frequencies of lasing on transitions with different V and J is fixed by the size of the vibrational anharmonicity and the value of the rotational constant of the carbon monoxide molecule. In contrast to carbon dioxide lasing, the cascade character of carbon monoxide lasing makes it impossible to transform the entire stored energy into radiation at a single frequency.

The total efficiency of the carbon monoxide laser over all the lasing lines may reach very high values. The laser can operate in the pulsed and cw modes. Using xenon as a buffer gas makes it possible to operate at room temperature and with sealed-off systems. The carbon monoxide laser is characterized by an absence of vibrational population inversion and the cascade character of lasing in the P-branch of the vibrational-rotational transitions.

The nitrogen and hydrogen lasers (Lecture 18). The nitrogen laser operates in the UV range, and the 337.1-nm radiation is the most important. Population inversion is created in a pulsed electric discharge on transitions between fairly high-lying excited electronic states of the N_2 molecule. In accordance with the Franck-Condon principle the discharge electrons participating in vertical transitions on the potential-energy vs. internuclear-distance diagram (see Figure

18.5) collisionally populate the state that lies higher on the energy scale, is fairly tight, and contains the upper laser levels. The looser state containing the lower laser levels and because of the looseness positioned to the right on the diagram and lower on the energy scale does not get populated. Lasing, also in accordance with the Franck-Condon principle, is achieved in radiative vertical transitions from the right turning points of the upper state to the lower state. A similar mechanism for creating population inversion operates in the hydrogen laser, which functions in the vacuum UV range, with wavelengths ranging from 116 to 126 nm.

The lower laser levels of the N_2 and H_2 molecules have a longer lifetime than the upper. Hence, the nitrogen and hydrogen lasers belong to the class of self-terminating lasers. Population inversion in these lasers lasts for a short time (3-10 ns for the nitrogen laser and less than 1 ns for the hydrogen laser). For this reason, population inversion is created by an excitation wave travelling along the laser's axis in synchrony with a light pulse. Because the time that population inversion exists is so short the nitrogen and similar lasers are super-fluorescence lasers.

A characteristic feature of the nitrogen and hydrogen lasers is the separation of the excitation and emission channels in accordance with the Franck-Condon principle for transitions between the electronic states of a molecule.

Excimer lasers (Lecture 18). Lasing takes place on transitions from bound electronic states (the upper state) to the repulsive electronic state (the lower state) of what is known as excimer molecules, that is, molecules that are stable only in an excited electronic state. The repulsive state in such molecules corresponds to the ground electronic state. Examples of excimer molecules are dimers of noble-gas atoms and halides of noble-gas atoms. The presence of excimer molecules is equivalent to population inversion. Hence, inversion is achieved by creating excimer molecules, which is done by employing a high-energy electron beam or in conditions of a gas discharge. Depopulation of the lower laser level takes place automatically during decay of the molecules that have returned to the ground state, that is, molecules that find themselves in the repulsive state as a result of radiative transitions.

Dimers of noble-gas atoms are formed during the excitation and ionization of the atoms by a beam of high-energy electrons in triple collisions with unexcited atoms, which requires high pressures (exceeding 10 atm). Monohalides of noble gases are formed during harpooning reactions between an excited atom of the noble gas and halogen, reactions realized in binary collisions, which proceed at fairly low pressures. This makes possible gas-discharge excitation of such lasers.

The lasing wavelengths of excimer lasers range from the optical region to vacuum UV. The most important applications have been established with XeF-, XeCl-, KrF-, and KrCl-lasers, operating at 352, 308, 249, and 222 nm, respectively.

A characteristic feature of excimer lasers is the fact that the lower level is completely depopulated automatically.

The ruby laser (Lectures 19 and 20). Lasing is achieved primarily in the pulsed mode on transitions between the metastable excited and ground states of Cr^{3+} ions (ground state $^4F_{3/2}$) isomorphically implanted in α-corundum Al_2O_3, at a wavelength of about 0.69 μm (the luminescence R-line of ruby). Population inversion is achieved through the three-level optical-pumping scheme (see Figure 19.2). The radiation emitted by a nonmonochromatic source of the gas-discharge flashlamp type is effectively absorbed on transitions from the ground state of chromium, $^4A_2$, to the broad $^4F_2$ and $^4F_1$ bands (the wavelengths of the pumping radiation are approximately 410 and 550 nm).

Excitation energy is not stored on the resonance levels of $^4F_2$ and $^4F_1$ but quickly transferred nonradiatively to the metastable state (the ^2E doublet), whose population grows owing to the small rate of decay of this state. As particles gather on the metastable levels, steady-state population inversion with respect to the ground state sets in:

$$n_3 - n_1 = N \, \frac{W - w_{31}}{W + w_{31}} , \qquad (19.7)$$

which remains valid as long as the upper level of the pumping transition becomes populated faster than the upper level of the laser transition becomes nonradiatively depopulated ($W > w_{31}$).

Creating conditions for population inversion requires an expenditure of energy at the initial stage because at least half of the particles must be transferred from the ground state to the metastable levels by the three-level pumping scheme.

Ruby lasers have high energy parameters. In the pumping bands absorption is 2-3 cm^{-1}. The threshold value of the pumping energy density in the green band is approximately 3 J/cm^3. When the threshold is considerably exceeded in free-running lasing pulses of about 1 ms, the radiant energy density is about 0.2-0.25 J/cm^3. The linear gain attained with ruby rods reaches 0.2-0.25 cm^{-1}. Pulsed-periodic and (for high-quality crystals) cw modes of operation are possible. Ruby lasers with Q-switching and mode-locking have been built.

A characteristic feature of ruby lasers is the three-level optical-pumping scheme.

The neodymium laser (Lectures 19 and 20). Lasing is carried out in the pulsed and cw modes on transitions between the metastable excited states of Nd^{3+} ions (ground state $^4I_{9/2}$) isomorphically implanted in crystals or glass. Population inversion is achieved through the four-level optical-pumping scheme (see Figure 19.4). The nonmonochromatic pumping radiation is effectively absorbed on transitions from the ground state to a set of several fairly narrow bands from which the excitation energy is rapidly transferred to the metastable level $^4F_{3/2}$. This level serves as the initial level for laser transitions.

The $^4F_{3/2} \to {}^4I_{11/2}$ transition has the highest probability. The laser wavelength is 1.06 μm. The energy gap between the $^4I_{11/2}$ and $^4I_{9/2}$ states (see Figure 20.3) ensures that the four-level optical-pumping scheme is realized in this laser. At a high rate of nonradiative energy transfer from the pumping bands to the upper laser level the steady-state population inversion condition is realized, that is,

$$n_3 - n_4 = \frac{NW(w_{41} - w_{34})}{w_{23}w_{34}}, \tag{19.20}$$

which sets this laser apart from the ruby laser.

The population inversion is positive if $w_{41} > w_{34}$, meaning that the lower laser level becomes depopulated faster owing to nonradiative transitions to the ground states than it becomes populated owing to transitions from the upper laser level. The pumping threshold for population inversion is low and practically absent if the conditions for the realization of the four-level optical-pumping scheme are met, that is, when the lower level in the laser transition lies above the ground state by $\Delta E \gg kT$.

Neodymium-doped laser glasses are known to have a high concentration of active centers, a strong inhomogeneous amplification-line broadening, and the possibility of obtaining the active media in large volumes. A typical operational mode of neodymium-doped glass lasers is the high-energy single-pulse mode. Crystals, primarily of the YAG type, are activated by neodymium to lower concentrations, the amplification lines in them are broadened to a much lesser extent, and the volume of the active medium is significantly limited by the technical difficulties of growing large homogeneous crystals. Neodymium-garnet lasers easily operate in the cw and pulsed-periodic modes.

The characteristic features of neodymium lasers are the four-level optical-pumping scheme and high energy parameters.

Dye lasers (Lecture 22). Lasing is carried out in the pulsed and cw modes on transitions between the levels of the excited and ground singlet states of the complex molecules of organic dyes. Usually the dyes are used in the form of highly diluted liquid solutions. Population inversion is achieved by employing the four-level optical-pumping scheme (see Figures 22.2 and 22.3). In the process of singlet-singlet absorption the pumping radiation populates the vibrational-rotational states of the singlet state being excited. In accordance with the Franck-Condon principle, for the case of shifted equilibrium configurations the uppermost vibrational levels are excited.

Within the excited singlet state fast nonradiative relaxation occurs and the excitation energy is transferred to the lower vibrational levels of this state. From the lower levels the molecule, again in accordance with the Franck-Condon principle, performs a radiative transition to the upper vibrational levels of the ground singlet state. The energy of the emitted photon is lower than the energy of the absorbed photon from the pumping radiation (the Stokes shift). The excess energy of the lower laser levels relaxes in the process of intrastate maxwellianization. The Franck-Condon principle and the rapid (0.1-1 ps) intrastate relaxation ensure the four-level character of optical pumping in dye lasers.

The continuous-electronic-state spectrum is the result of superposition of many adjacent vibrational states of the heavy polyatomic molecule of the organic dye and corresponds to the inhomogeneous broadening of the spectral lines representing the transitions between states. When the cavity is tuned to a definite frequency within the amplification line, the excited state is radiatively depopulated at that frequency as a result of positive feedback. The laser frequency changes as the cavity is tuned to other frequencies. Owing to the high rate of intrastate relaxation the entire energy stored by the excited state (minus Stokes losses) is pumped into single-frequency radiation.

Dye lasers operate at wavelengths ranging from the near IR to the near UV. Continuous tuning of the laser wavelength is achieved in bands several dozen nanometers wide with a monochromaticity up to several megahertz.

A characteristic feature of dye lasers is the continuous tuning of the lasing wavelength combined with the possibility of operating in the cw mode and the pulsed mode down to the subpicosecond range.

Color-center lasers (Lecture 23). The level diagram and the method used to obtain population inversion are similar to those of dye lasers. The active substance consists of point defects in the structure that serve as color centers. Their luminescence spectrum lies in the near IR range, where lasing in the pulsed and cw modes has been achieved with continuous tuning in a frequency band approximately 1000 cm^{-1} wide. The width of the amplification lines is due to the intense interaction of the electrons localized at defects with nearest-neighbor ions in the crystal lattice, and this leads to the formation of vibrational energy levels in the electronic states. There is a great diversity of point defects of the anion-vacancy/localized-electron type and their aggregates, which are color centers in various transparent crystals. Examples of color centers effectively used in lasers are F_2^+-centers (0.8-1.1 μm) and F_2^--centers (1.1-1.3 μm) in LiF crystals.

A characteristic feature of color-center lasers is the use, as impurity centers immersed in the crystalline matrix, of point defects of the matrix proper, whose spectrum allows for the four-level optical-pumping cycle in the broad luminescence lines of the near IR range.

Semiconductor lasers (Lectures 24 and 25). Lasing is maintained in the pulsed and cw modes on band-to-band recombination transitions of direct-gap semiconductors. Population inversion is achieved by creating nonequilibrium charge carriers in the conduction and valence bands. For population inversion to take place the quasi-Fermi levels of the carriers must lie within the respective allowed bands, that is,

$$F_n - F_p > E_g. \tag{24.13}$$

In injection semiconductor lasers the nonequilibrium electron-hole pairs are created by a direct current that injects the carriers into the p-n junction of the semiconductor diode. Creation of population inversion is greatly facilitated when the laser diodes use heavily doped semiconductors, whose electron and hole gases in the n- and p-region are strongly degenerate. Here the high density of states in the bands leads to high gain values. The width of the spectral region where amplification is possible is specified by the frequency condition

$$E_g < \hbar\omega < F_n - F_p. \tag{24.14}$$

Injection lasers have been realized in many one-valley direct-gap semiconductors. One of the best is the gallium arsenide laser ($\lambda = 0.84$ μm). Manufacturing the p-n junctions of laser diodes in the form of so-called heterostructures reduces the threshold injection current and makes possible cw lasing at room temperature. The dimensions of the active region of diode lasers do not exceed several micrometers. Hence, the power output of a single laser is moderate and in the cw mode no higher than several watts. The wavelength range of such lasers is determined by the band gap of the semiconductor employed. Crystals of varying composition make it possible to span a large range. Depending on its composition, the AlGaAs system emits radiation in the 0.63-0.90 μm range, the AlGaSb in the 1.2-1.80 μm range, and the GaInAs in the 0.9-3.4 μm range. The solid solutions PbSnTe and CbHgTe enable constructing lasers operating in the 30-40 μm range.

The radiation's wavelength of a diode laser can be continuously tuned by changing the temperature of the crystal, by applying hydrostatic pressure, or by applying a magnetic field.

It is characteristic of injection lasers that population inversion is obtained by generating nonequilibrium charge carriers in the p- and n-region of the semiconductor diode by means of direct current pumping. This implies compactness, high efficiency, low inertia, broad frequency band, and the possibility of fine tuning the laser radiation.

Free-electron lasers (Lecture 26). Radiation is observed when a relativisitc electron beam passes through a spatially periodic external field. Most often the relativistic electrons are made to propagate along the axis of a device known as undulator, that is, along the axis of a magnet whose field is constant in time and varies periodically in space. Stimulated undulator emission occurs when parallel to the electron beam there propagates along the axis an external electromagnetic wave whose frequency is somewhat lower than the resonance interaction frequency ω_0 (see Eq. (26.7)).

The radiation is amplified in the region where an electron packet is localized as the packet passes through the undulator. The mirrors of the cavity, into which the undulator is placed, confine the train of amplified radiation to the laser until the next electron packet arrives. Amplification is then repeated. Multiple repetition of this process results in pulsed lasing, with the lasing wavelength determined by the undulator period Λ and the relativistic factor γ as follows:

$$\lambda = \Lambda/2\gamma^2. \tag{26.2}$$

In real devices an electron energy of 25-50 MeV corresponds to radiation with a wavelength in the center of the IR range.

A characteristic feature of free-electron lasers is the use as an active medium of a system of particles that can be considered in the classical framework, that is, not necessarily a quantum system. One result of this is continuity of the spectrum of possible lasing frequencies, which implies that it is easy to tune the lasing wavelength, determined in such lasers by the energy of the electrons in the relativistic beam.

* * *

Essentially, these lectures have set out the physical basics of quantum electronics and explained how the effect of stimulated emission of radiation can be used in systems with discrete energy levels to amplify and lase electromagnetic waves. As has been said, quantum electronics is almost exclusively the electronics of bound states, electronics in which monochromatic radiation is lased primarily at fixed frequencies.

The possibility of tuning the frequency of laser radiation corresponding to transitions between bound states appears as the effect of external agents, which broaden the transition lines, increases. Examples are high-pressure molecular lasers, dye lasers, and color-center lasers. But when the electrons become less bound and instead of discrete energy levels broadened by external agents we get broad energy bands of allowed states, tuning the radiation's wavelength becomes much easier, as the example of semiconductor diode lasers demonstrates. The next step consists in going over to free electrons, which possess a continuous spectrum and, therefore, make it possible to tune the radiation's wavelength in a simple manner by varying the energy of the accelerated electrons.

CONCLUSION

Development Trends

New wavelengths of laser radiation. The IR range. The optical range. Methods of nonlinear optics, harmonic generation, and difference-frequency generation. The SRS laser. The far UV and X-ray ranges. Gamma-ray lasers. Applications of lasers

One of the main goals of quantum elecronics is to broaden the range of wavelengths of laser radiation, primarily in spectral regions that have yet to be conquered by laser methods. To conclude this course of lectures let us briefly examine the prospects for achieving this. The most general method of mastering the entire optical range, from the far IR to the far UV, by quantum electronics is to develop the appropriate free-electron lasers (see Lecture 26).

Although these lasers are still at the stage of theoretical and preliminary experimental studies, there is a clear picture of their potential. Free-electron lasers are quickly mastering the UV and even far UV ranges. But there is still much to do in the IR range to improve the energy characteristics of these lasers, raise their efficiency and gain, and increase the monochromaticity of their radiation. Special accelerators can also be expected to be designed for these lasers, since the possibility of building free-electron lasers has already been demonstrated in principle with the help of accelerating devices that can hardly be considered optimal.

Nevertheless, it must be clearly understood that free-electron lasers are complicated technically and differ from the idea of a laser accepted in quantum electronics. Hence, the great importance of developing new lasers based on the traditional methods of quantum electronics.

In the IR range, including the far IR (submillimeter) range, optical pumping of molecular gases has proved to be promising. Owing to the narrowness of the absorption lines (in contrast to the case of solids), pumping in gases must be resonant, that is, laser pumping. And since the vibrational-rotational spectra of molecules are rich, population inversion can be attained in many schemes of optical pumping of the vibrational levels. Theoretically, population inversion can be attained on transitions between excited levels belonging to different vibrational modes of a molecule (i.e. on composite vibrations with only one mode being pumped), between the excited levels of a single vibrational mode with only one harmonic being pumped, and between the rotational levels of the ground and/or an excited vibrational state with the ground vibrational-rotational transition being pumped. In the latter case lasing is usually observed in the submillimeter range.

The most promising have proved to be the methods of creating population inversion on excited vibrational levels with a composite vibration being pumped (the CF_4 laser, $\nu \approx 600$ cm^{-1}), and on the excited vibrational levels belonging to the band in which pumping takes place, with a defect in rotational quanta (the NH_3 laser, $\nu \approx 800$ cm^{-1}).

Submillimeter rotational-rotational lasers, and also NH_3 and CF_4 lasers and similar vibrational-vibrational lasers, operate under pumping by radiation from carbon dioxide lasers (see Lectures 15 and 16). The interest in pumping molecular lasers by radiation from carbon dioxide lasers stems from the possibility of drastically broadening the wavelength range mastered by quantum electronics into the 10-1000 μm region and constructing tunable sources of radiation for this region. It is important that the unique properties of the carbon dioxide laser can be effectively carried over to longer waves.

A class of promising methods of generating new lasing lines in the visible and adjoining spectral ranges is based on creating new and more complex excimer molecules, such as exciplexes (excited complexes) ArF_2^*, $XeCl_2^*$, and $XeBr_2^*$ and excimers that are noble gas oxides. There is also a future for the photodissociation of vapors of complex molecules, which leads to the appearance of excited dimers (e.g. dissociation of dihalides of cadmium, zinc, and mercury, which yields the CdI*, ZnI*, HgI*, HgBr*, and HgCl* dimers), and the dissociation of complex halogen-containing compounds, which leads to the appearance of excited interhalides of the IF* type. Of interest, too, is utilization of the recombination luminescence of N_2, O_2, and Cl_2 molecules. Apparently, the gas lasers of the noted types are extremely promising for mastering practically all lasing wavelengths in the entire visible range. This is corroborated by the fact that new excimer and photodissociation lasers have shown to lase radiation at 310, 440, 475, 490, 503, 520, 555, and 656 nm.

The potential of solid-state lasers in the visible and IR ranges is far from exhausted. This is true both of lasers that use crystals with transition-element ion impurities and of color-center lasers. Not so long ago the possibility of lasing on electronic-vibrational transitions of the $^4T_2 \rightarrow \, ^4A_2$ band of the Cr^{3+} ion was realized in a gadolinium-scandium-gallium garnet. Further search for promising matrices and impurity ions and ways of controlling energy migration and sensitization should lead to a considerable broadening of the spectral region mastered by lasers of this type.

The review article "Crystals of rare-earth gallium garnets with chromium as the active media of solid-state lasers" by E. V. Zharikov, V. V. Osiko,

A. M. Prokhorov, and I. A. Shcherbakov (*Izv. Akad. Nauk SSSR, Ser. Fiz.* English translation: *Bull. Acad. Sci. USSR, Phys. Ser.* 1, vol. 48, No. 7, pp. 1330-42, 1984) shows the universality of this new class of laser crystals with excellent mechanical, thermal, and optical properties, inherent in oxide crystals with the garnet crystal. The possibility of doping these crystals not only with chromium ions but also with ions of many transition elements, notably, rare-earth element ions, opens up new horizons in creating new active media for solid-state lasers with diverse modes of operation and a broad spectrum of properties.

Similarly, we may expect the appearance of an ever growing number of new color-center lasers that will enable moving into ranges where there are as yet no tunable lasers with a worthwhile output power. Examples are lasers for the 2-3 μm range based on color centers in KCl or RbCl crystals with lithium.

A special place in quantum electronics is occupied by the methods of nonlinear optics now being rapidly developed for, among other things, the transformation of laser frequencies. These methods are of a sufficiently general nature and should be treated in a special course of lectures. However, in speaking of the possibility of broadening the spectrum of laser frequencies we cannot ignore the methods used in nonlinear optics. In Lecture 3, following S. V. Vavilov, we introduced the concept of nonlinear optics in discussing the decrease in light absorption by a medium with increasing irradiation intensity (the saturation effect). The propagation of waves in a medium is described by linear equations only when the optical parameters of the medium are independent of the wave's field. The optics corresponding to this case is known as linear. But when the optical parameters begin to depend on the strength of the field of the light wave propagating in the medium, the equations become nonlinear and we are dealing with nonlinear optics.

The methods based on nonlinear processes of transforming the frequencies of monochromatic electromagnetic radiation, such as detection and the generation of harmonics, difference frequencies, and sum frequencies, have been well-developed in classical electronics. The high intensity, directivity, and monochromaticity of laser light have made it possible to employ these radio engineering methods extensively in the optical range. A great contribution to developing the methods of nonlinear optics, primarily in the field of transforming laser frequencies, was made by Academician R. V. Khokhlov (1926-1977), and the Nobel Prize Winner Nocolaas Bloembergen.

For the sake of brevity we will mention only two basic propositions on which the nonlinear-optics methods of transforming laser frequencies rest.

First, a strong field makes the medium's susceptibility nonlinear, which means that in the expansion of the medium's polarization in powers of the electric field strength in the light field,

$$P = \chi^{(1)}E + \chi^{(2)}E^2 + \chi^{(3)}E^3 + \ldots,$$

where the $\chi^{(n)}$ ($n > 1$) are known as the nonlinear susceptibilities (of order n), the higher-order terms become important. This implies that under a strong monochromatic field the polarization, that is, dipole moment per unit volume of medium, oscillates not only at the fundamental frequency but also at harmonics. If the radiation contains more than one frequency, there also emerge oscillations at sum and difference frequencies. As a result the radiation field propagating in a nonlinear medium contains harmonic components and components at sum and difference frequencies.

Second, a necessary condition for effectively transforming a frequency is the presence of phase synchronism between the waves of the initial and required frequencies, a synchronism that ensures build-up of the transformation effect along the entire length of the nonlinear material (phase matching). The reason is that in optics, in contrast to the radio range, the region of nonlinear interaction is usually much greater in size than the wavelength and the interaction takes place in the travelling-wave mode.

It is also important to choose the nonlinear material to meet the needs of pushing the laser frequency into the required range. The best results in generating harmonics in the visible range have been achieved with such crystals as ADP (ammonium dihydrophosphate), KDP (potassium dihydrophosphate), lithium iodate and lithium niobate, and sodium-barium niobate. For the generation of difference frequencies to enable continuous spanning of the IR region up to 20 μm, such crystals as gallium selenide, silver thiogallate, zinc germanium-phosphide, and cadmium arsenogermanate have found application.

Note that highly effective transformations have been achieved in the generation of harmonics for the visible and near UV regions. The intensities obtained in this way coincide in order of magnitude with the initial irradiances attained with ruby, neodymium, and dye lasers. However, in the IR range the radiation at difference frequencies is still weak, reaching fractions of a millijoule in the pulsed mode and fractions of a milliwatt at high repetition frequencies. There is still much to be done in this area.

The use of solids as nonlinear materials is restricted by their spectral transparency range. Atoms and molecules in the gaseous phase have attracted great interest for moving farther into the UV range. The decrease in the medium's density in gases can be compensated for by the increase in nonlinear suscepti-bilities resulting from the proximity to a resonance of the frequencies of the interacting fields. Since the resonance lines in gases are narrow and outside resonances gases are transparent, gaseous media are suitable for nonlinear in-teractions from the IR range to the far UV range and soft X-ray radiation.

One great advantage of gases is the possibility of controlling the phase synchronism by mixing gases that have different wavelength dependencies of the refractive index. At 10^{14}-10^{15} W/cm^2 intensities of the fundamental radia-tion of the fifth, seventh, and ninth harmonics are generated in alkali metal vapors and in noble gases, producing radiation in the 100-40 nm range. The transformation efficiency is usually very low. At present the best transforma-tion efficiency in the 100-nm range is only 10^{-5}. The generated power reaches several hundred watts in picosecond pulses. This, of course, is a great achieve-ment, but it is still too early to speak of mastering the far UV range by sources of laser radiation.

Among the nonlinear effects that can be successfully used for transforming laser frequencies is the Raman effect. This effect is most often used to scatter light on molecular vibrations. The energy of interaction of molecules with a light wave is determined by the square of the wave's field strength. At high intensities of the incident radiation the overall effect of the electric fields of the incident and the scattered light results in a situation in which the force acting on the molecules contains a sizable component at the difference fre-quency of these fields. But according to the Raman effect this freqency is equal to a frequency of natural vibrations of a molecule. What we have here is a resonance buildup of molecular vibrations, which leads to an increase in the scattering intensity.

This growth in the intensity of the scattered light further stimulates vibra-tion of the molecules of the scattering medium, and feedback results. What is observed is stimulated Raman scattering (SRS). By placing in a laser cavity the medium that scatters light in such a manner we get the SRS laser, or simply the Raman laser. The shift in frequency of the radiation emitted by the Raman laser with respect to the frequency of the pumping laser radiation is a multiple of the frequency of natural vibrations of the molecules of the scattering medi-um that are active in Raman scattering. Raman lasers can operate on the rota-tions of molecules as well as on the vibrations. The efficiency of transforming the pumping radiation into the required radiation may be very high and is restricted, theoretically, only by Stokes losses.

Raman lasers provide a promising way of transforming the laser frequency into a given spectral region and are widely used in combination with such lasers as the excimer, the chemical, the neodymium, and the carbon dioxide. Also quite obvious is the often used potential of this laser for transforming the frequency step-by-step when the Raman effect is used in cascade processes.

The methods of nonlinear optics have proved highly productive, but their use is limited by the possibilities of the existing lasers and in the far UV and X-ray regions is not very effective. To really master new spectral regions with ever decreasing wavelengths it is necessary to develop direct methods of creating population inversion.

So far we have touched on ways of broadening the range of wavelengths of laser radiation that have left the R&D stage and are moving into the everyday practice of quantum electronics. We will now examine the less developed but also promising avenues of laser research.

First we consider the far UV region, which is commonly understood to encompass radiation with wavelengths ranging from 1 to 50 nm. In its shortwave section the far UV region joins the soft X-ray region.

As noted in Lecture 18, the difficulties in mastering the UV range are fundamental. As (18.6) shows, gain decreases like v^{-5} as we move into the region of ever shorter waves. Such a sharp drop in gain with increasing frequency requires a drastic increase in the pumping intensity, which requires an excessive expenditure of energy to create population inversion.

We will consider the main aspects of this problem step by step drawing on the article "On the prospects of amplifying the light of the far UV range" by F. V. Bunkin, V. I. Derzhiev, and S. I. Yakovlenko (*Kvantov*. Elektron. [English translation: *Sov. J. Quantum Electron.*], vol. 8, No. 8, pp. 1621-49, 1981). We note first that an analysis of the possibilities of building lasers for the far UV range has shown the expedience of restricting the discussion to the problems of creating population inversion that will lead to strong superradiance. It is too early to discuss problems related to feedback in this extremely short-wave range. Cavities in their classical double-mirror form can hardly be utilized in the far UV range, although distributed feedback circuits could, apparently, be used. Of course, the most important problem is how to create population inversion.

Let us now turn to the choice of the proper active medium. We recall that a wavelength of 1 nm corresponds to a photon with an energy of 1240 eV. At the same time the greatest variation in the energy of an outer (optical) electron of a neutral atom on transitions between discrete levels cannot exceed the energy needed to tear the electron away from the atom. Of all the elements in the Periodic Table helium has the highest suitable ionization potential (24.6 eV). Transitions involving energies of hundreds of electronvolts can be achieved only between energy levels of the inner electrons in multielectron atoms. However, the spectroscopy of multielectron configurations is poorly developed, the kinetics of the relaxation processes is complicated, and the presence of many radiative decay channels seriously lowers the reliability of a theoretical analysis and hinders experimental study of the excitation and relaxation of such systems. Therefore, the main interest has been focused on an analysis of the possibilities of amplification on transitions in the outer shells of multiply charged ions. In the case of hydrogenlike ions of iron, copper, and zinc, for instance, the only remaining electron is "pressed" to the nucleus so hard that transitions between its energy levels correspond to the short-wave part of the far UV range. A similar situation is encountered in helium- and neon-like ions of atoms heavier than helium and neon, respectively. Modern technical means, such as laser-produced breakdown and electron beams, make it possible to create plasma containing the required multiply charged ions.

There can be many methods for creating population inversion on transitions in the outer shell of a multiply charged ion. The most promising is the one that uses recombination cascade population of the upper operating state accompanied by radiative depopulation of the lower operating state. Practically for all types of collisional recombination the electron first finds itself in a highly excited state and then, after performing a series of cascade transitions, appears in the ground state. If one of the excited states in this cascade decays faster than an upper-lying state, population inversion sets in in the recombining plasma and exists as long as the collisional recombination flux dominates over ionization and excitation. Such nonequilibrium recombining plasma must be supercooled, that is, the electron temperature must be fairly low. To maintain the nonequilibrium character of the plasma the cooling of the plasma must proceed fairly rapidly, that is, the cooling time must be shorter than the recombination time.

Note that there is a certain analogy here with the gasdynamic method of creating population inversion (see Lecture 16). Under cooling the plasma cylinder disperses, and the cooling time can be estimated by assuming it equal, in order of magnitude, to the ratio of the initial cylinder radius to the rate of boundary movement. For a cylinder radius of 10^{-2} cm and an expansion rate of 10^6-10^7 cm/s, the cooling time proves equal to 1-10 ns. The recombination time is determined by the chosen level diagram, the type of ion, the mechanism of depopulation of the lower operating state, the plasma parameters, and the sought lasing wavelength. In the most promising case of radiative depopulation of the lower level of a hydrogenlike ion in the creation of population inversion on the $n = 4 \rightarrow n = 3$ transition, where n is the principal quantum number, the requirement that the plasma be recombination-nonequilibrium limits the lasing wavelength from below to 2-3.5 nm.

More important, however, is that the possibility of moving into the 1-nm region is limited not by the cooling time but by the necessary energy deposit. In the scheme considered here we find that the threshold energy input grows so rapidly as the wavelength decreases that in going over from $\lambda = 10$ nm to $\lambda = 1$ nm the value of the required energy input increases from 3 kJ/cm^3 to 0.3 GJ/cm^3.

We will not consider the difficulties encountered in forming a plasmoid with the parameters required for the lasing of far UV radiation. We only note that lasing in a nonequilibrium recombining plasma of a noble gas combined with metal vapor has been achieved in the optical range and adjacent regions of the IR and UV ranges on many transition lines (about 40) of singly charged ions of Be, Mg, Al, Ca, Sr, Sn, Ba, and Pb, to name only some. It has also been found that the lower operating states are effectively depopulated (radiatively). There are also reports that population inversion has been observed in the recombining plasma of a laser-produced breakdown on transitions of multiply charged ions in the far UV range.

Apparently, building effective sources of stimulated radiation in the far UV range has proved feasible.

We have considered only one of the possible approaches to solving this important problem of quantum electronics. The recombination instability of a plasma can be created and maintained in a steady state by intense ionizing radiation from an external source. If the source is sufficiently strong (10^{15}-10^{16} W/cm^2), there is no need to separate in time the processes of energy deposit into the plasma and cooling of the electron gas. This lifts the restriction imposed on the lasing wavelength from below by the smallness of the cooling time, and lasing of X-ray radiation in the 0.1-1 nm wavelength range on transitions between hydrogenlike states of multiply charged ions of high-Z atoms becomes possible.

We note once more that building sources of laser light in the far UV and X-ray ranges is one of the most important problems of quantum electronics. The problem is far from solution, although we have the necessary theoretical prerequisites and the first experimental proof that a solution does exist.

For example, in a laser-induced explosion of 75-nm thin foil manufactured from selenium and yttrium a (5.5 ± 0.5)-cm^{-1} gain has been attained at a wavelength of approximately 21 nm. The experiment was carried out at the Lawrence Livermore National Laboratory (USA) using the Novetta laser facility, developed for the laser fusion program. Two-sided irradiation of a selenium or yttrium film sputtered onto a transparent polymer support 110-nm thick was employed. The second harmonic of the radiation of the Novetta laser (0.53 nm) was focused into a strip 2.2-cm long and 0.02-cm wide as homogeneous as possible on the film. The radiation pulse was 450 ps long and the intensity 5×10^{13} W/cm^2. The result was the formation of a plasma cylinder containing neon-like ions of selenium (or yttrium). Population inversion forms between the $2s^2 2p^5 3p$ and $2s^2 2p^5 3s$ configurations, the latter rapidly decaying into the ground state $2s^2 2p^6$. The dynamics of population inversion formation leads to the assumption that the mechanism of this formation is electron-collisional. In the case of selenium, gain was observed at 20.9 and 20.6 nm and in the case of yttrium at 15.5 nm.

Actually, what was demonstrated in these experiments was the amplification of spontaneous emission in the near X-ray spectral region (60-80 eV photons). The exponential growth of the luminance with the length of the irradiated surface and the angular anisotropy of the luminance attest to the laser nature of the observed effect. The equivalent brightness temperature of the 20.9-nm amplified radiation in the direction of greatest brightness reached 5.5×10^8 K. The use of heavier neon-like ions in the same method should lead to a laser effect at $\lambda = 8$ nm.

A much more complicated question is whether it is possible to build gamma-ray lasers to lase monochromatic electromagnetic radiation with photon energies ranging from several dozen kiloelectronvolts to hundreds of kiloelectronvolts (wavelengths ranging from tenths of a nanometer to several thousandths). Analysis of the potential of gamma-ray lasers encounters problems referring to many fields of science. Besides quantum electronics, these include optical and nuclear spectroscopy, chemistry, crystallography, solid-state physics, and neutron physics.

Many schemes for building gamma-ray lasers exist, many theoretical studies of suggested operational modes have been carried out, many contradictions have been resolved, many technical difficulties have been pointed out, and many ways of surmounting these difficulties have been examined. However, there is still no report on the results of experimental studies that could be considered a path to creating a gamma-ray laser.

Since the possible avenues of application of monochromatic radiations at wavelengths comparable to, or even smaller than, atomic dimensions appear to be practically boundless in both fundamental studies and applications, the search for ways to realize the idea of the gamma-ray laser is being actively pursued.

In conclusion it must be stressed that although quantum electronics is based on the R&D and commercial production of lasers, its content goes far beyond the basics of laser physics or lasers alone. Quantum electronics made possible in the IR, optical, and UV ranges very high values of radiative energy density in space, time, and a frequency interval and led to the appearance and rapid development of entirely new directions in science and technology, each of which deserves a separate course of lectures.

Here are some of these new directions.

In nonlinear optics, which was discussed earlier, methods of generating harmonics and difference frequencies are being developed, parametric amplifiers and lasers are being created, induced scattering of light is being studied, wavefront reversal (phase conjugation) is being realized, and the possibility of constructing nonlinear adaptive optical systems is being studied. A large share of the nonlinear optics studies deals with the self-interaction of intense light, during which the optical characteristics of the medium change and influence the conditions of propagation of light in the medium; for instance, self-focusing of light may occur.

Nonlinear spectroscopy investigates coherent processes of photon echo and light nutation and their spectroscopic applications, studies saturation and intra-Doppler effects, examines aspects of lasing frequency stabilization, develops the applications of coherent anti-Stokes scattering, and inspects processes of multiphoton light absorption, multiphoton ionization of atoms, and multiphoton dissociation of molecules.

Nonlinear laser spectroscopy is closely related to the study of resonance interactions of strong laser light with matter, which lead to laser-induced isotope separation, laser control of chemical reactions (including biochemical reactions), and the possibility of detecting single atoms and molecules by laser light.

Nonresonance processes of interaction of high-power laser radiation with matter include the optical breakdown of gases, laser-induced heating of breakdown plasma, vaporization of metals by laser light, laser cutting, welding, and hardening of metals, laser-assisted generation of sound, laser thermochemistry, and the destruction of transparent insulators by laser light. Great interest has been shown in the possibility of using lasers to obtain high-temperature plasma and inertia confined thermonuclear fusion.

In holography, integrated optics, and optical fiber communications, all of which are fields of optics born of quantum electronics, the use of lasers has opened up new horizons for transmitting, processing, and storing information.

Other fields rapidly developing are laser biophysics, laser biochemistry, and medical applications of lasers.

For many laser applications the useful effect arises from the possibility of concentrating energy at a given time in the interaction region. The number of latent applications of this potential is extremely high. The main property of laser light directly used here is the high spatial coherence and, hence, high directivity.

An entirely new possibility, that of a strong resonance action on matter, usually absent from electron-beam, plasma, explosion, and other methods of intense action, is provided by monochromatic laser light. The entire set of properties of laser light manifests itself in resonance interactions.

Above all, the monochromaticity and tunability of laser light have opened new areas in spectroscopy by greatly increasing the sensitivity, resolving power, and operation speed and developing remote control features in spectrometers that employ lasers. The dye, color-center, and semiconductor lasers have found wide application in these areas.

Also, since laser light is not only monochromatic but intense, this intensity finds its use in nonlinear laser spectroscopy. An example is coherent anti-Stokes Raman spectroscopy (CARS).

When intense laser light interacts resonantly with matter, the saturation effect becomes important. For instance, when the homogeneous component of an inhomogeneously broadened Doppler line is saturated, a narrow dip forms in the line. This effect is used in nonlinear intra-Doppler spectroscopy, whose resolving power for gases at low pressures may reach 10^4-10^6 Hz.

However, high-resolution spectroscopy, including nonlinear laser spectroscopy, does not exhaust the field of intense resonance interactions. The high spectral brightness of monochromatic laser light may lead to a situation in which the interaction of the resonance radiation with matter proceeds selectively.

In spectroscopic studies one assumes that light has no noticeable irreversible effect on the medium. The laser light used in laser spectroscopy neither destructs nor noticeably perturbs the medium investigated. However, the possibility of high energy concentrations in laser light ensures that lasers act energetically on matter and essentially change the state of the object on the macroscopic scale. The possible changes are extremely varied. They may be changes in the coordinates and velocities of the particles of the irradiated substance, changes in structure, phase transitions, and the discrimination of certain components. When these or other similar changes occur due to monochromatic laser light and the result of the interaction is found to be spectrally dependent, we are dealing with the resonant interaction of laser light with matter. The interaction is intense if the redistribution of population caused by the resonant absorption of light is noticeable and manifests itself in macroscopic changes in the properties or behavior of the system.

It is worth noting that the resonant excitation of a noticeable fraction of particles constituting the irradiated substance takes the system out of equilibrium considerably. Relaxation processes tend to return it to equilibrium. These processes, which in the final analysis lead to heating of the substance, hinder the selective action of laser light. For this reason the main problem in studying and applying intense resonant interactions is to determine the conditions for conserving the selectivity of the resonant action and choose the proper methods for creating such conditions.

The overall idea of realizing selective photoprocesses lies in breaking down such a process into at least two stages. The first stage is the resonant excitation of a noticeable quantity of microparticles, which can easily be made selective. The second stage must lead to irreversible changes in the physical properties of the selectively excited particles, in other words, to fixing the excitation. Loss of selectivity occurs primarily at this stage. Hence, it is this stage that undergoes the most thorough investigation. In such selective processes as laser-induced isotope separation, laser purification of gases, and laser photochemistry the two- and multistage ionization and dissociation have found the main application. The second stage in these processes is reduced to physical separation or chemical fixation of fragments of the atoms or molecules selectively obtained under intense laser irradiation. There is also the possibility of selective laser control of processes taking place at the boundary between substances in essentially different states of aggregation.

Owing to the high intensity of laser light, the interaction of light and matter cannot always be considered by first-order perturbation theory. Then, if the energy of the interaction of the radiation field with an atom or molecule becomes comparable to a characteristic internal energy or if the transition from state to state corresponds to the absorption of several photons, so-called multiphoton processes become essential. In such processes the absorption (emission) of several photons is synchronized, and the total energy of the photons is equal to the transition energy. Multiphoton processes considerably broaden the possibilities of effective resonance action of light on matter.

Further development of quantum electronics, the creation of new lasers, the broadening of lasing wavelength ranges and ranges of continuous tuning of laser frequencies in the IR, optical, and UV sections of the spectrum will greatly widen the field of laser applications and increase laser efficiency.

* * *

In conclusion we give approximate formulas and estimates of the radiation frequencies and photon energies in units characteristic of various application ranges:

$$\nu = \frac{3}{\lambda} \times 10^{14} \text{ Hz}, \quad \frac{\nu}{c} = \frac{10}{\lambda} \times 10^3 \text{ cm}^{-1},$$

$$h\nu = \frac{1.986}{\lambda} \times 10^{-12} \text{ erg} = \frac{1.986}{\lambda} \times 10^{-19} \text{ J} = \frac{5.52}{\lambda} \times 10^{-24} \text{ kW·h},$$

$$N_A h\nu = \frac{28.5}{\lambda} \text{ kcal/mol}, \quad \frac{h\nu}{k} = \frac{14.3}{\lambda} \times 10^3 \text{ K}, \quad h\nu = \frac{1.24}{\lambda} \text{ eV}.$$

Here $c = 2.997\,924\,58 \times 10^{10}$ cm/s is the speed of light, λ the wavelength in micrometers, $h = 6.626\,176 \times 10^{-34}$ J/Hz Planck's constant, $N_A = 6.022\,045 \times 10^{23}$ mol^{-1} Avogadro's number, and $k = 1.380\,662 \times 10^{-23}$ J/K the Boltzmann constant, and the relevant conversion factors are

$$1 \text{ cal} = 4.1868 \text{ J}, \quad 1 \text{ eV} = 1.602\,19 \times 10^{-19} \text{ J}.$$

Answers to Problems

1.2. Approximately 2×10^{12} times.

1.3 and **1.4.** $N_2/N_1 = \exp(-\Delta E/kT)$.

T, K	N_2/N_1		
	Maser	CO_2 laser	Ar laser
300	0.996	0.044	0
77	0.985	0	0
4.2	0.759	0	0

Thus, in contrast to masers, where even at helium temperatures the upper level is highly populated, in a laser it is almost always empty. This is one of the basic differences between lasers and masers.

1.5. $(N_1 - N_2)/N = \{1 - \exp(-\Delta E/kT)\}/\{1 + \exp(-\Delta E/kT)\}$.

ΔE, eV	10	1	0.1	0.01	0.001
λ, μm	0.124	1.24	12.4	124	1240
ν, GHz	2.4×10^{14}	2.4×10^{14}	2.4×10^{13}	2.4×10^{12}	2.4×10^{11}
$(N_1 - N_2)/N$	1	1	0.96	0.2	0.02

1.6. It is not necessary to know Planck's formula to derive Eq. (1.10). Suffice it to analyze the behavior of Eq. (1.8) under variations of temperature T. As $T \to \infty$, the radiative energy density tends to infinity as well. This is possible only if

$$\frac{g_1 B_{12}}{g_2 B_{21}} \exp \frac{E_2 - E_1}{kT} - 1 = 0.$$

Since

$$\lim_{T \to \infty} \exp \frac{E_2 - E_1}{kT} = 1,$$

we find that $g_{21} B_{12} = g_{21} B_{21}$.

1.7. The population of discrete states for particles obeying the Bose-Einstein statistics is given by the following formula: $n = \left[\exp\left(-\dfrac{E - \mu}{kT}\right) - 1\right]^{-1}$, where E is the energy of a single particle, and μ the electrochemical potential. Since photons are bosons for which $E = h\nu$ and $\mu = 0$, we have $n = \left[\exp\left(-\dfrac{h\nu}{kT}\right) - 1\right]^{-1}$. The average energy per one type of oscillation, that is, per mode, in the frequency interval $\Delta\nu$ is

$$E_{\nu,\nu+\Delta\nu} = \sum_{\nu}^{\nu+\Delta\nu} n(\nu)h\nu.$$

Going over from summation to integration and allowing for the two directions of polarization, we arrive at Planck's formula

$$\rho_\nu d\nu = 8\pi\nu^2 c^{-3} h\nu \left[\exp\frac{h\nu}{kT} - 1\right]^{-1} d\nu,$$

where $8\pi\nu^2 c^{-3} d\nu$ gives the number of modes per unit volume and in the frequency interval from ν to $\nu + d\nu$ for radiation contained in a volume whose linear dimensions are much greater than the radiation's wavelength λ.

2.1. $\Delta\nu_{\text{nat}} = \Delta\nu_0/2\pi = 1/2\pi\tau_0 \approx 15$ MHz.

2.2. $\Delta\nu_{\text{X-ray}} \approx \Delta\nu_{\text{vis}} \dfrac{\nu_{\text{X-ray}}^3}{\nu_{\text{vis}}^3} \approx 2000$ GHz.

2.3. $\tau_0 = 3$ s.

2.4. Homogeneous broadening of lines is predominant in cubic crystals, and inhomogeneous broadening in glasses. In an impurity crystal at low temperature the basic reason for inhomogeneous broadening may be the nonuniform inner stresses caused by the low symmetry of the crystal (the closer the structure of a real lattice is to the perfect cubic lattice the smaller the contribution of inhomogeneous broadening is to the linewidth), while at high temperatures the basic reason may be the electron-phonon interactions.

2.5. $T_2 = 0.02$ ps.

2.6.

$$\Delta\nu = \frac{1}{2\pi}\left(\frac{1}{\tau_{01}} + \frac{1}{\tau_{02}}\right),$$

$$\Delta\nu_{01} = \frac{1}{2\pi\tau_{01}} = \Delta\nu - \frac{1}{2\pi\tau_{02}} \approx \Delta\nu.$$

We see that the lasing linewidth (9000 GHz) is generally determined by the width of the upper laser level.

2.7. The R line of chromium ions in this case is homogeneously broadened.

2.8. According to formula (2.28), $\Delta\nu_D = 540$ MHz.

2.9. About 10 MHz.

2.10. The collision-broadened linewidth is inversely proportional to the square root of the temperature, while the Doppler linewidth is directly proportional to the square root of the temperature. For this reason the gas must be heated to $T = 2T_{room}$ for the Doppler linewidth to become twice the collision-broadened linewidth.

2.11. The mean free path l equals $u_0 \tau_{col}$ or, according to formulas (2.27), (2.28), (2.31), and (2.32),

$$ l = \frac{\Delta\nu_D}{4\pi\sqrt{\ln 2}\ \Delta\nu_{col}}\ \lambda = 0.16\lambda, \quad \text{i.e.} \quad l < \lambda. $$

Thus, the molecular velocity changes often and its mean projection on the direction of observation becomes smaller than the projection's value in the absence of collisions. Hence, the Doppler frequency shift diminishes, which leads to a decrease in linewidth known as Dicke narrowing.

3.3. According to formula (3.14), $\sigma \approx 2 \times 10^{-18}$ cm^2.

3.4. $\tau_0 = 3$ s.

3.5. Using the Bougier-Lambert-Beer law, we find that $\sigma = 4 \times 10^{-18}$ cm^2.

3.6. According to formula (3.28), $\tau = 3$ ms. Note that here we are speaking of saturation intensity of amplification rather than of saturation intensity of absorption because in a four-level system the $2 \to 1$ transition amplifies the radiation rather than absorbs it.

4.1. In deriving the formula we implicitly used the condition $t \gg 1/(\omega + \omega_0)$ in averaging (4.15) and the condition $t \ll 1/|\omega - \omega_0|$ in carrying $b(t)$ outside the integral in (4.11).

4.2. When deriving formula (4.46) for the transition probability, we lifted the restriction $t \ll 1/|\omega - \omega_0|$ and the condition $\langle \mu E \rangle / \hbar \ll 1$ necessary to ensure the smallness of $b(t)$. However, it can be directly shown that (4.18) is the particular case of (4.46) for $\Omega_R \ll \delta$.

4.3. See formula (4.20) in which the value of t is determined by the physical process that limits the interaction time.

4.4. The quantity $1/t$ determines the transition's linewidth. When the distribution of ϱ_ν is broad, the transition probability incorporates the entire area of the line. In contrast, for a very narrow ϱ_ν (coherent interaction), formula (4.22) transforms into

$$\Pi \approx \frac{8\pi}{3} \frac{\langle\mu^2\rangle}{\hbar^2} \int_{-\infty}^{\infty} \varrho_\nu t^2 d\nu,$$

which, allowing for the condition $t \ll 1/|\omega - \omega_0|$ (see Problem 4.1), is the generalization of (4.52) for small t's.

4.5. Since the interaction must be coherent, (a) the emission linewidth must be much smaller than the detuning from resonance, δ, which, in turn, must be smaller than the inverse of the transition lifetime (see Problem 4.4); (b) the coherence time of the radiation must exceed $1/\Omega_R$; and (c) the coherence length must be larger than the sample's dimensions.

4.6. The coherence time is greater than 0.01 ns, the detuning δ is less than $1/t \approx 10$ MHz, and the linewidth $\Delta\nu$ of the exciting radiation is smaller than 10 MHz.

4.7. The minimal Rabi frequency must exceed the transition linewidth $\Delta\nu$. Assuming that $\Delta\nu \approx \Delta\nu_{nat}$, we have $\Omega_R > 10$ MHz, whence the intensity $I > 10^{-2}$ W/cm^2.

5.1. Transforming Eq. (5.5), we get

$$\Delta\nu = \Delta\nu_{line}\sqrt{\ln 2}\,(n\sigma l - \ln 2)^{-1/2} \approx 120 \text{ GHz}.$$

5.2. Yes, it does. The effect of gain saturation leads to a situation in which the gain at the center of the line begins to drop and the effective linewidth grows. The effect manifests itself when $I_{out} \approx I_{sat}$. For the given amplifier $I_{in} = I_{sat}/G_0 \approx 4$ W/cm^2.

5.3. For a four-level system $n_1 \ll n_2$. Then formula (5.15) for P_{in}^{eff} assumes the form

$$P_{in}^{eff} = h\nu(G_0 - 1)/G_0.$$

For this reason a quantum amplifier based on the three-level scheme has an $n_2/(n_2 - n_1)$-fold higher effective noise level than a four-level amplifier does.

5.5. Equation (5.17) with $\beta = 0$ yields the following maximum value of the power output produced by a unit volume of the active medium:

$$P_{max} = \alpha I_{sat} = 22 \text{ kW/cm}^3.$$

5.6. With all laser active media with $\Delta\nu_{line} \gg 15$ MHz. For instance, with the active media of solid-state lasers, whose linewidths amount to hundreds of gigahertz.

5.7. An Nd-glass laser.

6.1. Let $l = 20$ cm. Then $Q_{opt} = 2.5 \times 10^8$ in the optical range and $Q_{mcw} = 10^4$ in the microwave range, with the result that $Q_{opt}/Q_{mcw} = 2.5 \times 10^4$. The bandwidths of the chosen cavities in the optical and microwave ranges are

$$\delta\nu_{opt} = \delta\nu_{mcw} = 2.4 \text{ MHz.}$$

6.2. The bandwidth of the cavity will become narrower by a factor of five under regeneration (see (6.4)). For generation to appear the gain must be higher than $\sqrt{2}$ (see (6.13)).

6.3. Using formulas (6.15) and (6.16), we find that $l_{max} = 15$ cm.

6.4. 10 J/cm^3 (see (6.58)).

6.5. 0.6 J.

6.6. (a) $N_{max} = 2 \times 10^{19}$ cm^{-3}; (b) No, it cannot since the population inversion density cannot exceed the number density of the active particles; (c) $l_{max} = 50$ cm.

6.7. $\tau_{pulse} \sim E/P$, where E is the energy and P the power output of the radiation. Using (6.58), we get

$$\tau_{pulse} = \pi d^2 lNh\nu/8P = 20 \text{ ns.}$$

6.8. From Eqs. (6.46) and (6.47) it follows that the reflectivity of the output mirror R is equal to 0.9.

7.1. Formula (7.7) has been derived for oblique beams whose Q-factors exceed half of the Q-factor of a normal beam. At $R = 1$ the Q-factor of normal beams is infinite (under the assumptions made in the problem). At $R = 1$ the above is meaningless because the Q-factor of an oblique beam is finite.

7.2. $N_F = (a/2l) \div (\lambda/2a)$, where $a/2l$ is the angle at which one mirror is seen from the center of the other, and $\lambda/2a$ the angle of diffraction divergence.

7.3. After the parallel beam with a diffraction angle $\lambda/2a$ has propagated over a distance l (see the previous problem) its diameter increases by $l\lambda/a$. Assuming this increment to be small, we get $\lambda l/a^2 \ll 1$.

7.4. $N = 1.5 \times 10^{12}$; the number of modes decreases 25-fold.

7.5. $\tau_{eff} = 1/\alpha c \approx 66$ ns and $Q = \omega\tau_{eff} = 2.2 \times 10^8$.

7.6. $(1 - R)^3 < 1/6\pi^2 N_F^2$, that is, $R > 97.7\%$.

8.1. $w_0 = \sqrt{l\lambda/4\pi} = 0.2$ mm.

8.2. $w^2 = w_0^2 + (z/kw_0)^2$; substituting $z = l/2$ and $z = l/2 + 1$, we get $2w = 0.8$ mm and $2w = 2.5$ mm, respectively, and $\theta = 1/kw_0 = 4 \times 10^{-4}$ rad.

8.3. Differentiating (8.10) with respect to z, we get $R_{max} = l$ at $z = l/2$, that is, at the mirrors.

8.4. We use the formula for the focal length of a thin lens[8]

$$F = \frac{n_m}{n_1 - n_m} \frac{1}{1/R_2 - 1/R_1}.$$

In our case $F = R(n - 1)$. Using (8.12), we get $r = R/n$. Knowing r and the spot size $w_0\sqrt{2}$, we solve Eqs. (8.4) and (8.10) simultaneously, and substituting the result into (8.11a), we find that the divergence angle increases $\sqrt{(1 + n^2)/2}$-fold.

8.5. $\Delta x = v_0 = \lambda/2\pi$.

8.6. Assuming that maximum focusing is achieved, we have a

$$\left(\frac{w_0}{\lambda/2\pi}\right)^2 = 6 \times 10^6\text{-fold}$$

increase in intensity.

8.7. $x = \dfrac{(R_2 - L)L}{R_1 + R_2 - L}$, where x is the distance from the neck to the mirror with the curvature radius R_1.

8.8. $w_0 = \sqrt{\lambda L_{equiv}/2\pi}$, where L_{equiv} is the length of the equivalent confocal cavity, $L_{equiv}^2 = (2R - L)L$, and

$$w_{1,2} = \sqrt{\frac{\lambda e}{2\pi}} \left(\frac{4R^2}{L(2R - L)}\right)^{1/4}.$$

8.9. $P \propto E^2$; substituting this into (8.8), we get

$$P \propto \exp[-(x^2 + y^2)/w^2], \quad \text{i.e.} \quad w_P = w_E/\sqrt{2}.$$

---

[8] M. Born and E. Wolf, *Principles of Optics: Electromagnetic Theory of Propagation, Interference and Diffraction of Light*, 5th ed., Oxford: Pergamon Press, 1975.

8.10. Integrating the result of Problem 8.9 with respect to x and y, we find that

$$P_0 = I_0 \pi w_E^2.$$

8.11. In the far zone, that is, for $z \gg l/2$, (8.8) becomes

$$E(x, y) = 2\pi w_0^2 \frac{E_0}{z} \exp\left(-\frac{x^2 + y^2}{2w^2}\right) \cos(\omega t - kz),$$

which links the electric field in the wave to the intensity. Integrating with respect to x and y and equating the result with P_0, we find the value of E_0, which yields

$$I(x, y) = \frac{1}{\pi} P_0 \frac{w_0^2 k^2}{z^2} \exp\left(-\frac{x^2 + y^2}{z^2/k^2 w_0^2}\right).$$

Since the total angular divergence, θ_0, is equal to 2θ, with θ given by (8.11a), we have

$$I(x, y) = \frac{4}{\pi} P_0 \frac{1}{z^2 \theta_0^2} \exp\left[-\frac{4(x^2 + y^2)}{z^2 \theta_0^2}\right].$$

9.1. The neck lies on the line connecting the point of intersection of two circles with diameters R_1 and R_2 centered at the focal points of mirrors 1 and 2, respectively.

9.2. If the circles do not intersect, the cavity is unstable. If they touch, coincide, or are two parallel lines, the cavity is on the verge of stability. Prove this independently using the results of Problem 8.7.

9.3. The neck coincides with the flat mirror.

9.4. $L_{max} = 3$ m, $L_{min} = 0$.

9.5. $L_{max} = 2$ m, $L_{min} = 1$ m.

10.1. For the radiation to fill the laser tube in an optimal manner, the mirror diameter must be greater than the tube diameter.

10.2. Using formulas (10.17) and (10.18), we get

$$R_1 = 2.67 \text{ m}, \quad R_2 = 0.67 \text{ m}, \quad 2a_1 > 2 \text{ cm}, \quad 2a_2 = 5 \text{ mm}.$$

10.3. Using formula (10.20), we find that $N_F{}^{equiv} = 18.36$; suppression is satisfactory.

10.4. Using formulas (10.17), (10.18), and (10.20), we get

$$2a_2 = 8.47 \text{ cm}, \quad 2a_1 > 10 \text{ cm}, \quad R_2 = 31 \text{ m}, \quad R_1 = 37 \text{ m}.$$

10.5. For the efficiency to be high the tube diameter must be slightly greater than 10 cm (see Problems 10.1 and 10.2).

11.3. $\tau \approx E/P_{max} = 2\tau_{ph}$.

11.5. $\tau \approx 1$ ns.

11.6. Since $n_{thr} = 1/c\sigma\tau_{ph}$ (see Eqs. (3.13) and (11.13)), we find that, as $\tau_{ph} \to 0$, n_{thr} grows and the condition $n_0 > n_{thr}$ ceases to be met, say nothing of the approximation $n_0 \gg n_{thr}$.

11.7. $Q = \omega\tau_{ph} \approx 10^8$, $\Delta\nu_{min} = 1/2\tau_{ph} \approx 10$ MHz.

11.8. By a factor of two and three, respectively.

13.1. $\Delta\nu_{line} \approx \Delta\nu_{nat} \approx 20$ MHz.

13.2. Yes. What glows is the helium. There is five times more helium than neon.

13.3. $\eta < E_{pump}/h\nu_{em} \approx 8\%$ ($\lambda = 0.63\mu$m), 4% ($\lambda = 1.15$ μm), and 2% ($\lambda = 3.39$ μm).

13.4. The radiation at 0.63 μm is completely linearly polarized, and so is the radiation at 1.15 μm, since the losses for the radiation polarized in the plane normal to the plane of incidence on the Brewster windows are much higher than amplification.

13.5. It is best to place the cell inside the cavity and reduce pumping almost to the threshold level.

13.6. The homogeneous linewidth must be reduced as much as possible, $\Delta\nu_{col} \approx n u_0 \sigma_{col} \approx n\sqrt{kT}$ (Lecture 2). Since the necessary value of n is determined by the gain of the system, it is clear that the methane must be cooled in such a way that $\Delta\nu_D$ remains greater than $\Delta\nu_{col}$. One must also bear in mind the need to reduce the transit linewidth (2.33).

14.1. A powerful solenoid coaxial with the tube; this device is used in all designs of argon lasers.

14.2. Equation (14.10) shows that under the given condition the difference $n_2 - n_1$ is much smaller than N, which means that the population difference constitutes only a small fraction of the number of active particles.

14.3. Yes, there will because the separation between the cavity modes, $c/2l = 125$ MHz, is smaller than the homogeneous linewidth.

14.4. 460-800 MHz; see the previous problem.

14.5. $v_{dr} = \Delta\nu\lambda/2 \approx 100$ m/s.

14.6. $n_i = n_e = J/2ev_{dr} = 3 \times 10^{14}$ cm^{-3} and $n = p/kT =$

2×10^{21} cm^{-3}, whence $n_i/n \approx 10^{-7}$. Do not confuse this result with the condition $n_2 - n_1 \ll N$ in Problem 14.2; the active particles are the ionized ones, that is, $N = n_i$.

14.7. For the pressure value given in the lecture, $\Delta\nu_{\text{hom}} \approx 1/2\pi\tau = 150$ MHz, where τ is the smallest of the two lifetimes of the P- and D-states.

14.8. According to (2.28), $\Delta\nu_D \approx 20$ GHz.

15.1. Level 01^10 is actually the lower laser level. Hence, its population must be as low as possible. Assuming the population of this level to be one-tenth of that of the ground level, we find that

$$N_{01^10}/N_{00^00} = \exp\left(-\Delta E/kT\right) = 0.1 \quad \text{at} \quad T = 450 \text{ K.}$$

15.2. Using the explanatory text to Figure 15.5, we find that

$$P_{\max} = \frac{\alpha_0}{\beta} \, I_{\text{sat}} \frac{\pi D^2}{4} = 150\text{-}300 \text{ W.}$$

15.3. $\eta = P_{\text{out}}/P_{\text{el}} = P_{\text{out}}/(E/p)plI = 15\text{-}30\%$.

15.4. According to (2.28), $\Delta\nu_D = 50$ MHz.

15.5. Substituting the data into Eq. (15.18), we get $J_{\max} = 18$.

16.1. To achieve continuous tuning we must ensure that gain exceeds the threshold value in the entire spectral range of the P- and R-bands, that is, the widths of separate rotational transitions must be roughly equal to the distances between the lines corresponding to these transitions. However, $\Delta\nu_{\text{col}} = 760$ torr $\times$ 6 MHz/torr $= 0.15$ cm$^{-1} \ll 1\text{-}2$ cm^{-1}.

16.2. Homogeneous collision, except in the case of extremely low pressures.

16.3. The gasdynamic laser operates according to the self-terminating three-level scheme and the others according to the four-level.

16.4. In this pumping channel one excitation quantum of level 00^02 is exchanged for two laser quanta with no loss of energy (the quantum yield is two).

16.5. When mode locking with the narrowest pulse possible must be achieved.

16.6. By using a laser pulse that is shorter than the time interval in which energy is exchanged between rotational sublevels, say, by generating such a pulse via active mode locking.

17.1. At the leading edge of the pulse the spectrum consists of all the rotational branches allowed by the selection rules, but as the V-V relaxation time passes, the spectrum narrows to the width of a part of the P-branch (partial inversion).

17.2. The two-level scheme operates in the case of complete vibrational inversion and the four-level in the case of partial inversion.

17.3. In the case of complete vibrational inversion by removing the reaction products by pumping through and in the case of partial inversion by the same mechanism plus an additional depopulation channel through R-T relaxation.

17.4. No, since the lower laser levels must be depopulated and the reaction products removed.

17.5. Theoretically, yes, but the channel along which the active substance circulates must incorporate a chemical plant.

18.1. The energy level diagram shows that the carbon monoxide laser emits in a cascade process with low Stokes' losses:

$$\eta_q = \frac{\Sigma h\nu_{em}}{E_{pump}} \approx 1.$$

18.2. The carbon monoxide laser operates on a large number of successive laser transitions, while the carbon dioxide laser operates on a single transition but the excitation energy is exchanged for two photons, that is, also without any additional losses.

18.3. Because an oscillator (a quantum oscillator included) spends most of its time at the turning points.

18.4. $\eta_q = h\nu/E_{pump} \approx 30\%$.

18.5. The disappearance of population inversion is unimportant for single-pass lasers.

18.6. There is no standing wave in these types of lasers and no Lamb dip either.

19.1.

$$\frac{dn_3}{dt} = Wn_1 - n_3(w_{31} + w_{34}), \quad \frac{dn_4}{dt} = n_3 w_{34} - n_4 w_{41},$$

$$n_2 w_{23} = n_1 W, \quad n_1 + n_3 + n_4 = N.$$

19.2. Formulas (19.2) and (19.20) show that the statement of the problem is valid if

$$W^{(3)} = \frac{w_{31}(w_{23} + w_{21})}{w_{23} - 4w_{31}}$$

in the three-level system and

$$W^{(4)} = \frac{w_{23} w_{34}}{w_{41} - 4w_{34}}$$

in the four-level, that is, $W^{(3)} = W^{(4)}$, and $W \to \infty$ as $w_{23} \to 4w_{31}$ and $w_{41} \to 4w_{34}$.

19.4. No, there will not, since at the liquid nitrogen temperature $kT = 54$ cm^{-1} and $\Delta E \gg kT$. Energy transfer will be effective at $T > 470$ K.

19.5. *Hint*: Read the material after (19.21) carefully.

19.6. 1/3.

19.7. This condition has a negative effect on the efficiency and the rate of nonradiative depopulation of the levels.

20.1. Employing Eq. (7.16) and formula (11.24) for the peak output power, we find that

$$P_{\max} = 6.75 \text{ MW/cm}^3.$$

The minimum rate of pumping of the upper laser level is $\Lambda = 2.5 \times 10^{20}$ s$^{-1} \cdot$cm^{-3}.

20.2. Using the system of equations obtained in Problem 19.1 and allowing for the fact that $w_{34} \ll w_{41}$, we find that

$$n_3 - n_4 \approx n_3 = \frac{WN}{W + w_{31} + w_{34}}.$$

Equations (3.13) and (7.16) yield

$$n_{\text{thr}} = \frac{n}{c\sigma\tau_{\text{eff}}} = 3.3 \times 10^{16} \text{ cm}^{-3}.$$

Substituting $w_{31} + w_{34} = 230$ μs and allowing for the fact that the standard concentration of Nd^{3+} ions is 6×10^{19} cm^{-3}, we obtain

$$W_{\text{thr}} = \frac{n_3(w_{34} + w_{31})}{N - n_3} = \frac{n_{\text{thr}}}{N} \frac{1}{\tau} = 2.4 \text{ s}^{-1}.$$

Then, ignoring the pumping losses and recalling that the pumping rate is greater than the lasing rate by 2000-3000 cm^{-1} (see the lecture), we find that

$$P_{\text{thr}} > Wh\Omega_{\text{pump}}N \approx 40 \text{ W/cm}^3.$$

20.3. $\tau_{\text{pulse}} \approx h\nu\Lambda\tau/2P_{\text{max}} = 10$ ns.

20.4. The strong inhomogeneous broadening is caused by the considerable inhomogeneity of the intrinsic electric field in the glass.

20.5. $P = n_{\text{thr}}h\nu V/0.04\tau = 7.5$ kW.

20.6. No, it cannot, since the maximum length of the active element must not exceed $l = c/2\Delta\nu = 50$ μm. The minimum pulse length τ in mode-locked lasing is approximately $1/\Delta\nu = 0.3$ ps.

21.1. $p = 6000/750 = 8$.

21.2. From Eq. (21.6) it follows that the sought effect emerges when the term $\exp(-\hbar\omega/kT)$ ceases to be negligible. Replacing $(1 - \exp(-\hbar\omega/kT))^p$ with $1 - p\exp(-\hbar\omega/kT)$ and assuming that $p\exp(-\hbar\omega/kT)$ is close to unity, we get $T > \hbar\omega/k \ln p \approx 250$ °C.

21.3. Expanding the exponential in Eq. (21.6) in a power series yields

$$W(\Delta E, T) \propto T^p.$$

21.4. (1) The transfer must be of a resonance nature, (2) the interaction must be weak, (3) the interaction must be dipole-dipole, (4) the transfer must be irreversible, (5) the spectra must be broad and continuous, and (6) both donor and acceptor must be in a medium with a refractive index n.

21.5. Using the expression for the quantity γ introduced in formula (21.27), we find that $C_{DA} = 9 \times 10^{-40}$ cm^6/s for the YAG crystal and $C_{DA} = 2.2 \times 10^{-38}$ cm^6/s for the GSGG crystal.

22.1. The appearance of the factor $\exp[-h(\nu_{\text{el}} - \nu)/kT]$ is due to the four-level nature of the operation of a dye laser. Actually this is the equilibrium (fractional) population of the lower laser level.

22.2. Using the result of the previous problem, we find that the "gain" in threshold pumping at room temperature is $2.5\text{-}20 \times 10^3$ times.

22.3. (a) Since the threshold of a three-level system is much higher than that of a four-level, the efficiency of such a system may be lower by an unlimited number of times provided that the length of the lasing pulse is longer than the time it takes to depopulate the lower laser level of the four-level system. (b) The efficiency of the four-level system is twice as high as that of the three-level because for a population inversion density close to the active particle density in the three-level system only half of the stored energy will transform into laser energy.

22.4. $P = 10^9$ s^{-1}.

22.5. $F \approx 0.03$ J/cm^2.

22.6. Using the result obtained in the previous problem, we can find the condition for calculating the size of the spot:

$$10F = \frac{P_{av}}{f_{rep}} \frac{4}{\pi d^2},$$

which yields $d \approx 0.3$ mm for the spot diameter.

22.7. A cavity can be assumed optimal if the neck size is equal to the pumping spot size. Then, if we put $\lambda \approx 600$ nm, we can calculate the length of the cavity:

$$l = nd^2/\lambda \approx 50 \text{ cm.}$$

22.8. (a) $T = 2l/\tau = 3$ ns; $\tau \approx 6$ ps. (b) For the laser to operate with passive mode locking the pulses must pass several times through a saturating absorber. This requires making the cavity shorter, introducing into it, if necessary, additional lenses for focusing the beam, and raising the repetition frequency by placing the absorber at the cavity's center, for instance.

23.2. By a factor of four.

23.3. $\tau = 1/2\pi\Delta\nu = 1$ ps.

23.5. We denote the population numbers of levels $^4T_2$ and ^2E by n_2 and n_1, respectively. We then have

$$\frac{dn_2}{dt} = -g_2 n_2 A_2 + w_{12} n_1 - w_{21} n_2,$$

$$\frac{dn_2}{dt} = Q - g_1 n_1 A_1 - w_{12} n_1 + w_{21} n_2.$$

This yields

$$\frac{d}{dt}(n_1 + n_2) = Q - g_2 n_2 A_2 - g_1 n_1 A_1 = 0.$$

Allowing for (1.4), we get

$$\frac{I}{h\nu} = g_2 A_2 n_2 = Q \left[1 + (g_1 + g_2)^2 \frac{A_1}{A_2} \exp\left(\frac{\Delta E}{kT}\right) \right]^{-1}.$$

Substituting numerical values yields $I/h\nu = Q$ at $T = 300$ K and $I/h\nu = 10^{-5}Q$ at $T = 77$ K.

23.6. No, it cannot. See the previous problem.

24.1. Yes, we can, because the bands are not continuous but consist of a large number of levels. Hence, population inversion can be created in any pair of levels.

24.3. In all cases lowering the temperature raises the efficiency (see Eq. (24.7) or Figure 24.2).

25.2. The upper boundary is determined by overheating, the lower by the threshold.

25.3. It decreases (cf. Problem 24.3).

25.5. Lowering the temperature makes the short-wave edge of the spectrum sharper.

25.6. Even if we assume that the cavity is confocal, the spot size is approximately 50 μm. The width of the active region is much less: $2d_{\text{dif}} \approx 2$ μm.

25.8. In identical conditions the amplification and emission spectra coincide and the lasing spectrum has a linear mode structure.

25.9. $\Omega \approx 2\lambda/d$. In the junction plane $\Omega \approx 3'$ and in the normal plane it is roughly $3°$.

References

1. M. Betolotti, *Masers and Lasers. A Historical Approach*, Adam Hilger Ltd., 1983.
2. Y. R. Shen, *The Principles of Nonlinear Optics*, John Wiley, Chichester, 1984.
3. D. Bäuerle, *Chemical Processing with Lasers*, Springer Verlag, Berlin, 1986.
4. W. T. Witteman, *The CO_2-laser*, Springer Verlag, Berlin, 1986.
5. M. von Allmen, *Laser-Beam Interactions with Material. Physical Principles and Applications*, Springer Verlag, Berlin, 1987.
6. I. W. Boyd, *Laser Processing of Thin Films and Microstructures. Oxidation, Deposition and Etching of Insulators*, Springer Verlag, Berlin, 1988.
7. M. D. Levenson and S. S. Kano, *Introduction to Nonlinear Laser Spectroscopy*, Academic Press, London, 1988.
8. D. R. Hall and P. E. Jackson, Ed., *Physics and Technology of Laser Resonators*, Adam Hilger, Bristol, UK, 1989.
9. G. P. Agrawal, *Nonlinear Fibre Optics*, Academic Press, London, 1989.
10. S. Ungar, *Fibre Optics, Theory and Applications*, John Wiley, Chichester, U. K., 1990.
11. Chai Yeh, *Handbook of Fibre Optics, Theory and Applications*, Academic Press, London, 1990.
12. F. J. Duarte and I. W. Hillman, *Dye Laser Principles. With Applications*, Academic Press, London, 1990.
13. R. Elton, *X-Ray Lasers*, Academic Press, London, 1990.
14. M. J. Adams and I. D. Henning, *Optical Fibres and Sources for Communications*, Plenum Press, 1990.
15. V. M. Akulin and N. V. Karlov, *Intense Resonant Interactions in Quantum Electronics*, Springer Verlag, Berlin, 1991.
16. J. N. Dodd, *Atoms and Light Interactions*, Plenum Press, 1991.
17. P. Meystre and M. Sargent III, *Elements of Quantum Optics*, Springer Verlag, Berlin, 1992.
18. W. M. Steen, *Laser Material Processing*, Springer Verlag, Berlin, 1992.
19. Franc J. Duarte, *High-Power Dye Lasers*, Springer Verlag, Berlin, 1992.
20. M. Young, *Optics and Lasers*, Springer Verlag, Berlin, 1992.
21. W. Koechner, *Solid-State Laser Engineering*, 3rd ed., Springer-Verlag, Berlin, 1992.

INDEX